Ruminant Physiology: Digestion, Metabolism

Ruminant Physiology: Digestion, Metabolism

Editor
Julio Mendes

scitus
academics

Ruminant Physiology: Digestion, Metabolism

Edited by **Julio Mendes**

ISBN: 978-1-68117-626-0
Library of Congress Control Number: 2016947446

SCITUS Academics LLC,
616, Corporate Way, Suite 2, 4766,
Valley Cottage, NY 10989

www.scitusacademics.com

Printed in United States of America on Acid Free Paper

Preface

Ruminating animals have various physiological features that enable them to survive in nature. One feature of ruminants is their continuously growing teeth. During grazing, the silica content in forage causes abrasion of the teeth. This abrasion is compensated for by continuous tooth growth throughout the ruminant's life, as opposed to humans or other non-ruminants, whose teeth stop growing after a particular age. Most ruminants do not have upper incisors; instead, they have a thick dental pad to thoroughly chew plant-based food. Ruminants are mammals that digest plant based food by processing it in a series of chambers in their stomachs. There are about 150 species of ruminants, including both domestic and wild species. Ruminating mammals include cattle, goats, sheep, giraffes, bison, moose, elk, yaks, water buffalo, deer, camels, alpacas, llamas, and antelope. Ruminants are distinguished from other animals by having a four-compartment stomach. The four compartments are called the rumen, the reticulum, the omasum, and the abomasum. The rumen and the reticulum are connected and work in concert and are therefore sometimes called the "reticulorumen".

Ruminant Physiology: Digestion, Metabolism is a valuable tool of information for researchers, nutritionists, advisors, and advanced graduate students who want to have up-to-date and concise information on ruminant digestive system.

Table of Contents

CHAPTER 1

The Roles of Syncytin-Like Proteins in Ruminant Placentation

Yuki Nakaya [1,2,*] and Takayuki Miyazawa [2,*]

[1] Department of Infectious Diseases, Kyoto Prefectural University of Medicine, 465 Kawaramachi-hirokoji-Kajiicho, Kamigyo-ku, Kyoto 602-8566, Japan

[2] Laboratory of Signal Transduction, Department of Cell Biology, Institute for Virus Research, Kyoto University, 53 Shogoin-Kawaharacho, Sakyo-ku, Kyoto 606-8507, Japan

ABSTRACT

Recent developments in genome sequencing techniques have led to the identification of huge numbers of endogenous retroviruses (ERV) in various mammals. ERVs, which occupy 8%–13% of mammalian genomes, are believed to affect mammalian evolution and biological diversity. Although the functional significance of most ERVs remains to be elucidated, several ERVs are thought to have pivotal roles in host physiology. We and other groups recently identified ERV envelope proteins (e.g., Fematrin-1, Syncytin-Rum1, endogenous Jaagsiekte sheep retrovirus Env) that may determine the morphogenesis of the unique fused trophoblast cells, termed trinucleate cells and syncytial plaques, found in ruminant placentas; however, there are still a number of outstanding issues with regard to the role of ERVs that remain to be resolved. Here, we review what is known about how these ERVs have contributed to the development of ruminant-specific trophoblast cells.

KEYWORDS

ruminants; endogenous retrovirus; envelope glycoprotein; fematrin-1; syncytin-rum1; placenta; cell-to-cell fusion

1. INTRODUCTION

Endogenous retroviruses (ERVs) occupy 8%–13% of mammalian genomes, while only 2% of those genomes are protein-coding genes [1]. "Endogenization" occurs when exogenous retroviruses integrate into the genomes of germ cells; these ERVs are then inherited by offspring according to Mendelian genetics [2]. The detailed mechanism of by which endogenization occurs is not clearly understood, due to the lack of animal models with which to monitor the endogenization process. As an exceptional case, Tarlinton *et al.* reported that the exogenous koala retrovirus (KoRV) is currently colonizing the koala genome; KoRV is, thus, expected to serve as a good model for understanding endogenization [3]. As with exogenous retroviruses, proviruses of ERVs consist of *gag*, *pol* and *env* genes flanked by long terminal repeats (LTRs) at both ends. Most ERVs are considered to be junk DNA, due to the many deletions and substitutions in their coding sequences; however, recent studies have revealed that several ERVs still retain intact coding sequences and that the encoded proteins are necessary for their host's physiology [2].

The placenta is a transient organ that appears only during pregnancy and plays essential roles in maintaining the fetus throughout gestation. Recent studies have provided us with evidence that ERVs are involved in placentation in various species [4,5,6,7,8,9,10,11,12]. In this paper, we review recent findings regarding ERVs and placentation, especially in ruminants.

2. DIVERSITY OF MAMMALIAN PLACENTA

All placentas consist of fetal trophoblast cells and maternal uterine cells; however, placental morphologies differ between species [13]. The gross anatomical morphology of the placenta is divided into four groups:

discoidal placenta (e.g., human, mouse), zonary placenta (e.g., cat, dog), cotyledonary placenta (e.g., cow, sheep) and diffuse placenta (e.g., horse, pig) (Figure 1). Each type of placenta has a unique microscopic anatomical feature, namely, hemochorial placenta (discoidal), endotheliochorial placenta (zonary), synepitheliochorial placenta (cotyledonary) and epitheliochorial placenta (diffuse) (Figure 2). The histological classification is defined by the morphology of the fetomaternal interface (Table 1).

In both hemochorial and endotheliochorial placentas, the trophoblasts deeply invade maternal tissues to allow the efficient exchange of nutrients, gases and hormones. They also form specialized cells called syncytiotrophoblasts, which may prevent rejection of the fetus, the result of maternal immune responses. Syncytiotrophoblasts are formed by continuous cell-to-cell fusion of cytotrophoblasts. Maternal endometrial epithelial cells are completely diminished in these placentas (Figure 2).

The most characteristic feature of synepitheliochorial placentas is that they develop only at projections on the uterine wall, termed caruncles. These placentas are characterized by the presence of about 100 placentomes, which are derived from trophectodermal villi and caruncles, respectively (Figure 1). In synepitheliochorial placentas, the trophoblast cells do not invade the maternal uterine tissue as deeply as occurs in hemochorial and endotheliochorial placentas (Figure 2). At the periimplantation period, the trophoblast cells produce several hormones and proteases that degrade the epithelia of uterine caruncles [14,15,16]. The trophectoderm subsequently adheres to the caruncles and develops the cotyledonary placentomes. The uterine epithelium might be restored after implantation. Some of the trophectodermal cells, termed trophoblast giant cells or binucleate cells (BNCs), fuse with the endometrial cells to form fetomaternal hybrid cells throughout the different stages of pregnancy, including implantation (Figure 2) [17,18]. This fusion is believed to enhance implantation and transport of various substances [17]. This fetomaternal cell-to-cell fusion is the origin of the term “syn” epitheliochorial placenta.

In contrast, no fused cells appear in epitheliochorial placenta and the trophoblasts do not invade maternal tissues (Figure 2). The fetal and maternal tissues are completely separable in this placenta.

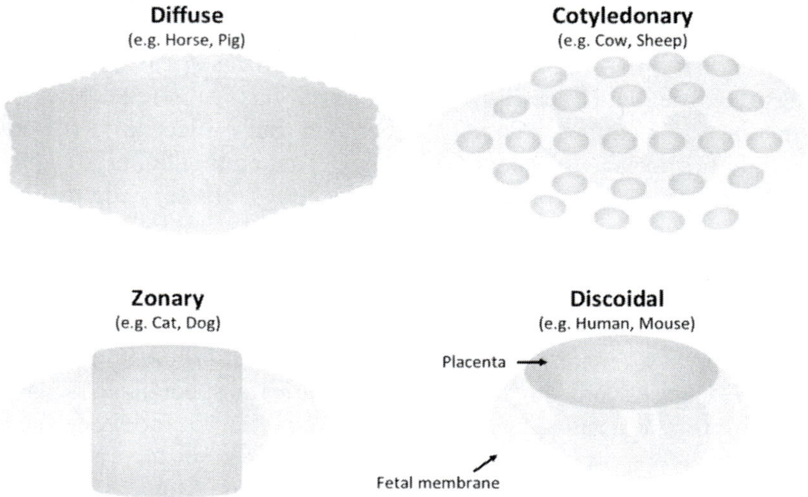

Figure 1. Gross morphology of placentas in different species. The fetus is surrounded by a fetal membrane (light pink). The placenta (light brown) is formed on the fetal membrane and its morphology varies by species.

3. ERVS INVOLVED IN PLACENTATION

3.1. Retroviral Envelope Glycoproteins

The main functions of retroviral envelope glycoproteins (Env) are binding to the entry receptor on the host membrane and the induction of viral-cellular membrane fusion. Retroviral Envs have two subunits: surface (SU) and transmembrane (TM) subunits. The translated Env polyprotein is cleaved into each subunit by cellular proteases, after which SU and TM assemble to establish mature Env trimers of SU-TM heterodimers [19,20,21,22,23,24]. SU recognizes the entry receptors by its receptor binding domains [25,26,27,28,29,30]. The binding of SU leads to conformational changes in Env that expose the fusion peptide at the N-terminus of TM, which then penetrates into the host cell membrane to initiate membrane fusion [31]. Env expression on the cell surface can result in cell-to-cell fusion, as well as viral-cellular membrane fusion; however, the degree of cell-to-cell fusion varies among different retroviral Envs [4,5,6,11,12,32,33].

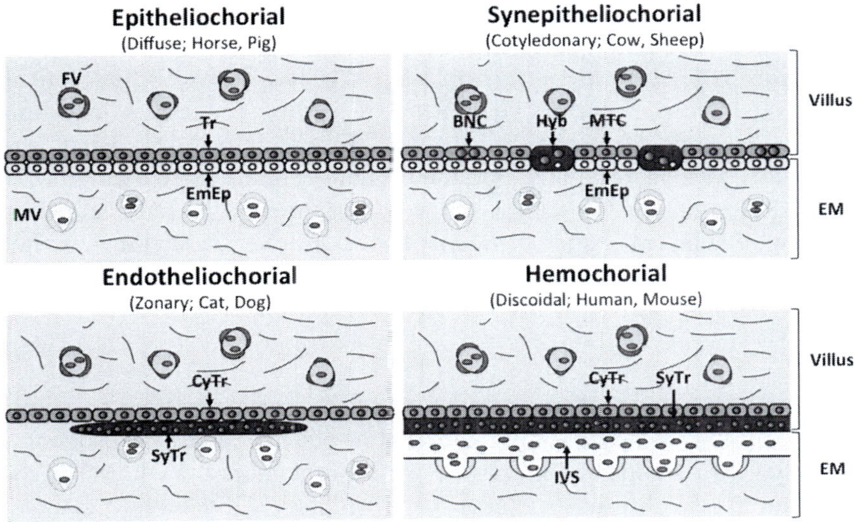

Figure 2. Structure of different types of placentas. The fetomaternal interfaces of the placentas are represented. The endometrial epithelium is retained in epitheliochorial and synepitheliochorial placentas, while it is degraded in endotheliochorial and hemochorial placentas. Abbreviations: FV; Fetal blood vessel, MV; Maternal blood vessel, Tr; Trophoblast, EmEp; Endometrial epithelium, BNC; Binucleate cell, Hyb; Hybrid cell, MTC; Mononucleate trophoblast cell, CyTr; Cytotrophoblast, SyTr; Syncytiotrophoblast, IVS; Intervillous space, EM; Endometrium.

3.2. Discovery of Envs Derived from ERVs in Human Placenta

How syncytiotrophoblast morphogenesis occurs was controversial until syncytin-1, the Env of human endogenous retrovirus (HERV) W at 7q21 (ERVW-1), was reported to be involved in the process [4,34]. Although other ERVs (e.g., ERV-3) have also been identified as candidates that might play a role in forming human syncytiotrophoblasts, definitive evidence for their function has not yet been established [35,36,37,38,39]. Syncytin-1 is preferentially expressed in placenta, especially cytotrophoblasts and syncytiotrophoblasts, and has strong fusogenic activity with human trophoblastic cells [4,34]. Syncytin-1 integrated into the common ancestor of hominoids and Old World monkeys (OWMs) between 20 and 30 million years ago (MYA) [40]. While syncytin-1 is believed to contribute to syncytiotrophoblast formation in hominoid placenta, it is inactivated in OWMs that are close relatives of hominoids [40]. Esnault *et al.* recently suggested that OWMs instead use the EnvV syncytin to form syncytiotrophoblasts [41]. They showed that both the *envV1*and *envV2* genes have intact Env coding sequences in primate genomes; however, only EnvV2 had fusogenicity [41]. Both *envV syncytins* were incorporated into the primate genome more than 40 MYA [41]. Even though the *envV2* syncytin is expressed in both OWM and hominoid placenta, the fusogenicity of EnvV2 syncytin has been maintained only in OWMs and not in hominoids [41]. Thus, hominoids and OWMs utilize different ERV Envs for syncytiotrophoblast morphogenesis.

After the discovery of syncytin-1, syncytin-2 was identified as another ERV Env that is specifically expressed in placenta and has fusogenicity [5]. Syncytin-2 appeared in the primate genome more than 40 MYA, as it is found in most primates except for prosimians [5]. While syncytin-1 is preferentially expressed in most syncytiotrophoblast cells, syncytin-2 expression is limited to a few cytotrophoblasts [42,43]. Conversely, the receptors for syncytin-1 (ASCT-2) and syncytin-2 (MFSD2A) are expressed in cytotrophoblasts and syncytiotrophoblasts, respectively [44,45]. Additionally, the fusion activity of syncytin-2 appears to be considerably lower than that of syncytin-1 [33]. Therefore, syncytin-1 and syncytin-2 might have different roles in placentation.

Retroviral Envs also have another important function, namely, immunosuppression. Although the molecular mechanism is not fully elucidated, the leading model suggests that retrovirus-related immunosuppression can be attributed to an immunosuppressive domain (ISD) found in TM [46]. ISD function was first confirmed using a synthetic 17 amino acid-long peptide CKS-17 (LQNRRGL DLLFLKEGGL). This peptide sequence is highly conserved in many retroviruses, including human T-lymphotropic viruses, type C-related HERV, Mason-Pfizer monkey virus, and Syncytins [33,46,47]. Both syncytin-1 and syncytin-2 have ISD motifs; however, only syncytin-2 was able to suppress the immune response *in vivo*, while syncytin-1 was reported to have immunosuppressive activity *in vitro* [33,48,49]. This indicates that syncytin-2 might be an essential piece of the strategy that the placenta adopts to maintain maternal immune tolerance during pregnancy. Taken together, humans and related species might use syncytin-1/EnvV2 and syncytin-2 mainly for forming syncytiotrophoblasts and immunosuppression, respectively.

Table 1. Classification of placenta.

Gross Morphology	Microscopic Structure	Species	Fetomaternal Interface	Type of Fused Cells
Diffuse	Epitheliochorial	Horse, Pig	Trophoblast-Epithelial	None
Cotyledonary	Synepitheliochorial	Cow, Sheep	Trophoblast-Epithelial (partially fused)	Fetomaternal hybrid
Zonary	Endotheliochorial	Cat, Dog	Trophoblast-Endothelial	Syncytiotrophoblast
Discoid	Hemochorial	Human, Mouse	Trophoblast-blood	Syncytiotrophoblast

A number of syncytin-like elements have also been identified in additional species, including rodents (syncytin-A and syncytin-B), lagomorphs (syncytin-Ory1) and carnivores (syncytin-Car1) [7,8,9]. These syncytin-like elements are also believed to form syncytiotrophoblasts. More details regarding these syncytin-like elements are reviewed elsewhere [50].

3.3. ERV Envs Expressed in Ruminant Placenta

Three types of trophoblast cells exist in ruminant placenta: mononucleate trophoblast cells (MTCs), BNCs and trinucleate cells (TNCs)/syncytial plaques (SyPs) [17]. Although the exact differentiation process of these cells has not been completely elucidated, a leading model has emerged [18,51]. In this model, BNCs appear as the result of MTC endoreduplication, the replication of genomic DNA without cellular mitosis. The genes for many pregnancy-associated molecules, which have important roles in maintaining pregnancy, are specifically expressed in BNCs and are modulated by epigenetic modification [17,51,52]. The BNCs fuse with maternal endometrial cells to efficiently transfer these molecules to the maternal blood stream [17]. Consequently, fetomaternal hybrid cells appear at the interface of the fetal and the maternal placentomes (Figure 2) [17]. These hybrid cells are categorized as TNCs or SyPs, depending on the species (e.g., Bovinae; TNCs, Caprinae; SyPs) [13,53]. TNCs are composed of a BNC and an endometrial cell, while SyPs are composed of multiple BNCs and an endometrial cell [17,54]. The population of hybrid cells is small and they disappear by apoptosis immediately after exocytosis of the pregnancy-associated molecules [17].

For the past decade, it has been suggested that differentiation of ruminant trophoblast cells is driven by ERVs [9,10,11,12,55]. Endogenous Jaagsiekte sheep retroviruses (enJSRVs) are derived from exogenous JSRV and enzootic nasal tumor virus (ENTV), both of which cause tumors in the ovine respiratory tract [56,57,58]. The ovine genome harbors at least 27 enJSRVs and five of them are still intact [57,58]. Some of the enJSRVs are believed to function as restriction factors because they impede the replication of JSRV and ENTV [59,60,61]. In addition, *env* mRNA of enJSRVs named enJS5F16 and enJS56A1 was detected in the ovine placentome, consisting of BNCs, SyPs, and maternal endometrial tissue [9,62]. Dunlap *et al.* demonstrated that inhibition of enJSRV Env expression by morpholino antisense oligos led to death of the conceptus during the periimplantation period due to a failure of BNC development [9]. Even though enJSRV Env is essential for the appearance of BNCs, it remains to be determined whether it is required for the fetomaternal cell-to-cell fusion found in ovine placenta.

Recently, a large number of bovine endogenous retroviruses (BERVs) were identified by various sequencing methods [10,63,64,65,66]. We searched the bovine genome database to identify potential ERV Envs, which could play roles in the fetomaternal cell-to-cell fusion. We identified two candidates, named BERV-K1 and BERV-K2 [10]. Both BERV-K1 and BERV-K2 are categorized as betaretroviruses, which also include JSRV and mouse mammary tumor virus [10,11]. While the BERV-K2 has all of the coding sequences including *gag*, *pro-pol* and *env*, BERV-K1 only retains the*env* coding sequence [10]. We examined the expression of both BERV-K1 and BERV-K2 *env* in bovine organs and found that BERV-K1 *env* but not BERV-K2 *env* was preferentially expressed in bovine placenta [10,11]. This result was further supported by *in situ* hybridization and immunohistochemistry, further demonstrating that BERV-K1 Env alone was detected specifically in BNCs (Figure 3) [11]. We also conducted *in vitro* fusion assays using BERV-K1 Env and primary bovine endometrial cells and found that BERV-K1 Env had high fusogenic activity [11]. This result suggests that BERV-K1 Env induces fetomaternal cell-to-cell fusion at the fetomaternal interface. We also conducted fusion assays using BERV-K2 Env; however, BERV-K2 Env was not able to induce cell-to-cell fusion with endometrial cells [11]. This phenomenon was attributed to a failure of Env glycoprotein maturation [67].

BERV-K1 is located in intron 13 of the bovine FAT tumor suppressor homolog 2 (*bFAT2*) gene, which is highly conserved evolutionarily in eukaryotes [11]. We examined other mammalian species for the presence of BERV-K1 by genomic PCR and found BERV-K1 in *Bovinae* (*Bos taurus*,*Bos javanicus*, *Bubalus bubalis*, and *Tragelaphus spekii*) but not other species, including *Caprinae* (*Ovis aries*, *Capra hircus*), human, mouse, dog and cat [11]. From this, we estimate that BERV-K1 infected the common ancestor of *Bovinae* 20 MYA after the separation of *Bovinae* and *Caprinae* [11]. Moreover, BERV-K1 shows evidence of purifying selection (dN/dS < 1) [11]. Thus, we suggested that BERV-K1 Env has a role in forming TNCs in *Bovinae* placenta, although further studies are required to determine its precise function *in vivo*. The BERV-K1 Env was named Fematrin-1 (Fetomaternal trinucleate cell inducer 1) because it is genetically different from syncytins found in the ERVs of other species [11]. We also found that *bFAT2* expression levels were higher in placenta than in other tissues [11]. Additionally, not only *bFAT2* but also ovine *FAT2* was preferentially expressed in placenta [11]. This suggests that the placenta-specific expression of Fematrin-1 is due to its integration into a locus (*bFAT2*), which has high transcriptional activity in ruminant placenta.

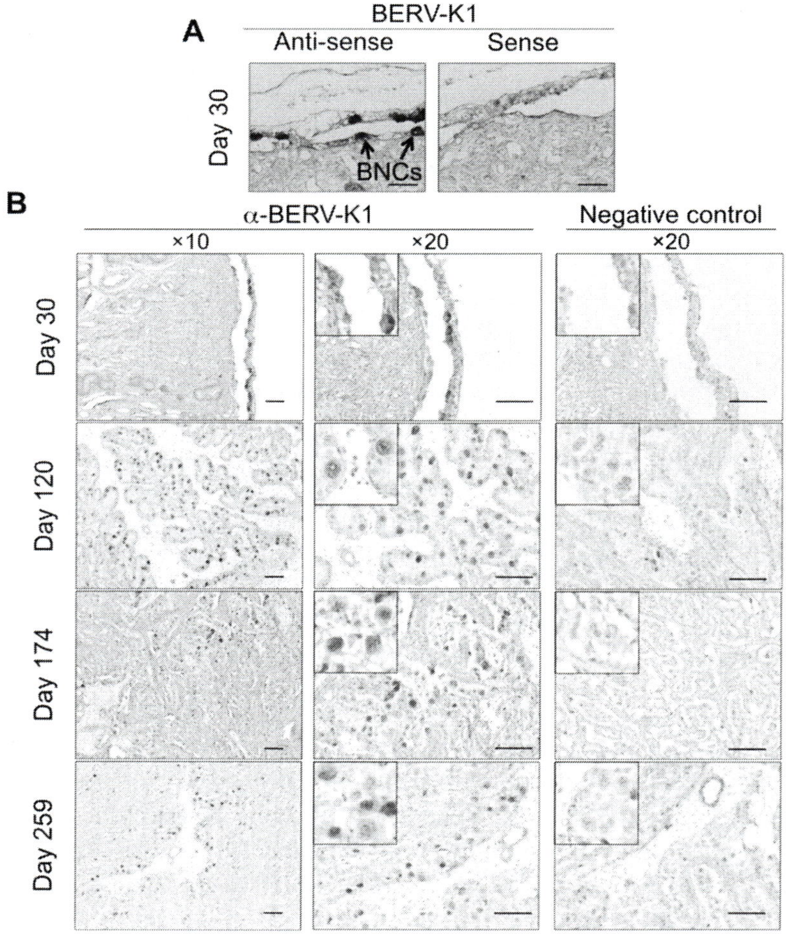

Figure 3. Expression of BERV-K1 envelope in bovine placental tissue. (**A**) *In situ* hybridization of BERV-K1 *env* mRNA on day 30 of gestation in the bovine fetal membrane and maternal endometrium. Scale bars represent 100 μm. DIG-labeled antisense probe for BERV-K1 *env* was used for the hybridization. BNCs express BERV-K1 *env* are shown as blue foci and as indicated by arrows; (**B**) Immunohistochemistry of bovine uterus and fetal membranes on days 30, 120, 174 and 259 gestation. Scale bars represent 100 μm. BNCs are enlarged in small panels. BERV-K1 Env was detected by anti-BERV-K1 Env and appear as brown foci.

Just prior to our publication describing Fematrin-1, Cornelis *et al.* identified Syncytin-Rum1, also derived from an ERV Env, as a potential factor involved in fetomaternal cell fusion [12]. Syncytin-Rum1 integration occurred over 30 MYA in the common ancestor of ruminants [12]. Syncytin-Rum1 has been subjected to purifying selection and is conserved among most ruminants, with the exception of *Tragulidae* (mouse-deer), whose ruminant ancestor diverged from the others about 50 MYA [12]. Syncytin-Rum1 is expressed in both bovine and ovine placenta and like *Fematrin-1*, was specifically detected in BNCs [12]. Syncytin-Rum1 has the unique characteristic that it is able to exert fusogenicity only under acidic (pH 5.0) but not neutral (pH 7.0) conditions [12]. Hence, the authors suggest that the boundary region of the fetomaternal placenta might be of acidic pH [12]. Further studies regarding whether acidic conditions exist in the ruminant placenta are required to support this hypothesis. Syncytin-Rum1 transcript levels are higher in ovine placenta than bovine placenta throughout gestation [12]. This difference in expression levels between sheep and cattle may be one of the causative agents for the morphological difference in fetomaternal hybrid cells between the two species [12].

bERVE-A and BERV-P *envs* have also been identified as transcripts in bovine placenta and trophectoderm [55,63]. The amino acid sequence of bERVE-A is similar to syncytin-1 and is expressed preferentially in BNCs [17]. However, bERVE-A might be the result of BNC appearance rather than an inducer of cell-to-cell fusion because it lacks an intact coding sequence [51]. BERV-P*env* was identified in bovine conceptuses, especially trophectoderm, in the periimplantation period [63]. The amino acid sequence of BERV-P Env is similar to syncytin-Car1 and has both a fusion peptide and an ISD in the TM subunit [63]. However, the functional significance of these elements has not been determined. Table 2 summarizes the placenta-related endogenous retroviral *env* genes that have been identified in ruminants thus far.

4. PERSPECTIVE ON THE RELATIONSHIP BETWEEN FEMATRIN-1 AND SYNCYTIN-RUM1

As described above, various mammals utilize at least two ERV Envs to induce cell-to-cell fusion in their placenta. However, why mammals utilize multiple ERVs in placental development remains an enigma. Mangeny et al reported that each syncytin has a different primary role, namely either cell-to-cell fusion or immunosuppression, in both human and mouse placentation [33]. In other words, syncytin-1 and syncytin-A have higher fusogenic activity than syncytin-2 and syncytin-B, respectively, while syncytin-2 and syncytin-B have higher immunosuppressive activity [33]. Nakamura and Imakawa previously proposed the "baton pass" hypothesis, in which the roles of either the original host genes or older ERVs have been transferred to newer ERVs [68]. For example, in the case of human syncytins, syncytin-2 probably played roles in both cell-to-cell fusion and immunosuppression at the beginning of endogenization. However, its fusogenic activity would have been gradually attenuated after syncytin-1 integration and ultimately, the main roles of syncytin-1 and syncytin-2 became distinct.

Likewise, a similar situation might be happening in ruminant placenta, at least in the case of the*Bovinae* placenta. Syncytin-Rum1 has both fusogenic and ISD, although the immunosuppressive activity has not been proved [12]. Moreover, syncytin-Rum1 was incorporated into the ruminant genome immediately after the appearance of the ruminant ancestor [12]. In contrast, Fematrin-1 does not contain an ISD, possesses only fusogenic activity and is specific for *Bovinae* [10,11]. We compared the fusogenicity of syncytin-Rum1 and Fematrin-1 in neutral acidic conditions and showed that Fematrin-1 possesses higher activity than syncytin-Rum1 [11]. Thus, we propose that the relationship between syncytin-Rum1 and Fematrin-1 is quite similar to that of human syncytin-1 and syncytin-2: that the role of syncytin-Rum1 (syncytin-2) in cell-to-cell fusion has been being replaced by Fematrin-1 (syncytin-1) in *Bovinae*. Even though no ERVs other than syncytin-Rum1 are known to be involved in cell-to-cell fusion in other ruminants, some as-of-yet unidentified ERV Env(s) or Fematrin-1 could potentially replace, or be in the process of replacing, the cell-to-cell fusion function of syncytin-Rum1. This hypothesis is supported by the observation that syncytin-Rum1 from Giraffidae, Cervidae and some of Bovidae (especially Antilopini) did not show fusogenicity [12]. This suggests that syncytin-Rum1 contributed, in an evolutionary sense, to the appearance of synepitheliochorial placenta and that subsequently, newly acquired ERVs, such as Fematrin-1, further engendered morphological variations in fetomaternal hybrid cells (TNCs or SyPs) in *Bovinae*. Unfortunately, it is difficult to reproduce such ERV-driven ruminant placental evolution, due to the lack of animal models. However, we are able to get an insight into such placental evolution by comparative analysis of ruminant placentas. *Tragulidae* species, like the mouse-deer, might be good animals to study ERV-driven ruminant placental evolution, particularly because their placentas are considered to be a primitive form of the ruminant placenta. Moreover, they are the unique as ruminants, since they lack syncytin-Rum1 [11,12,69].

Table 2. Endogenous retrovirus envelope genes supposed to play roles in placentation in ruminant.

Name of ERV	Species	Expressed Tissues	Expressed Period	Appearance in Host Genome
Syncytin-Rum1	Most ruminants (except for Tragulidae)	Placenta (BNCs)	Through gestation	Over 30 MYA
BERV-K1 (Fematrin-1)	*Subfamily Bovinae*	Placenta (BNCs)	Through gestation	25.4 to 18.3 MYA
bERVE-A	*Bos taurus*	Placenta (BNCs)	Through gestation	Not examined
BERV-P	*Genus Bos*	Trophectoderm	Periimplantation	16.9 to 7 MYA
enJSRV	*Subfamily Caprinae*	Placenta (BNCs), Endometrium	Through gestation	Within 7 MYA

It is really interesting that most mammals utilize ERV Envs to form fused cells in placentas even though the morphology of their placentas differ. Mammals also seem to have evolved the "baton pass" to employ different ERVs in placentation. This could be considered a kind of convergent evolution among mammals. Comparative studies and mechanistic investigations into ERV-host co-evolution using animal models will bring us new insights into various fields, such as virology and evolutional, reproductive and cell biology.

ACKNOWLEDGMENTS

We are grateful to Susan R. Ross for her helpful critical reading. We are also grateful to Kazuyoshi Hashizume and Katsuo Koshi for their cooperation in this study. This work was supported by the Grant-in-Aid for the Japan Society for the Promotion of Science Fellows (for Y.N., research project number 13J05281) and the Programme for Promotion of Basic and Applied Researches for Innovations Bio-oriented Industry of the Ministry of Agriculture, Forestry, and Fisheries of Japan.

AUTHOR CONTRIBUTIONS

Both Y.N. and T.M. wrote this manuscript.

CONFLICTS OF INTEREST

The authors declare no conflict of interest.

REFERENCES

1. Gifford, R.; Tristem, M. The evolution, distribution and diversity of endogenous retroviruses. Virus Genes 2003, 26, 291–315.

2. Stoye, J.P. Studies of endogenous retroviruses reveal a continuing evolutionary saga. Nat. Rev. Microbiol. 2012, 10, 395–406.

3. Tarlinton, R.E.; Meers, J.; Young, P.R. Retroviral invasion of the koala genome. Nature 2006,442, 79–81.

4. Mi, S.; Lee, X.; Li, X.; Veldman, G.M.; Finnerty, H.; Racie, L.; LaVallie, E.; Tang, X.Y.; Edouard, P.; Howes, S.; et al. Syncytin is a captive retroviral envelope protein involved in human placental morphogenesis. Nature 2000, 403, 785–789.

5. Blaise, S.; de Parseval, N.; Bénit, L.; Heidmann, T. Genomewide screening for fusogenic human endogenous retrovirus envelopes identifies syncytin 2, a gene conserved on primate evolution. Proc. Natl. Acad. Sci. USA 2003, 100, 13013–13018.

6. Dupressoir, A.; Marceau, G.; Vernochet, C.; Bénit, L.; Kanellopoulos, C.; Sapin, V.; Heidmann, T. Syncytin-A and syncytin-B, two fusogenic placenta-specific murine envelope genes of retroviral origin conserved in Muridae. Proc. Natl. Acad. Sci. USA 2005, 102, 725–730.

7. Heidmann, O.; Vernochet, C.; Dupressoir, A.; Heidmann, T. Identification of an endogenous retroviral envelope gene with fusogenic activity and placenta-specific expression in the rabbit: A new "syncytin" in a third order of mammals. Retrovirology 2009, 6, e107.

8. Cornelis, G.; Heidmann, O.; Bernard-Stoecklin, S.; Reynaud, K.; Véron, G.; Mulot, B.; Dupressoir, A.; Heidmann, T. Ancestral capture of syncytin-Car1, a fusogenic endogenous retroviral envelope gene involved in placentation and conserved in Carnivora. Proc. Natl. Acad. Sci. USA 2012, 109, E432–E441.

9. Dunlap, K.A.; Palmarini, M.; Varela, M.; Burghardt, R.C.; Hayashi, K.; Farmer, J.L.; Spencer, T.E. Endogenous retroviruses regulate periimplantation placental growth and differentiation. Proc. Natl. Acad. Sci. USA 2006, 103, 14390–14395.

10. Baba, K.; Nakaya, Y.; Shojima, T.; Muroi, Y.; Kizaki, K.; Hashizume, K.; Imakawa, K.; Miyazawa, T. Identification of novel endogenous betaretroviruses which are transcribed in the bovine placenta. J. Virol. 2011, 85, 1237–1245.

11. Nakaya, Y.; Koshi, K.; Nakagawa, S.; Hashizume, K.; Miyazawa, T. Fematrin-1 is involved in fetomaternal cell-to-cell fusion in Bovinae placenta and has contributed to diversity of ruminant placentation. J. Virol. 2013, 87, 10563–10572.

12. Cornelis, G.; Heidmann, O.; Degrelle, S.A.; Vernochet, C.; Lavialle, C.; Letzelter, C.; Bernard-Stoecklin, S.; Hassanin, A.; Mulot, B.; Guillomot, M.; et al. Captured retroviral envelope syncytin gene associated with the unique placental structure of higher ruminants. Proc. Natl. Acad. Sci. USA 2013, 110, E828–E837.

13. Wildman, D.E.; Chen, C.; Erez, O.; Grossman, L.I.; Goodman, M.; Romero, R. Evolution of the mammalian placenta revealed by phylogenetic analysis. Proc. Natl. Acad. Sci. USA 2006,103, 3203–3208.

14. Yamada, O.; Todoroki, J.; Kizaki, K.; Takahashi, T.; Imai, K.; Patel, O.V.; Schuler, L.A.; Hashizume, K. Expression of prolactin-related protein I at the fetomaternal interface during the implantation period in cows. Reproduction 2002, 124, 427–437.

15. Yamada, O.; Todoroki, J.; Takahashi, T.; Hashizume, K. The dynamic expression of extracellular matrix in the bovine endometrium at implantation. J. Vet. Med. Sci. 2002, 64, 207–214.

16. Kizaki, K.; Ushizawa, K.; Takahashi, T.; Yamada, O.; Todoroki, J.; Sato, T.; Ito, A.; Hashizume, K. Gelatinase (MMP-2 and -9) expression profiles during gestation in the bovine endometrium. Reprod. Biol. Endocrinol. 2008, 6, e66.

17. Klisch, K.; Pfarrer, C.; Schuler, G.; Hoffmann, B.; Leiser, R. Tripolar acytokinetic mitosis and formation of feto-maternal syncytia in the bovine placentome: Different modes of the generation of multinuclear cells. Anat. Embryol. 1999, 200, 229–237.

18. Klisch, K.; Hecht, W.; Pfarrer, C.; Schuler, G.; Hoffmann, B.; Leiser, R. DNA content and ploidy level of bovine placentomal trophoblast giant cells. Placenta 1999, 20, 451–458.

19. Hallenberger, S.; Bosch, V.; Angliker, H.; Shaw, E.; Klenk, H.D.; Garten, W. Inhibition of furin-mediated cleavage activation of HIV-1 glycoprotein gp160. Nature 1992, 360, 358–361.

20. Schultz, A.; Rein, A. Different recombinant murine leukemia viruses use different cell surface receptors. Virology 1985, 145, 335–339.

21. Chang, C.; Chen, P.T.; Chang, G.D.; Huang, C.J.; Chen, H. Functional characterization of the placental fusogenic membrane protein syncytin. Biol. Reprod. 2004, 71, 1956–1962.

22. Chen, C.P.; Chen, L.F.; Yang, S.R.; Chen, C.Y.; Ko, C.C.; Chang, G.D.; Chen, H. Functional characterization of the human placental fusogenic membrane protein syncytin 2. Biol. Reprod. 2008, 79, 815–823.

23. Wallin, M.; Ekström, M.; Garoff, H. Isomerization of the intersubunit disulphide-bond in ENV controls retrovirus fusion. EMBO J. 2004, 23, 54–65.

24. Henzy, J.E.; Coffin, J.M. Betaretroviral Envelope Subunits Are Noncovalently Associated and Restricted to the Mammalian Class. J. Virol. 2013, 87, 1937–1946.

25. Cheynet, V.; Oriol, G.; Mallet, F. Identification of the hASCT2-binding domain of the ENV ERVWE1/syncytin-1 fusogenic glycoprotein. Retrovirology 2006, 3, e41.

26. Watanabe, R.; Miyazawa, T.; Matsuura, Y. Cell-binding properties of the envelope proteins of porcine endogenous retroviruses. Microbes Infect. 2005, 7, 658–665.

27. Argaw, T.; Figueroa, M.; Salomon, D.R.; Wilson, C.A. Identification of residues outside of the receptor binding domain that influence the infectivity and tropism of porcine endogenous retrovirus. J. Virol. 2008, 82, 7483–7491.

28. Kim, F.J.; Manel, N.; Garrido, E.N.; Valle, C.; Sitbon, M.; Battini, J.L. HTLV-1 and -2 envelope SU subdomains and critical determinants in receptor binding. Retrovirology 2004, 1, e41.

29. Jones, K.S.; Lambert, S.; Bouttier, M.; Bénit, L.; Ruscetti, F.W.; Hermine, O.; Pique, C. Molecular aspects of HTLV-1 entry: Functional domains of the HTLV-1 surface subunit (SU) and their relationships to the entry receptors. Viruses 2011, 3, 794–810.

30. Zhang, Y.; Rassa, J.C.; deObaldia, M.E.; Albritton, L.M.; Ross, S.R. Identification of the receptor binding domain of the mouse mammary tumor virus envelope protein. J. Virol.2003, 77, 10468–10478.

31. Checkley, M.A.; Luttge, B.G.; Freed, E.O. HIV-1 envelope glycoprotein biosynthesis, trafficking, and incorporation. J. Mol. Biol. 2011, 410, 582–608.

32. Bobkova, M.; Stitz, J.; Engelstädter, M.; Cichutek, K.; Buchholz, C.J. Identification of R-peptides in envelope proteins of C-type retroviruses. J. Gen. Virol. 2002, 83, 2241–2246. [PubMed]

33. Mangeney, M.; Renard, M.; Schlecht-Louf, G.; Bouallaga, I.; Heidmann, O.; Letzelter, C.; Richaud, A.; Ducos, B.; Heidmann, T. Placental syncytins: Genetic disjunction between the fusogenic and immunosuppressive activity of retroviral envelope proteins. Proc. Natl. Acad. Sci. USA 2007, 104, 20534–20539.

34. Blond, J.L.; Lavillette, D.; Cheynet, V.; Bouton, O.; Oriol, G.; Chapel-Fernandes, S.; Mandrand, B.; Mallet, F.; Cosset, F.L. An envelope glycoprotein of the human endogenous retrovirus HERV-W is expressed in the human placenta and fuses cells expressing the type D mammalian retrovirus receptor. J. Virol. 2000, 74, 3321–3329.

35. Lyden, T.W.; Johnson, P.M.; Mwenda, J.M.; Rote, N.S. Expression of endogenous HIV-1 crossreactive antigens within normal human extravillous trophoblast cells. J. Reprod. Immunol. 1995, 28, 233–245.

36. Lin, L.; Xu, B.; Rote, N.S. Expression of endogenous retrovirus ERV-3 induces differentiation in BeWo, a choriocarcinoma model of human placental trophoblast. Placenta 1999, 20, 109–118.

37. Boyd, M.T.; Bax, C.M.; Bax, B.E.; Bloxam, D.L.; Weiss, R.A. The human endogenous retrovirus ERV-3 is upregulated in differentiating placental trophoblast cells. Virology 1993,196, 905–909.

38. Lin, L.; Xu, B.; Rote, N.S. The cellular mechanism by which the human endogenous retrovirus ERV-3 env gene affects proliferation and differentiation in a human placental trophoblast model, BeWo. Placenta 2000, 21, 73–78.

39. Langat, D.K.; Johnson, P.M.; Rote, N.S.; Wango, E.O.; Owiti, G.O.; Isahakia, M.A.; Mwenda, J.M. Characterization of antigens expressed in normal baboon trophoblast and cross-reactive with HIV/SIV antibodies. J. Reprod. Immunol. 1999, 42, 41–58.

40. Cáceres, M.; Thomas, J.W. The gene of retroviral origin Syncytin 1 is specific to hominoids and is inactive in Old World monkeys. J. Hered. 2006, 97, 100–106.

41. Esnault, C.; Cornelis, G.; Heidmann, O.; Heidmann, T. Differential evolutionary fate of an ancestral primate endogenous retrovirus envelope gene, the EnvV syncytin, captured for a function in placentation. PLoS Genet. 2013, 9, e1003400.

42. Lee, X.; Keith, J.C., Jr.; Stumm, N.; Moutsatsos, I.; McCoy, J.M.; Crum, C.P.; Genest, D.; Chin, D.; Ehrenfels, C.; Pijnenborg, R.; et al. Downregulation of placental syncytin expression and abnormal protein localization in pre-eclampsia. Placenta 2001, 22, 808–812.

43. Malassiné, A.; Blaise, S.; Handschuh, K.; Lalucque, H.; Dupressoir, A.; Evain-Brion, D.; Heidmann, T. Expression of the fusogenic HERV-FRD Env glycoprotein (syncytin 2) in human placenta is restricted to villous cytotrophoblastic cells. Placenta 2007, 28, 185–191.

44. Hayward, M.D.; Pötgens, A.J.; Drewlo, S.; Kaufmann, P.; Rasko, J.E. Distribution of human endogenous retrovirus type W receptor in normal human villous placenta. Pathology 2007,39, 406–412.

45. Esnault, C.; Priet, S.; Ribet, D.; Vernochet, C.; Bruls, T.; Lavialle, C.; Weissenbach, J.; Heidmann, T. A placenta-specific receptor for the fusogenic, endogenous retrovirus-derived, human syncytin-2. Proc. Natl. Acad. Sci. USA 2008, 105, 17532–17537.

46. Cianciolo, G.J.; Copeland, T.D.; Oroszlan, S.; Snyderman, R. Inhibition of lymphocyte proliferation by a synthetic peptide homologous to retroviral envelope proteins. Science1985, 230, 453–455.

47. Sonigo, P.; Barker, C.; Hunter, E.; Wain-Hobson, S. Nucleotide sequence of Mason-Pfizer monkey virus: An immunosuppressive D-type retrovirus. Cell 1986, 45, 375–385.

48. Tolosa, J.M.; Parsons, K.S.; Hansbro, P.M.; Smith, R.; Wark, P.A. The placental protein syncytin-1 impairs antiviral responses and exaggerates inflammatory responses to influenza.PLoS ONE 2015, 10, e0118629.

49. Tolosa, J.M.; Schjenken, J.E.; Clifton, V.L.; Vargas, A.; Barbeau, B.; Lowry, P.; Maiti, K.; Smith, R. The endogenous retroviral envelope protein syncytin-1 inhibits LPS/PHA-stimulated cytokine responses in human blood and is sorted into placental exosomes.Placenta 2012, 33, 933–941.

50. Lavialle, C.; Cornelis, G.; Dupressoir, A.; Esnault, C.; Heidmann, O.; Vernochet, C.; Heidmann, T. Paleovirology of “syncytins”, retroviral ENV genes exapted for a role in placentation. Philos. Trans. R. Soc. Lond. B Biol. Sci. 2013, 368, 20120507.

51. Nakaya, Y.; Kizaki, K.; Takahashi, T.; Patel, O.V.; Hashizume, K. The characterization of DNA methylation-mediated regulation of bovine placental lactogen and bovine prolactin-related protein-1 genes. BMC Mol. Biol. 2009, 10, e19.

52. Su, J.; Wang, Y.; Xing, X.; Liu, J.; Zhang, Y. Genome-wide analysis of DNA methylation in bovine placentas. BMC Genomics 2014, 15, e12.

53. Klisch, K.; Mess, A. Evolutionary differentiation of cetartiodactyl placentae in the light of the viviparity-driven conflict hypothesis. Placenta 2007, 28, 353–360.

54. Wooding, F.B.; Hobbs, T.; Morgan, G.; Heap, R.B.; Flint, A.P. Cellular dynamics of growth in sheep and goat synepitheliochorial placentomes: An autoradiographic study. J. Reprod. Fertil. 1993, 98, 275–283.

55. Koshi, K.; Suzuki, Y.; Nakaya, Y.; Imai, K.; Hosoe, M.; Takahashi, T.; Kizaki, K.; Miyazawa, T.; Hashizume, K. Bovine trophoblastic cell differentiation and binucleation involves enhanced endogenous retrovirus element expression. Reprod. Biol. Endocrinol. 2012, 10, e41.

56. York, D.F.; Vigne, R.; Verwoerd, D.W.; Querat, G. Nucleotide sequence of the jaagsiekte retrovirus, an exogenous and endogenous type D and B retrovirus of sheep and goats. J. Virol. 1992, 66, 4930–4939. [PubMed]

57. Carlson, J.; Lyon, M.; Bishop, J.; Vaiman, A.; Cribiu, E.; Mornex, J.F.; Brown, S.; Knudson, D.; DeMartini, J.; Leroux, C. Chromosomal distribution of endogenous Jaagsiekte sheep retrovirus

proviral sequences in the sheep genome. J. Virol. 2003, 77, 9662–9668.

58. Arnaud, F.; Caporale, M.; Varela, M.; Biek, R.; Chessa, B.; Alberti, A.; Golder, M.; Mura, M.; Zhang, Y.P.; Yu, L.; et al. A paradigm for virus-host coevolution: Sequential counter-adaptations between endogenous and exogenous retroviruses. PLoS Pathog. 2007, 3, e170.

59. Mura, M.; Murcia, P.; Caporale, M.; Spencer, T.E.; Nagashima, K.; Rein, A.; Palmarini, M. Late viral interference induced by transdominant Gag of an endogenous retrovirus. Proc. Natl. Acad. Sci. USA 2004, 101, 11117–11122.

60. Murcia, P.R.; Arnaud, F.; Palmarini, M. The transdominant endogenous retrovirus enJS56A1 associates with and blocks intracellular trafficking of Jaagsiekte sheep retrovirus Gag. J. Virol. 2007, 81, 1762–1772.

61. Arnaud, F.; Murcia, P.R.; Palmarini, M. Mechanisms of late restriction induced by an endogenous retrovirus. J. Virol. 2007, 81, 11441–11451.

62. Dunlap, K.A.; Palmarini, M.; Adelson, D.L.; Spencer, T.E. Sheep endogenous betaretroviruses (enJSRVs) and the hyaluronidase 2 (HYAL2) receptor in the ovine uterus and conceptus. Biol. Reprod. 2005, 73, 271–279.

63. Nakagawa, S.; Bai, H.; Sakurai, T.; Nakaya, Y.; Konno, T.; Miyazawa, T.; Gojobori, T.; Imakawa, K. Dynamic evolution of endogenous retrovirus-derived genes expressed in bovine conceptuses during the period of placentation. Genome. Biol. Evol. 2013, 5, 296–306.

64. Garcia-Etxebarria, K.; Jugo, B.M. Genome-wide detection and characterization of endogenous retroviruses in Bos taurus. J. Virol. 2010, 84, 10852–10862.

65. Garcia-Etxebarria, K.; Jugo, B.M. Evolutionary history of bovine endogenous retroviruses in the Bovidae family. BMC Evol. Biol. 2013, 13, e256.

66. Xiao, R.; Park, K.; Lee, H.; Kim, J.; Park, C. Identification and classification of endogenous retroviruses in cattle. J. Virol. 2008, 82, 582–587.

67. Nakaya, Y.; Miyazawa, T. Dysfunction of bovine endogenous retrovirus K2 envelope glycoprotein is related to unsuccessful intracellular trafficking. J. Virol. 2014, 88, 6896–6905.

68. Nakamura, Y.; Imakawa, K. Retroviral endogenization and its role in the genital tract during mammalian evolution. J. Mamm. Ova Res. 2011, 28, 203–218.

69. Wooding, F.B.; Kimura, J.; Fukuta, K.; Forhead, A.J. A light and electron microscopical study of the Tragulid (mouse deer) placenta. Placenta 2007, 28, 1039–1048.

CHAPTER 2

Water Stress in Small Ruminants

Lina Jaber[1], Mabelle Chedid[1] and Shadi Hamadeh[1]

[1] Department of Animal and Veterinary Sciences, Faculty of Agricultural and Food Sciences, American University of Beirut, Riad el Solh, Beirut, Lebanon

1. INTRODUCTION

Small ruminants are an integral part of farming systems in the marginal arid regions of the world. These areas are characterized by water scarcity and fluctuating precipitation; under the effect of global warming and unpredictable weather, rainfall is becoming even more irregular and water availability more limited. Along with water accessibility, feed and other resources will be markedly affected by climate change. Livestock that are able, in open range, to select high quality forage to maintain a relatively similar basal diet quality from season to season, will have their intake significantly reduced in extremely dry seasons when forage biomass and its quality are low [1]. Hence, selection of adapted animal breeds is very valuable for sustaining animal production under an increasingly challenging environment [2].

Breeds of ruminants which are well adapted to arid environments demonstrate a greater capability than non-desert breeds to endure the stressful environmental effects [3]. Although small ruminants in hot arid and semi-arid regions may survive up to one week with little or even no water, water deficiency is proved to affect animals' physiological homeostasis leading to loss of body weight, low reproductive rates and a decreased resistance to diseases [4]. In addition, under natural conditions, water scarcity often occurs at times of high environmental temperature and low feed quality and availability. Therefore, the effects of these three constraints are often confounded.

In this review, the effects of various degrees and forms of dehydration on small ruminants are presented. The findings are based on previous literature on the subject as well as the results of original research by the authors on water restriction in Awassi sheep. The major changes in physiological indicators and blood parameters are presented, in addition to the interaction of dehydration with physiological status. Finally, a brief overview of new approaches for water stress alleviation, through drugs, is exposed.

2. GENERAL CHARACTERISTICS OF SMALL RUMINANTS IN ARID AND SEMI-ARID REGIONS

Small ruminants in the Middle East and North Africa region are mostly reared under extensive and traditional pastoral farming systems, centuries-old, relying on natural pastures and mobility to secure water and feed year round. Giger-Reverdin and Gihad (1991) [5] reported that the water requirement for maintenance of goats in temperate climates is 107ml/kg $BW^{0.75}$; they also indicated that the water requirements under different ambient temperatures, based on previous work, range between 3.15 kg/kg DM (at 23°C) to 4.71 kg/kg DM (at 35°C). However, the effect of ambient temperature should be viewed in combination with the particular breed origin and adaptability, since wide differences in response to heat have been reported in different breeds [6].

Sheep breeds differ in their capacity to overcome water limitation; experiments show a variety of results: Yankasa sheep survived 5 days of water restriction but with several physiological changes [7,8]. Jaber et al. (2004) [9] concluded that Awassi females can withstand more than one month of watering every 2 days without significant changes, while a regime of watering once every five days causes important physiological perturbations. The Australian Merino sheep survived 10 days without water [10], and the desert bighorn sheep withstood water deprivation up to 15 days [11; 12], while the Barki sheep in Egypt did not endure 3 days without drinking [11].

Similarly, variations in water deprivation tolerance are observed in goat breeds. Ahmed and El Kheir (2001) [13] report that desert goats raised under traditional systems may be watered only once every 3-6 days, when water is scarce. The Black Bedouin and the Barmer goats are another example of adapted breeds that can live on a once every four days watering regime [14; 15].

Review papers are numerous on this subject, indicating differences between adapted and non-adapted sheep and goat breeds in tolerating water deprivation and the general arid conditions [6, 16].

Indigenous small ruminants are able to thrive despite extreme temperatures and limited water through their behavioral adaptations in combination with both morphological and physiological adaptations [17].

2.1. Behavioral Adaptations

Feeding behavior is affected by environmental constraints. Nocturnal feeding has been documented in bighorn sheep [18] in order to avoid high temperatures during the day. Similar behavior is also reported in goats [16]. The author also indicated that feeding frequency is modified in some adapted goats which resort to more frequent and shorter meals in order to reduce heat production associated with rumen fermentation. Langhans et al. (1991) [19] further observed that feed intake is less affected by water deprivation in adapted breeds such as pygmy goats as compared to non-adapted breeds. Furthermore, Lechner-doll et al. (1995) [20] observed that adapted goats can select high quality feed during the dry season while sheep showed less selectivity to high quality feed. Drinking behavior is also affected by water restriction whereby water deprived sheep and goats tend to drink large volumes of water in one bout upon watering. This capacity is more pronounced in goats than in sheep [5].

Timing of reproduction is another adaptive feature of sheep living in semi-arid areas: parturition is timed to a favorable period of the year to ensure offspring survival [21]. Seasonality in small ruminants is commonly observed. Reproductive cycles are thought to be regulated by environmental cues, most importantly photoperiod. Research has demonstrated that nutrition and the general body score are also important factors for normal cyclicity. Under-nutrition (below 40-60% of maintenance requirement) is reported to cause an immediate retardation in follicular growth [22]. In addition, prolonged under-nutrition induces a delay in estrous behavior that lasts for a shorter period, as compared to normally fed animals. Similarly, fasting was reported to cause major changes in the concentrations of reproductive hormones and ovulation rate [23, 24]. Since water deprivation is often accompanied by feed intake reduction, the same effects on ruminant reproductive cycles can be expected. Indigenous sheep living in the tropical and sub-tropical regions tend to breed throughout the year; however, their sexual activity may be limited, to a degree, during the summer season when the environmental temperature is elevated and feed is lacking [25]. In arid and semi-arid regions, where differences in daylight, as well as in food and water availability are well marked, the breeding season usually spans from June to November [26, 27]. Consequently, kidding and lambing mostly occur between February and April, when food and climate become more hospitable for the newborns.

2.2. Morphological Adaptations

Ruminants are usually classified as grazers, browsers or intermediate feeders. Sheep are usually classified as grazers feeding mainly on grasses while goats are intermediate feeders which can use grasses as well as shrubs [28]. These observed preferences in feed selection have been linked to underlying morphological and physiological digestive differences. These include larger rumen and long feed passage time for grazers to allow them to digest their high fiber diet, while browsers have simple and smaller digestive system with profuse saliva production to effectively process their feed high in cell solubles [28]. However, other scientists have argued that the observed differences in feed selection are related more to body size than to actual differences in the digestive anatomy and physiology [29]. Other morphological differences are also noted in relation to the mouth anatomy with goats having a mobile upper lip, while sheep are characterized by a cleft upper lip, features that allow them to best use the available vegetation. In addition, goats have the capacity to assume a bipedal position thus they are capable of browsing higher vegetation that is beyond the reach of sheep [30]. This area of research relating morphology, diet selection and utilization and response to changes in the available vegetation is still in need of further exploration [29, 31] as it is affected by the climatic changes and drought spells that put pressure on the vegetation cover and the animals that feed on it.

Morphological characteristics such as body shape and size help reducing heat loads and minimizing water losses; it is noted that goat breeds of arid and semi-arid regions are relatively smaller than their European counterparts [6]. Smaller animals benefit from a relatively larger surface area which allows them to better dissipate heat to the environment. Fleece is another feature that plays a major role in controlling body temperature, serving as a thermal barrier that reduces the effects of the ambient temperatures through the formation of a milder microclimate within the fleece [32]. In addition, fleece and hair color of small ruminants play a role in reflecting solar radiation with light colors absorbing less heat that the darker ones thus leaving the underneath skin relatively cooler [33]. Therefore, thermostability could be maintained without directly resorting to evaporative cooling (panting) which leads to high water loss [34]. Location of body fat also affects heat dissipation rates: arid-adapted sheep exhibit highly localized fat storage [35], such as in the fat-tail, as opposed to high subcutaneous fat in non-adapted breeds; this again facilitates heat conductance to the periphery for dissipation [34]. Moreover, fat-tails are important energy reserves that help in buffering long-term dietary shortfalls to maintain survival and productivity [36, 37]. The Awassi sheep, as a representative breed adapted to arid regions, presents a medium body size, with a large fat-tail and carpet type fleece. The carpet-type wool of the Awassi allows convective heat loss from the skin to the environment [34]. Their large ears are another anatomical adaptation that is thought to help in convective heat loss [34]. Finally, as mentioned above, the localization of the fat stores in the fat-tail facilitates body heat dissipation and serves as energy reservoir for times of scarcity.

2.3. Physiological Adaptations

Physiologically, ruminant breeds of arid regions show many adapted mechanisms to conserve water in times of heat and drought. Adapted breeds resort to reduction of urine volume and fecal moisture. The production of more concentrated urine is related to the length of Henlé loops located in the medulla of the kidney [38]. The thickness of the medulla is relative to kidney size, and is frequently used as an index of kidney concentrating ability [38, 39]. For instance, the desert bighorn sheep exhibits a medulla nearly twice thicker than that of other domestic sheep and thus produces highly concentrated urine of 3900 mOsm/liter H_2O [40, 41]. The Awassi sheep demonstrated a similar ability to highly concentrate urine (up to 3244 mOsm/Kg H_2O) under dehydration, and to drink large volumes upon rehydration without disrupting their homeostasis [42].

Urea renal retention is similarly increased under dehydration leading to increased urea concentration in the blood, on the other hand, urea recycling from the blood in to the gut is often observed under these conditions and is thought to contribute as a nitrogen source in times when the quality of the offered feed is low in protein [33]. The rumen is another organ that plays an important role in maintaining homeostasis under dehydration in adapted ruminants, particularly goats. Due to its relatively large volume, it acts as an important water reservoir providing most of the water lost during prolonged dehydration to maintain blood volume. It also allows the intake of large volumes of water upon rehydration which is temporarily sequestered in the rumen. Through efficient and well-coordinated mechanisms of saliva recycling and high water and Na^+ retention in the kidneys, slow rehydration is achieved without causing water toxicity and with minimal water losses. These processes are detailed in [15]. Small increases in body temperature are also observed during the hottest parts of the day, followed by body cooling at night through conduction and radiation. The capacity to tolerate this increase in temperature means that less water is needed for evaporative cooling [33].

3. PHYSIOLOGICAL CHANGES IN RESPONSE TO WATER STRESS

3.1. Effect On Feed Intake and Body Weight

Studies show a close relation between water intake and feed consumption [3, 43, 44]. Ruminant feeding behavior can be affected by the changes in osmolality of body fluids [45]. Feed intake causes hypovolemia and hyperosmolality due to the secretion of saliva and gastric juices. These mechanisms can urge ruminants, as well as other animals, to drink while eating, or alternatively not to eat when severely dehydrated [45]. Moreover, an adequate level of water intake is necessary for proper digestive functions [46]. On the other hand, Kay (1997) [33] states that drinking water is not needed for swallowing and moistening feed, since water can be circulated from the blood to maintain high salivation. However, water is needed to replace the inevitable water loss by excretion and evaporation. A possible explanation for the physiological mechanism behind the reduction in feed intake under water restriction, mainly through the reduction in meal size, may be attributed to the postprandial hyperosmolality of the ruminal fluid [19].

The co-occurrence of decrease in feed intake along with water restriction renders the differentiation between water versus feed shortage related effects difficult. Previous work has shown that Awassi sheep under 3-4 days intermittent watering regime reduced their voluntary feed intake to approximately 60% of the control [9, 47]. Similar rates are reported in [48] in different domestic ruminants subjected to dehydration, especially when combined with heat stress. The drop in feed intake under dehydration is also dependent on the type of feed that is available to the animals. Water restricted goats reduced their feed intake by 18.8% when offered legume hay compared to 21.21% when offered grass hay with lower crude protein content [49]. Therefore the negative effect of water restriction is more pronounced on low versus high quality forage [16]. This reduction in feed intake is partially compensated for by a slower feed movement and longer retention time in the digestive tract [46, 50, 51]. This is thought to lead to an increase in digestibility and nutrient utilization, as longer time is available for the microflora in the digestive tract to act on the feed [49, 52]. However, this hypothesis needs further research as reports seem inconclusive. Further drop in feed intake was recorded with increasing the degree of water restriction in South African indigenous goats [52]; but an improved nutrient utilization was also reported by the same author. Concomitantly, Ahmed Muna and El Shafei Ammar (2001) [49] reported an improvement in digestibility of Lucerne hay under water restriction in desert goats, and similarly, higher organic matter digestibility was observed in water-restricted dairy cows [53]. These adaptations allow the exploitation of grazing areas which are distant from water sources, and prevent erosion especially in regions where water is scarce and grazing pressure is high. In contrast, others [46, 48, 54] found no changes in feed digestibility in water-restricted sheep and goats. It was suggested that the elevated digestibilities usually observed are rather the result of dry matter accumulation rather than "a real increase in fermentation or digestion" [50].

The drop in feed intake puts an additional burden on the water stressed animal. In fact, in order to survive such regimes, adapted ruminants are thought to resort to lowering their metabolic rate in order to reach a new body condition with lower maintenance requirements [55]. Consequently, the effects of this decrease in dietary intake should be considered along with the effects of dehydration. Feed restriction of 50% for only a 3-day period is enough to cause metabolic changes in lactating dairy Sarda ewes [56]. It has also been reported that the depleted body condition during periods of energy deficiency reduces heat tolerance [57], which in turn affects the reproductive potential of sheep [58,59].

As reported in [9, 47, 60-64] the most obvious physiological consequence of water restriction with the concomitant reduction in feed intake is weight loss. Many trials on dry and lactating Awassi ewes recorded a drop in weight ranging between 0.84% and 26% (Table 1). Besides the effect of the water regimen, other factors contribute to body weight variation such as the physiological status of the animal (lactating or dry), its age, and the prevailing climatic conditions during the experiment (ambient temperature). It is clear in Table 1 that watering every two days did not cause a mentionable weight loss in Awassi ewes even if the temperature reached up to 32°C. The highest weight loss (26.2%) was recorded in young sheep (2-year-old ewes) and in lactating animals. Reported results lead to one conclusion that dry Awassi have a high adaptation to dehydration, and can tolerate 3-day water restriction regime, up to one month with losing only 16.8% of their body weight (Chedid et al., unpublished).

Significant weight loss is documented in other breeds of sheep and goats subjected to feed and water stress [6, 65]. Part of this weight reduction is due to body water losses [9] while the other part is caused by the consequent mobilization of fat (and possibly muscle) used for energy metabolism to compensate the decrease in dietary intake [9, 66]. Furthermore it was observed that water restriction leads to more weight loss as compared to feed restriction alone [49, 62-63] although the difference was not always statistically significant.

The following table (Table 2) presents the effect of water and feed restriction on body weight of adult dry Awassi ewes [62, 63].

Results in both studies are in compliance: water restricted animals lost more weight (approximately two fold) than those drinking every day but receiving less feed; however, large individual variations were recorded within each of the experimental groups (N=8). Therefore, further research is needed for conclusive results on the difference between water and feed restriction impact on weight loss.

Table 1. Effect of water restriction on body weight of Awassi sheep.

Physiological status	Water restriction regime	Drop in weight (%)	Age	Ambient temp. (°C)	Reference
Non-lactating	2-day-restriction	0.84	mature	15-32	[9]
	3-day-restriction	9.98	mature	27-30	[62]
		16.8	mature	23-28	[63]
		16.7	mature	27-31	[47]
		26.2	2 years	30-31	[64]
		10.4	mature	18-21	[64]
	4-day-restriction	3.32	mature	15-32	[9]
	1L on day 4 and 3L on day 8 of 12-day water restriction	22.13	mature	25-35	[61]
	1L on day 4 of 7-day water restriction	16.8	mature	23-33	Chedid et al. (unpublished)
Lactating	3-day-restriction	26.2	mature	27-31	[47]

Table 2. Effect of water restriction versus feed restriction on body weight in Awassi ewes.

	Treatment*		Reference
	Water restriction	Feed restriction	
Weight loss (%)	9.98	5.7	[62]
	17.9	8.2	[63]

*Water restricted animals received water every 3 days;

Feed restricted animals had free access to water but received 60% feed of the ad libitum intake.

3.2. Fat Metabolism: Fat Cell Diameter, Cholesterol, Glucose, Fatty Acids, Leptin And Insulin

The fat deposited during the periods of pasture abundance is mobilized and utilized for maintaining the body and sustaining production during periods of scarcity [67, 68]. The specialized fat depot represented in the fat-tail of many indigenous sheep, serves as a readily available source of energy to circumvent variation in dietary energy intake. This was well described in the Barbarine sheep subjected to long periods of undernutrition [37] as well as in the Awassi which showed a reduction in the fat-tail adipocyte diameter following an intermittent watering regime [64]. Ermias et al. (2002) [68] highlighted the importance of the location of fat depots as adaptive features to periodic fluctuations in nutrition. They noted that the rump and fat-tail depots are the most responsive under such conditions. On the other hand, Atti et al. (2004) [37] noted that subcutaneous fat is the first energy depot to be mobilized when energy intake is deficient. This is true for the fat-tailed sheep such as the Barbarine on which they conducted their study, as well as other ruminants. However, the fat-tail provided an adaptive advantage by being slowly mobilized when undernutrion is extended over a long period thus allowing long-term survival by using up this important energy store.

Fat mobilization under water restriction is further denoted by high levels of cholesterol [69] and Free Fatty Acids (FFA) in the blood [61, 64]. The increase in plasma cholesterol following water scarcity is attributed to the decrease in energy intake, along with fat metabolism [70]. These results were recorded in different ruminant species submitted to water deprivation such as Awassi [9, 47] and Yankasa ewes [8]. Furthermore, the high level of FFA outlined in water restricted Awassi [61, 64] and the Sudanese desert sheep [71] reflects lipid mobilization within the adipocyte, thus permitting lipid stores to be used as fuel when feed intake is limited [72]. Similarly, FFA levels in the blood of lactating goats correlated positively with fat mobilization in times of under nutrition [73].

Reports about the changes in glucose levels in water restricted sheep are contradicting: while no significant change was recorded by some authors [8, 9, 61, 74], a decrease in plasma glucose level was observed in Merino sheep after 24 and 48 hours of fasting [75], and in intermittently watered Awassi, although the change was not significant [47]. Glucose metabolism decreases due to the decline of propionate (the major precursor for gluconeogenesis) production in the rumen caused by low feed intake [76]. Although ruminants derive most of their energy requirements from volatile fatty acids resulting from rumen fermentation of carbohydrates, they still have an absolute need for glucose, necessitating a good homeostatic control of this compound [77].

Fat mobilization seems to be coordinated by underlying changes in the levels of key hormones. In dry Awassi ewes, subjected to an intermittent watering once every four days, FFA were negatively correlated with insulin and leptin as highlighted in [64]. The decrease in plasma insulin is probably caused by the decline in feed intake, insulin secretion being accelerated by feeding [78]. Insulin levels are thought to remain low during fasting periods in order to facilitate lipolysis [79]. Similarly, the decrease in leptin levels in dehydrated and undernourished ruminants is well documented and explained by the decrease of the metabolic status which inhibits the adipose tissue from secreting leptin [80]. Moreover, a strong correlation between leptin concentration and fat-tail adipocyte diameter was noted in [64] highlighting the relation between the secretion of leptin by the adipose tissue and body fatness [81]. Chilliard et al. (2000) [36] proposed a model whereby insulin, cortisol and leptin interact in the process of adaptation to underfeeding and re-feeding in ruminants such as experienced also under intermittent watering. The drop in leptin following undernutrition leads to a chain of events that includes stimulation of re-feeding and decrease of energy expenditure and insulin-sensitivity, that serves to re-establish homeostasis by preventing excessive lipolysis that would yield toxic levels of fatty acids, and preserve body stores to prolong survival [36, 81].

3.3. Hematology: Pcv and Hemoglobin

Dehydration in warm weather conditions reduces plasma volume as water is taken up by the tissue [82]. Although some authors reported and agreed that increased PCV and Hb concentration are good indicators of dehydration [42, 61, 71, 83, 84] results on these two parameters have been inconclusive. Even though levels of hematocrit were found to increase in Awassi [42, 60] and Merinos [61] subjected to water stress in some experiments, no variation was remarked in Yankasa [7, 8], Awassi [9] and Australian whether sheep [85] under similar watering conditions. Similar contradictory results were reported regarding hemoglobin: whilst an elevation of hemoglobin level was attributed to a decrease in plasma volume due to water loss [47, 61, 85] others did not report any variation [7-9]. These undetermined results may be an indication that adapted sheep can maintain plasma volume [86] and redistribute body water after a long water deprivation period [87].

3.4. Blood Chemistry

3.4.1. Total Protein, Globulin and Albumin

According to Caldeira et al. (2007a, b) [88, 89], serum total protein, globulin and particularly albumin are good indicators for predicting the animal's protein status. A drop in serum albumin concentration is observed in ruminants with low dietary protein intake [88, 90], followed by a decrease in globulin concentration when this dietary insufficiency is prolonged [88]. Many authors reported an increase in blood albumin and globulin in sheep under water restriction [9, 61, 74, 91, 92]. The high protein concentration is explained by the reduced plasma volume due to dehydration [93]. However, reduction in total protein and albumin was noticed after 3 days of water deprivation in Awassi [47] and in Barki sheep under water stress [94] suggesting that low feed intake is behind this reduction and that circulating proteins are being used in order to compensate for the dietary shortfall. Accordingly, maximum values of total serum protein were recorded on the 8th day of water restriction in Awassi followed by a decline at the end of the experiment (12 days) [61]. Serum albumin is a major labile protein reservoir, but it is also very important in the maintenance of body osmoregulation. Consequently, some variations in serum albumin levels can occur, but the maintenance of normal levels has to be re-established as soon as amino acids from other sources like the skeletal muscle are available [95].

3.4.2. Creatinine and Urea

Urea is mainly synthesized in the liver using NH_4^+, the end product of protein catabolism, and is released to the blood [96]. Urea is excreted by the kidneys to rid the body of the excess N intake that was not used for maintenance or production [90], or it is recycled through saliva or by reabsorption into the rumen to be utilized by rumen microflora [96, 97]. Creatinine is produced in the muscles and excreted by the kidneys in proportion to the muscle mass and the rate of proteolysis [88, 97]. The transfer function of the kidney is altered under water stress [98] with slower glomerular filtration and higher urea re-absorption [6, 8, 99].

Water stress induces a decrease in urine output and the production of dry faeces under the action of vasopressin, and increased water reabsorption from the gastro-intestinal tract [100]. Urine volume dropped by 75% and fecal water output was 37% lower in desert sheep subjected to 5 days of water restriction [101]. Consequently, urea reabsorption by the kidney is also expected to increase as reflected by increased concentration in the blood [102], which was confirmed by several trials on Merinos [65], Yankasa [7, 8] and Awassi sheep [9, 42]. When Yankasa sheep were submitted to two consecutive periods of five-day water stress, an increase in urea and creatinine concentration was observed after the first period but only creatinine levels remained high after the second [8]. Thus the author suggested that urea is being re-circulated from the blood system into the digestive tract. This is consistent with the observations that urea conservation at the level of the kidneys and recycling into the gut is increased when dietary nitrogen intake is low [102].

On the other hand, creatinine levels in lambs were not affected by 48-hour water restriction [103], while others observed an increase of this parameter in water restricted animals [47, 91]. The creatinine concentration is influenced by the level of reliance on proteolysis and endogenous N sources [88, 98] as well as by higher kidney retention due to decreased glomerular filtration rate. In turn, these factors are related to the degree of protein/N intake deficiency that the animal is experiencing as well as the level of dehydration.

Figure 1, adapted from [47], illustrates the changes over repeated cycles of three days water restriction on major blood chemistry parameters of dry Awassi ewes. This figure shows that the response to the watering regime tends to decrease over time after it reaches a certain peak. This underlines the adaptation mechanisms that the animals activate in order to achieve homeostasis under the imposed treatment. The observed return to normal values in protein and albumin concentrations could indicate that these compounds are being used up from the blood in replacement of the deficient dietary intake, or this could also denote the mobilization of water from the extracellular fluid and the rumen into the blood stream [47]. Similarly, the decline in urea at the end of the treatment could be related to the above mentioned urea recycling mechanism, while creatinine concentrations remained relatively high as previously observed in [8]. The figure also illustrates that lactating and dry animals had similar responses to water restriction with no effect of lactation (2-3 months in lactation) on these blood parameters.

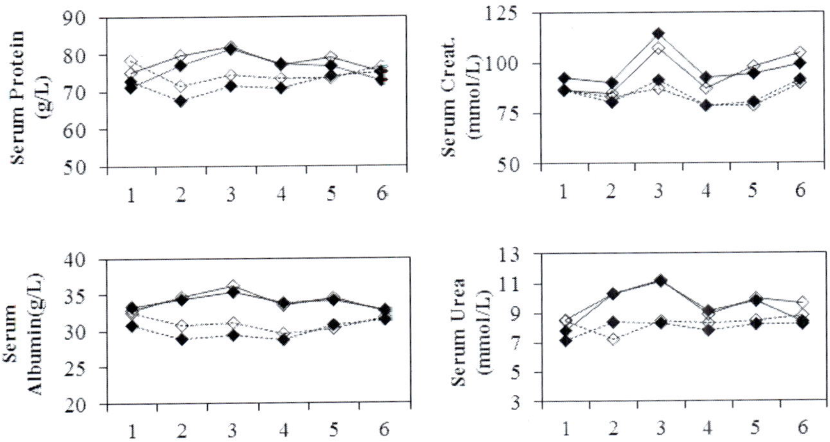

Figure 1. Changes in some serum chemistry indicators of dry (◊) and lactating (♦) Awassi ewes subjected to daily watering (—) or to water restriction (- - - -). (Adapted from [47])

3.4.3. Electrolytes and Osmolality

Analyzing osmolality is a good approach for monitoring the hydration status. It has been proved that osmolality and electrolytes levels are largely affected by water deprivation: the reduced plasma volume causes hyperosmolality inducing consequently an increase in electrolytes concentration [104] mainly sodium Na^+ and chloride Cl^- [60, 105, 106]. These results were reported in different breeds of sheep and goats [10, 63, 71, 91, 103, 107].

Increased renal retention of Na^+ is a physiological response to water restriction in different small ruminant breeds, which allows the maintenance of sodium balance in the body. Ashour and Benlemlih (2001) [87] attributed the increased renal retention to the influence of aldosterone, whereas McKinley et al. (2000) [107] added the effect of vasopressin secretion. Dehydration causes an increase in plasma vasopressin levels in both lactating goats [108-110] and non-lactating ones [108, 111]. According to Yesberg et al. (1970) [112] urinary vasopressin excretion rate is directly related to urinary osmolality and inversely related to urine flow rate. This explains why dehydrated goats and sheep decrease their urine volume while the osmolality and vasopressin levels augment. Olsson (2005) [113] noted that thermoregulation and fluid balance, as regulated by thirst control, vasopressin secretion, sodium balance and other osmotic and cardiovascular signals, seem to be centrally regulated at the hypothalamic level, particularly the preoptic and anterior hypothalamic neurons. Silanikove (1994) [15] presents a detailed account of the dynamics of major electrolytes under dehydration in connection with water conservation and homeostatic mechanisms in ruminants. In addition to the increased renal retention of water and Na^+, saliva secretion is decreased in dehydrated animals but its osmolality is increased. In parallel, the ruminants resort to utilize the large volume of water present in the gut through active transport of Na^+ across the rumen wall. This transport necessitates the presence of a minimal amount of volatile fatty acids in the rumen, hence the importance of sustaining some feed intake during dehydration. The hyperosmotic fluid absorbed from the rumen needs to be desalted the salivary flow to the rumen has to be maintained in order to preserve homeostasis.

Rehydration is an equally challenging situation for ruminants that can lead to hemolysis in non-adapted or severely dehydrated animals. Many studies reported the slow return to normal levels of blood volume, osmolality and other blood components after the rehydration of ruminants, although the animals drank large amounts of water at once [42, 92, 114]. The mechanisms allowing the slow release of the ingested water from the rumen into the blood stream are not fully understood. The production of hypotonic saliva is dramatically increased following rehydration [15] thus allowing the recycling of absorbed water and Na^+ from the blood back into the rumen to prevent a sudden drop in blood osmolality. In parallel, the kidney sustains its water/Na^+ conservation activity to prevent the loss of the ingested water which is vitally needed in anticipation of a future dehydration cycle. Rehydration also activates appetite and thermoregulatory mechanisms that allow the final restoration of homeostasis and normal functioning in up to 24 hours after rehydration or more.

Studies on different sheep breeds [115, 116] showed a negative correlation between Na^+ and K^+ in plasma. Concurrently, blood K^+ was reported to decrease in water-deprived sheep [9, 65] probably due to the intra-erythrocytic diffusion of K^+ or loss of these ions in urine in exchange of Na^+ re-absorption [8]. However, others observed an elevation in plasma K^+ under water restriction [7, 117], while [8, 47,61] did not report a variation in K^+ levels in Yankasa and Awassi sheep, respectively. These inconclusive results about K^+ alteration under water restriction do not make of potassium a reliable indicator of the hydration status; on the contrary they warrant further studies about the role played by this blood parameter during dehydration.

Chloride Cl^- is the major anion in extra cellular fluids (ECF). It functions primarily in transport processes integral to cation and water balance and as a conjugate anion in acid-base metabolism. Several findings reported that dehydration leads to an increase in plasma chloride levels in parallel to sodium levels [8, 9, 47, 74] as Cl^- is passively distributed in relation to the electrical gradients established by active Na+ transport [118]. This increase may be attributed to many phenomena such as the hemoconcentration resulting from a lower blood water level [74], and the increase in aldosterone and vasopressin concentrations [87] leading to increased renal retention.

Calcium plays an important role in regulating ion gating and as a co-factor for intermediary metabolism reactions. However, studies did not report variation in Ca^{++} under water deprivation in the Awassi [9,47] nor in Comisana sheep [74].

Finally, blood pH, a critical parameter for normal enzymatic and metabolic functions, seems to be well maintained in intermittently watered Awassi [9, 47, 61]. Increase in pH was only recorded in highly restricted Awassi following a once in a five days intermittent watering regime [9], this could be related to a combination of the high dehydration state and environmental heat which leads to hyperventilation and consequently to respiratory alkalosis as observed in other heat stressed animals due to the increased elimination of CO_2 [119].

Figure 2 is an illustration of the effect of a 12 days water restriction episode on Awassi dry ewes. The animals were offered 1 liter of water on day 4 and 3 liters on day 8 only. In this trial, the treatment had no significant effect on blood pH and K^+ concentration, while Na^+ and Cl^- were significantly increased under the restriction regime. The trial also included a group of water restricted animals that received 2.5g/d of vitamin C, the effect of which is discussed below (vitamin C section). It is worth noting here that the levels of Na^+ and Cl^- seem to have reached a peak after which they started to show a slight decline. This observation reinforces the previously proposed idea of adaptation to the restriction regime to reach a new homeostasis by activating water mobilization and conservation mechanisms to restore the blood volume and composition.

3.5. Cortisol and Other Hormones

Cortisol is a hormone secreted in order to deal with stress. It is released due to the activation of the hypothalamo-pituitary-adrenal axis by stress.

Although it plays a major role in maintaining the balance of water and electrolytes [120, 121], its mechanism is not very clear yet [122]. Dehydration had no effect on serum cortisol levels of Awassi [9, 61] and Clun forest sheep [123] which is consistent with the results obtained under laboratory conditions in sheep deprived of feed and water for 48h [124]. On the other hand, Kataria and Kataria (2004) [98] suggested that the increase in cortisol levels in dehydrated Marwari sheep (for 6 days) is a sign that the animals are under stress; they also reported that cortisol levels did not return to normal even after 72 hours of rehydration. Working on the same subject, Li et al. (2000) [85] reported plasma cortisol variation with fasting while water restriction had no additional effect on this parameter. A decline in serum cortisol was recorded in intermittently watered Awassi along the experimental period [47]. Concurring with [125], it is suggested that cortisol could be a good parameter in assessing acute stress response in small ruminants but not chronic stress such as dehydration.

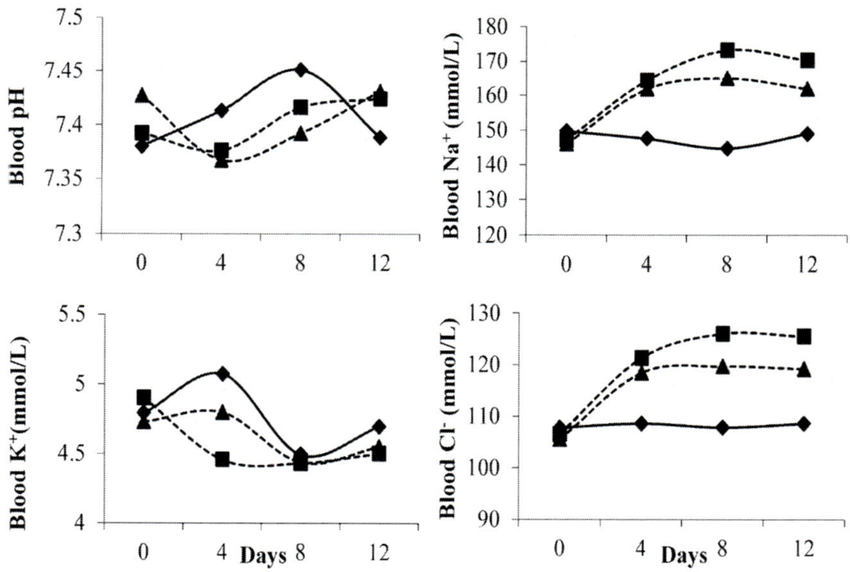

Figure 2. Effect of water restriction (- - -) on dry Awassi ewes with (▲) and without (■) vitamin C supplementation (Ghanem et al., 2008, unpublished data)

Thyroid hormones (Triiodothyronine T3 and Thyroxine T4) play a major role in many physiological events such as thermoregulation and metabolic homeostasis of energy and proteins [126-128]. T3 and T4 concentration is affected by many factors like reproductive status, climatic conditions [127, 129] nutrition, age and gender [130, 131]. Water restriction [64] and nutrient limitation [132] were found to lower the levels of T3 and T4 in dry Awassi ewes and pregnant Whiteface Western ewes, respectively. Similarly, Caldeira et al. (2007a) [88] noticed a decrease in these hormones with decreasing body score of ewes. They concluded that T3 is a good indicator of the metabolic state of the animal. On the other hand, variable T3 and T4 responses to seasonal variations and/or dehydration were reported in literature from various ruminants [133]. The concentrations of these two hormones were found to be strongly correlated (R^2=0.568; P=0.000) [64]. The authors suggested that this decline in T3 under water restriction reflected the declining metabolic state due to dehydration and decreased feed intake while the declining T4 concentration was probably a response to the thermal stress experienced by the animals in that experiment. Similarly, T4 concentrations were reported to vary with the season while T3 is affected by both the season and the physiological status of the sheep [134]. The reduction of thyroid hormone activity under dehydration is associated with the animal's attempt to minimize water losses by reducing general metabolism [135].

4. CHANGES IN RELATION WITH PHYSIOLOGICAL STATUS

As seen above, small ruminants in arid and semi-arid regions face many constraints related to fluctuating temperatures as well as shortages in feed and water sources. Pregnancy and lactation increase the needs for adaptive mechanisms due to the greater need for food, water and electrolytes in order to meet the requirements of the fetus and the mammary glands [113]. Water requirements could increase up to 50% by late pregnancy (around 160 ml/kg $BW^{0.75}$) while the requirement for milk production is stated as 165 ml/kg $BW^{0.75}$ for a milk production level of 148g milk/$kg^{0.75}$ [5]. Pregnant and lactating animals have 40-50% higher water turnover rates than dry animals [17]. However, Degen (1977) [35] recorded a small difference in water turnover between control and pregnant Awassi and Merinos sheep.

4.1. Pregnancy

The reported physiological changes vary according to the degree of water restriction and stage of pregnancy. Chokla pregnant ewes submitted to intermittent watering (every 72 and 96 hours) under semi-arid conditions showed significant hemocentration and reduction in extracellular fluid space as compared to pregnant Chokla receiving water daily [136]. Similarly, Olsson et al. (1982) [137] reported increases in plasma osmolality and Na^+ concentration in pregnant goats dehydrated for 30h accompanied by a decrease in glomerular filtration rate, while plasma protein and hematocrit did not change with dehydration. Interestingly, it was observed that pregnant goats [137] and sheep [138] have a lower capacity to concentrate urine in response to dehydration. Authors in both studies suggested that the apparent reason for this observation is a decreased sensitivity to Arginin-Vasopressin (AVP) which in turn, could be partially due to the effects of high prostaglandin concentrations which increase during late gestation. Research on the Matebele goat showed that a low nutrition level during late pregnancy had little effect on kid birth weight [139]. Similarly, a twice weekly watering regime imposed for a prolonged period had no effect on birth weight of desert adapted Magra and Marwari sheep [140]. On the other hand, pregnant Chokla ewes watered every 4 days gave birth to lambs of lower weight compared to ewes that were watered daily or every three day; however, at 12 weeks of age, lambs' weight were similar between the differently watered groups [136]. Working on goats, Mellado et al. (2006) [141] highlighted the importance of goat birth weight and weight gain at 25 days of age on their future reproductive performance under intensive conditions in hot arid environments. On the other hand, in a recent study prenatal feed restriction in the last trimester resulted in lower male offspring weight in goats, but had no effects on later behavior and growth [142]. Further studies are needed to assess the long term consequences of dehydration and/or intermittent watering during gestation on the growth and later performance of the offspring.

4.2. Lactation

Significant weight loss was recorded in lactating water-stressed Awassi [47] and Comisana ewes [74] as compared to control animals. Weight loss during lactation is due to body water loss caused by less water and feed intake combined with energy deficit that drives lactating animals to strongly depend on their body reserves [143].

Physiologically, lactating animals show lower hemoglobin concentration than dry ones [47]. According to [120], the decrease in hemoglobin concentration in lactating ewes is explained by the high water content and plasma volumes due to increased water mobilization to the mammary glands. It was found in [47] that lactation did not affect levels of serum glucose or those of cholesterol in Awassi ewes in their second to third month of lactation. Concurrently, no significant changes were recorded in glucose level caused by lactation beyond the first month [144]. Moreover, lactation did not affect blood total protein, albumin and globulin concentration of Awassi ewes in their mid-lactation [47]; no change in albumin was also reported in [145], however an increase in the gammaglobulin fraction was noticed. El-Sherif and Assad (2001) [94] observed a return to normal total protein levels on the fourth week of lactation most probably caused by a fall in globulin concentrations. The same authors reported that lactation significantly increased plasma albumin, albumin to globulin ratio and blood creatinine in Barki ewes under semi-arid conditions. No alteration was caused by lactation on serum urea and creatinine levels in Awassi ewes [47] and Corriedal [138]. Working with lactating Comisana subjected to 60% water deprivation, Casamassima et al. (2008) [74] reported significant elevations in serum concentration of triglycerides, albumin, total proteins and cholesterol. Intermittently watered Ethiopian Somali goats, exhibited similar physiological responses to those described in other breeds, namely increased osmolality, AVP secretion and blood protein concentration [110]. The authors also noted a significant capacity to fluctuate rectal temperature in response to heat stress reaching 5°C daily change in some animals; they also observed the activation of a water saving mechanism following the first cycle of water restriction, resulting in lower physiological changes in subsequent cycles. A similar trend was also noted in lactating Awassi ewes subjected to intermittent watering [47].

Concerning the effect of lactation on blood pH, Hamadeh et al. (2006) [47] reported significantly higher blood pH in water restricted lactating Awassi ewes as compared to dry ones. This increase in pH is correlated to the decrease in plasma Ca^{++} and K^{+} levels due to their need for milk production [47] and to the increase in plasma Na^{+} and Cl^{-} since Na^{+} is used for nutrient transport [146]. As for cortisol, lactating Awassi ewes tended to have high levels as compared to dry counterparts. However, more work is needed in order to elucidate the adaptive mechanisms of pregnant and lactating ewes to water restriction.

4.3. Milk Production and Composition

Water needs are reported to be the highest during lactation as compared to other physiological statuses [109]. Dehydration leads to a reduction in the blood flow to the mammary gland however, enough supply for milk production may still be achieved as supported by the sustained milk production in dehydrated goats [113] and sheep [74]. During dehydration, milk volume is generally decreased [109,110, 147], although the Bedouin goats could maintain their milk production when watered every second day [108]. Dahlborn et al. (1997) [148] suggested that the drop in milk volume observed in some dehydrated animals was mainly the result of lack of water as such and not a reflection of the consequent decrease in feed intake; furthermore the authors suggested that this drop could be related to the alteration in casein production observed under dehydration. In fact, Silanikove (2000) [6] showed that stress leads to a chain of events including increased cortisol secretion leading the activation of the plasmin system resulting in the release of a proteose peptone with channel blocking activity (PPCB) from β-casein and interfering with lactose secretion into the lumen of the mammary gland and consequently causing a drop in production. Previous studies have indicated an increase in milk osmolality, lactose and density following water deprivation for 48 hours [109, 149, 150]. Milk osmolality is strictly controlled to keep it isotonic with plasma; the increase in lactose under water restriction, being the major osmotic component of milk, is probably a response to the increase in serum osmolality under water restriction [149]. Nonetheless, milk can be less concentrated in fat and non-fat solids and more rich in water. Alamer (2009) [150] noted a decrease in fat content of milk of 25% water restricted goats but not in those that were 50% restricted. The increase in water content [151,152] is thought to be a form of adaptation allowing the offspring to receive the adequate quantity of water when water is not available.

5. STRESS ALLEVIATION DRUGS

Domestic animals are routinely faced with different stressors. Most stressful conditions, including diseases [153] and farming practices such as milking [143], isolation [154], introduction to a new flock [155], road transportation [156, 157] last for only a short period ranging between hours to days. On the other hand, harsh environmental factors are probably the stressors with the longest lasting effect since they may prevail for months. For example, heat stress and elevated ambient temperature are considered major risks affecting sheep performance [85; 158]. The negative impact of heat is translated into increased body temperature, higher respiration and heart rates followed by a drop in feed intake, redistribution in blood flow and alteration in endocrine function [158]. Low temperatures or cold leads to an equally stressful situation which affects sheep performance by increasing the metabolic rate [85,160]. Furthermore, different environmental constraints often come together such as the situation in arid and semi-arid areas during the dry season when heat stress is combined with water scarcity and low pasture quality.

Animal producers and researchers have looked for ways to alleviate the negative effects of common stressors. Stress alleviation strategies are numerous, and their availability to producers depend on the access to water and energy, the price they are able to pay and the adopted farming system [153]. These strategies vary from simple on-farm practices such as modifying the feeding pattern, feed composition [160], water management, cooling systems and environmental modifications like shading [153, 161], protection from solar radiation [153], the use of micro-sprinklers, spray jets and ventilation [162] to more scientific procedures like genetic selection [161] and others.

In this quest for stress alleviation in domestic animals, researchers have tested special drugs and/or nutritional supplements. Trials showed that a pre-transportation administration of ascorbic acid to goats facilitates the transition from depression to excitation; it exhibited potential depression amelioration after road trips [163] and significantly decreased weight loss caused by transportation under unfavorable thermal conditions [164]. Ali et al. (2005) [156] reported that a single dose of the anti-stressor xylazine administered to sheep and goats before road transportation considerably ameliorated the effects induced by the stressful stimulus; whereas a pretreatment with sodium betaine (a test compound) had no significant effects. Electrolyte therapy was also found effective in reducing stress of transportation in market cattle allowing a better meat quality and a reduction in live weight loss [157].

In this review we will focus on two compounds that appear to have a good potential for stress remediation in domestic animals: Vitamin C which is tolerated at high doses without apparent side effects [165] and aspirin which showed some potential advantages warranting further investigation [166-168].

5.1. Vitamin C

Although ruminants biosynthesize ascorbic acid under normal conditions and do not need any additional supplementation [45], and even though vitamin C administration is not a common practice in adult livestock nutrition [169] scientists however, decided to study the effect of Vitamin C administration to sheep under water-stress conditions [60, 61, 64, 92] and goats facing the stress of transportation [163, 164]. These trials were encouraged by promising results obtained on weaned pigs [170], Japanese quails [171], rabbits [172] and broilers [173] under stress.

Ascorbic acid is known for its function as an antioxidant mainly due to its redox properties; it acts as a free radical scavenger in numerous cellular oxidation processes [174] and has been demonstrated to be helpful for young ruminants in acclimatizing to cold stress [175]. Ascorbic acid plays an important role in modulating the immune response by enhancing neutrophil function and minimizing free radical damage [176] and by improving antibody response to antigen [177]. Concurrently, Minka and Ayo (2007) [164] reported that administering Vitamin C to Red Sokoto goats before transportation reduced the post-journey effect to a minimum or eliminated it completely; however the impaired animals' homeostasis was rapidly recovered after the trip.

A daily dose of Vitamin C was reported to decrease weight loss in adult female Awassi subjected to a water-restricted regime [61, 63] while this effect was not significant in other trials [63, 64, 92]. The alleviation of body weight loss can be explained by improved feed intake [92] and better feed conversion, also observed in other animals in stressful conditions and supplemented with vitamin C [164, 171, 178]. The ameliorated effect of ascorbic acid on weight loss during short-term transportation of goats in hot weather [164, 179] proves the advantage of Vitamin C supplementation in order to maintain an adequate live body weight for slaughter.

Vitamin C supplementation to sheep alleviated the effect of dehydration on PCV [61] but not on hemoglobin [61, 63]; while in goats submitted to transportation under unfavorable climate conditions, Vitamin C significantly decreased levels of both PCV and Hb [164]. It was also found that supplementing Vitamin C to deficient pigs increased hemoglobin levels probably due to increased iron absorption [180].

Lower serum protein concentrations were reported in Vitamin C administered water-restricted Awassi as compared to non-supplemented counterparts [61], while others found no significant differences in total protein and globulin levels due to Vit C [92]. The effect of vitamin C on albumin levels is inconclusive as well: although some [61] noted lower concentrations in treated ewes, others [47] reported higher values of albumin in treated animals. The daily Vitamin C dose also plays a considerable role on blood parameters: a daily dose of 5g significantly increased serum creatinine and urea concentrations as compared to 3g and to control [92]. This might be considered as an enhancement to the adaptive mechanisms of Awassi sheep to water restriction. However, in other experiments no effect of Vitamin C was observed on creatinine levels [63] warranting further investigations on the role of Vitamin C in urea and creatinine dynamics during dehydration.

In an experiment conducted in order to study the effect of Vitamin C on fat mobilization under water stress, Jaber et al. (2011) [64] reported no significant effect of this drug administration on adipocyte diameter, fat mobility and weight loss in water deprived sheep. Authors speculated that a daily dose of Vitamin C (3g or more) may be more helpful in increasing fat mobilization under water stress than single high dosages; they agreed that more work would be essential to confirm the observed trends.

A tendency for higher cholesterol levels was observed under Vitamin C administration [60, 61, 63]. Vitamin C interferes in norepinephrine formation, an important hormone that increases fat mobilization [172], it is also essential in carnitine formation, which upon reacting with acetyl CoA forms acetylcarnitine that transports fatty acids into the mitochondria to be oxidized [60].

Reports on the effect of Vitamin C on osmolality are scarce. However, Karnib (2009) [63] and Hanna (2006) [106] observed increased osmolality in water-stressed vitamin C supplemented Awassi ewes, but the mechanism of such phenomenon is still not well understood. Results obtained on the effect of Vitamin C on blood electrolytes are not very clear and need more elaborated work. Ghanem et al. (2008) [60] reported that Vitamin C administration alleviated the effect of water restriction as reflected in lower Na^+ and Cl^- (Figure 2). The authors attributed this observation to the role of vitamin C in norepinephrine formation, which affects the kidney function and therefore water and electrolytes dynamics. On the other hands, [72] reported an opposite result in one experiment while in a second experiment Na^+ and Cl^- were the same between supplemented and un-supplemented water restricted ewes.

Benefits of Vitamin C in decreasing stress hormones was reported by several authors: some [181] observed that ascorbic acid intake resulted in a drop in adrenal and plasma corticosterone levels, others showed that vitamin C eliminated the secretion of cortisol in animals subjected to stress [182]. Still others did not observe any effect of Vitamin C administration on cortisol concentrations in stressed animals [61, 122, 124]. Serum cortisol is a better marker for acute stress than for chronic stress [125]. While Vitamin C and cortisol interact, the anti-cortisol role of Vitamin C is still unclear and needs more research.

Concluding with Vitamin C, it has been shown that the most important parameter in highlighting the role of Vitamin C in counteracting the effect of water deprivation on sheep is the observed decrease in weight loss. Consequently, further work is warranted in order to elucidate the mechanisms of action of Vitamin C, and to determine the best dose recommendations that would increase the adaptive capacities of shepherds and flock keepers to the changing weather and increasing global warming.

5.2. Aspirin

Acetyl-salicylic acid ASA or aspirin has been used for ages as an antipyretic and analgesic agent [183]. New studies have emphasized the role of aspirin in the treatment of some types of cancer [168] and cardio-vascular diseases [184] although data on the effect of aspirin in animal production is contradictory. On one hand, supplementation of 20ppm of ASA to layer chicken under hot climates improved the number of eggs as well as their weight; it increased feed intake, improved fertility and hatchability [185]. On the other hand, chronic feeding of ASA had detrimental results to layer breeders with concerns for early hen livability and egg quality [186]. In mammals, aspirin reduced scouring and improved growth rate when supplemented to weanling pigs at a level of 125 or 250 ppm [166]. It also reduced plasma cholesterol levels in rats [167], and protected rats from colon cancer [168]. Recently, the role of aspirin in the protection from oxidative stress has been highlighted.

The role of aspirin in alleviating stress has been investigated in adult Awassi ewes subjected to water and feed restriction has been studied [62]. Treated animals with a daily dose of 100mg of ASA lost more weight than the untreated animals. However, the difference was not statistically considerable. Similar results had been reported in broilers [187] and weanling pigs [166].

Aspirin did not have any effect on rectal temperature; perhaps these results highlight the capacity of Awassi to remain thermostable even under dehydration [10, 35]. No changes were detected in PCV and hemoglobin levels [62] confirming that salicyclates do not usually alter these two parameters [188]. Moreover, results of the experiment revealed that aspirin has no effect on plasma concentrations of proteins, globulin and albumin. Similarly, no significant differences in the levels of urea, creatinine and osmolality were observed between treated and untreated animals. Furthermore, no significant effect of aspirin was observed on any fat mobilization indicator i.e. cholesterol, insulin, free fatty acids and leptin.

Other studies investigating the role of aspirin in water and feed stressed ruminants are not found. The available literature covers stress resulting from transportation and exposition to new environment [189] and physical pain caused by tail docking [190].

Further studies using different doses of aspirin could be done. Additional experiments might also help clarify the antioxidant property of aspirin and its effect on the ruminant immune system which might be compromised under stress.

6. CONCLUSION

This review helped in highlighting the adaptability of indigenous small ruminant breeds to water stress and the changes it induces under different physiological statuses. Most small ruminants respond to water stress by decreasing their feed intake, resulting in weight reduction due to water and body mass loss. The rumen plays an important role as water reservoir both in times of dehydration, to maintain blood volume, and upon rehydration to prevent hemolysis. Similarly, modulating saliva production and osmolality is an important mechanism for facing dehydration and rehydration cycles.

A ten-year research track on the Awassi demonstrated the key mechanisms that this breed activates in facing water stress. Strong water conservation is achieved at the level of the kidney, as reflected by a drop in urine output and increased blood Na^+, albumin and urea along with hyperosmolality. Furthermore, the Awassi seems to adapt to an intermittent watering regime after a couple of cycles by re-adjusting its blood volume and constituents to a new status tending toward control levels. Finally, the Awassi mobilizes its fat stores, including the fat-tail, to overcome the shortfall in dietary energy intake. This breed could be maintained, during the hot months and in times of severe water shortage, on short intermittent watering regimes. However, severe dehydration will ultimately result in detrimental effects on milk production, reproductive success, lambs' weight gain and disease resistance, particularly during gestation and peak lactation.

Stress alleviating supplements such as vitamin C show some promise in decreasing the effects of dehydration. Many tracks still need to be explored in future research such as feeding and nutritional manipulations to alleviate water stress, with special attention to the long-term effects of such approaches on overall productivity and welfare. The identification of breeds that show high adaptability to arid and semi-arid regions with an acceptable level of productivity is also important.

REFERENCES

1. Migongo-Bake W. Rumen dry-matter digestive efficiency of camels, cattle sheep and goats in a semi-arid environment in eastern Africa. Rome, Italy: FAO, The complementarity of feed resources for animal production in Africa; 1992.

2. Iñiguez, L. Small ruminant breeds in West Asia and North Africa. ICARDA Caravan 2005; Issue 22. http://www.icarda.org/Publications/Caravan/Caravan22/Focus_3.htm (accessed 15 August 2012).

3. Silanikove N. Effects of water scarcity and hot environment on appetite and digestion in ruminants: a review. Livestock Production Science 1992;30: 175-194.

4. Barbour E, Rawda N, Banat G, Jaber L, Sleiman FT, Hamadeh S. Comparison of immunosuppression in dry and lactating Awassi ewes due to water deprivation stress. Veterinary Research Communications 2005;29(1) 47-60.

5. Giger-Reverdin S, Gihad EA. Water metabolism and intake in goats. In: Morand-Fehr P. (ed.) EAAP 1991: Goat Nutrition: proceedings of EAAP, 24–26 April 1991, Pudoc Wageningen, Wageningen.

6. Silanikove N. The physiological basis of adaptation in goats to harsh environments. Small Ruminant Research 2000;35: 181– 193.

7. Aganga AA, Umunna NN, Oyedipe EO, Okoh PN. Influence of water restriction on some serum components in Yankasa ewes. Small Ruminant Research 1989;2: 19–26.

8. Igbokwe IO. Haemoconcentration in Yankasa sheep exposed to prolonged water deprivation. Small Rumin. Res. 1993;12: 99–105.

9. Jaber LS, Habre A, Rawda N, AbiSaid, M, Barbour EK, Hamadeh SK. The effect of water restriction on certain physiological parameters in Awassi sheep. Small Ruminant Research 2004;54: 115– 120.

10. MacFarlane W.V. Terrestrial animals in dry heat: ungulates. In: Dill D.B., Adolph E.F.A., Wilberg C.C. (eds) Handbook of Physiology, Section 4: Adaptation to the environment. American Physiology Society; 1964. p509-539.

11. Farid MF, Shawket SM, Abdel-Rahman MHA. Observations on the nutrition of camels and sheep under stress. In: Proceedings of workshop on camels, International Foundation of Science, Sweden, Kartoum, Sudan, 1979.

12. Turner JC. Osmotic fragility of desert bighorn sheep red blood cells. Comparative Biochemistry and Physiology 1979;A(64A) 167-175.

13. Ahmed MM, El Kheir IM. Thermoregulation and water balance as affected by water and food restrictions in Sudanese desert goats fed good-quality and poor-quality diets. Tropical Animal Health and Production 2004;36(2) 191-204.

14. Khan MS, Ghosh PK, Sasidharan TO. Effect of acute water restriction on plasma proteins and on blood and urinary electrolytes in barmer goats of the rajasthan desert. Journal of Agricultural Science Cambridge 1978;2(2) 395-398.

15. Silanikove N. The struggle to maintain hydration and osmoregulation in animals experiencing severe dehydration and rapid rehydration: the story of ruminants. Experimental Physiology 1994;79: 281-300.

16. Morand-fehr P. Recent developments in goat nutrition and application: A review. Small Ruminant Research 2005;60(1–2) 25–43.

17. Cain JW III, Krausman P, Rosenstock A, Turner J. Literature review and annotated bibliography: Water requirements of desert ungulates; 2005. Southwest Biological Science Center.

18. Miller GD, Cochran MH, Smith EL. Nighttime activity of desert bighorn sheep. Desert Bighorn Council Transactions 1984;28: 23-25.

19. Langhans W, Scharre E, Meyer AH. Changes in feeding behavior and plasma vasopressin concentration during water deprivation in goats. Journal Veterinary Medicine 1991;A(38) 11-20.

20. Lechner-Doll M, von Engelhardt W, Abbas HM, Mousa L, Luciano L, Reale E. Particularities in forestomach anatomy, physiology and biochemistry of camelids compared to ruminants. In: Tisserand JL (ed) Elevage et alimentation du dromadaire–Camel production and nutrition. Options Méditerranéennes CIHEAM, Paris 1995;B(13): 19–32.

21. Skinner JD, van Jaarsveld AS. Adaptive significance of restricted breeding in southern African ruminants. South African Journal of Science 1987;83: 657-663.

22. Blanc F, Bocquier F, Debus N, Agabriel J, D'hour P, Chilliard Y. La pérennité et la durabilité des élevages de ruminants dépendent des capacités adaptives des femelles. INRA Production Animale 2004;17: 287-302.

23. Al-Azraqi AA. Effect of fasting on luteal function, leptin and steroids concentration during oestrous cycle of the goat in natural photo-status. Animal Reproduction Science 2007;98: 343-349.

24. Kosior-Korzecka U, Bobowiec R, Lipecka C. Fasting-induced changes in ovulation rate, plasma leptin, gonadotropins, GH, IGF-I and insulin concentrations during oestrus in ewes. Journal of Veterinary Medicine Series A – Physiology, Pathology and Clinical Medicine 2006;53: 5–11.

25. Marai IFM, El- Darawany AA, Fadiel A, Abdel-Hafez MAM. Reproductive performance traits as affected by heat stress and its alleviation in sheep. Tropical and Subtropical Agroecosystems 2008; 8(3): 209-234.

26. Hamadeh SK, Shomo F, Nordblom T, Goodchild A, Gintzburger G. Small ruminant production in Lebanon's Beka'a Valley. Small Ruminant Research 1996;21(3) 173-180.

27. Amoah EA , Gelaye S, Guthrie P, Rexroad Jr. CE. Breeding season and aspects of reproduction of female goats. Journal of Animal Science 1996;74: 723-728.

28. Hofmann RR. Evolutionary steps of ecophysiological adaptation and diversification of ruminants: a comparative view of their digestive system. Oecologia 1989;78: 443-457.

29. Gordon IJ. Browsing and grazing ruminants: are they different beasts? Forest Ecology and Management 2003;181: 13–21

30. Jonsson H. *Foraging behaviour of cattle, sheep and goats on semi-arid pastures in Kenya. Veterinary Medicine thesis. Swedish University of Agricultural Sciences SLU; 2011. p.1-19.*

31. Shipley LA. Grazers and browsers: how digestive morphology affects diet selection. In: Launchbaugh K.L., Sanders K.D., Mosley J.C. (ed). Grazing behavior of livestock and wildlife. Idaho Forest, Wildlife and Range Experiment Station Bulletin No. 70; 1999. p.20-27.

32. Eyal E. Shorn and unshorn Awassi sheep. Body temperature. Journal of Agriculture Sciences 1963; 60: 159-176.

33. Kay RNB. Responses of African livestock and wild herbivores to Drought. Journal of Arid Environments 1997;37: 683–694.

34. Degen AA, Shkolnik A. Thermoregulation in fat-tailed Awassi, a desert sheep, and in German Mutton Merino, a mesic sheep. Physiological Zoology 1978;51: 333–339.

35. Degen AA. Fat-tailed Awassi and German Mutton Merino sheep under semi-arid conditions III: body temperature and panting rate. Journal of Agricultural Science 1977;89: 399-405.

36. Chilliard Y, Ferlay A, Faulconnier Y, Bonnet M, Rouel J, Bocquier. Adipose tissue metabolism and its role in adaptations to undernutrition in ruminants. Proceedings of the Nutrition Society 2000;59: 127-134.

37. Atti N, Bocquier F, Khaldi G. Performance of the fat-tailed Barbarine sheep in its environment: adaptive capacity to alternation of underfeeding and refeeding periods. A review. Animal Research. 2004;53: 165-176.

38. McNab BK. The physiological ecology of vertebrates. Comstock Publishing Associates, Cornell University Press, Ithaca, New York, USA; 2002.

39. Zervanos SM. Renal structural adaptations among three species of peccary. The Southwestern Naturalist 2002;47(4) 527-531.

40. Horst R, Langworthy M. Observations on the kidney of the desert bighorn sheep. Anatomical Record 1971;2: 343.

41. Turner JC. Water, energy and electrolyte balance in the desert bighorn sheep, Ovis Canadensis. Ph. D. Thesis. University of California, Riverside, California, USA; 1973. p.1-276.

42. Laden S, Nehmadi L, Yagil R. Dehydration tolerance in Awassi fat-tailed sheep. Canadian Journal of Zoology 1987;65: 363–367.

43. Forbes JM. The water intake in ewes. British Journal of Nutrition 1997;22: 33-34.

44. More T, Howard B, Siebert BD. Effect of level of water intake on water, energy and nitrogen balance and thyroxine secretion in sheep and goats. Australian Journal of Agricultural Research 1983;34: 441-446.

45. National Research Council of the National Academies. Nutrient requirements of small ruminants : sheep, goats, cervids, and New World camelids / Committee on Nutrient Requirements of Small Ruminants, Board on Agriculture and Natural Resources, Division on Earth and Life Studies, National Research Council of the National Academies; 2007.

46. Hadjigeorgiou I, Dardamani K, Goulas C, Zervas G. The effect of water availability on feed intake and digestion in sheep. Small Ruminant Research 2000;37(1/2) 147-150.

47. Hamadeh SK, Rawda N, Jaber LS, Habre A, Abi Said M, Barbour EK. Physiological responses to water restriction in dry and lactating Awassi ewes. Livestock Science 2006;101(1-3) 101-109.

48. Maloiy GMO, Kanui TI, Towett PK, Wambugu SN, Miaron JO, Wanyoike MM. Effects of dehydration and heat stress on food intake and dry matter digestibility in East African ruminants. Comparative Biochemistry and Physiology 2008;A(151): 185–190.

49. Ahmed Muna MM, El Shafei Ammar I. Effects of water and feed restriction on body weight change and nitrogen balance in desert goats fed high and low quality forages. Small Ruminant Research 2001;41(1; 1) 19-27.

50. Asplund JM, Pfander WH. Effects of water restriction on nutrient digestibility in sheep receiving fixed water: feed ratios. Journal of Animal Science 1992;6: 1271-1274.

51. Musimba NKR, Pieper RD, Wallace JD, Galyean ML. Influence of watering frequency on forage consumption and steer performance in Southeastern Kenya. Journal of Range Management 1987;40(5) 412-415.

52. Ajibola A. The effect of water deprivation and atropine administration on gastro-intestinal function in goats. M.Sc thesis. University of Pretoria; 2000.

53. Burgos MS, Senn M, Sutter F, Kreuzer M, Langhans W. Effect of water restriction on feeding and metabolism in dairy cows. *American Journal of Physiology: Regulatory, Integrative and Comparative Physiology* 2001;280: R418–R427.

54. Misra AK, Singh K. Effect of water deprivation on dry matter intake, nutrient utilization and metabolic water production in goats under semi-arid zone of India. Small Ruminant Research 2002;46(2;2) 159-165.

55. Gómez-Pastén M, Mora O, Pedraza-Chaverri, Shimida A. The effect of a long term feed restriction on metabolism and tissue composition of goats. The Journal of Agricultural Science 1999;132: 227-232.

56. Pulina G, Bomboi G, Mazzette FB, Dimauro C, Rassu SPG, Nudda A. Changes in metabolic and endocrine measurements during feed restriction in dairy ewes with different BCS. Journal of Animal Science 2007;85(supplement 1) 661.

57. Hooda OK, Naqvi SMK. Effect of thermal load and feed restriction on relative adaptability of Malpura and Avikalin sheep in semi-arid region. Indian Journal of Animal Science 1990;60: 608–611.

58. Rhind SM, Mc Neilly AS. Follicle populations, ovulation rates and plasma profiles of LH, FSH and prolactin in Scottish Blackface ewes in high and low levels of body condition. Animal Reproduction Science 1986;10: 105–115.

59. Maurya VP, Naqvi SMK, Mittal JP. Effect of dietary energy balance on physiological responses and reproductive performance of Malpura sheep in the hot semi-arid regions of India. Small Ruminant Research 2004;55(1-3) 117-122.

60. Ghanem AM, Barbour EK, Hamadeh SK, Jaber LS, Abi Said M. Physiological and chemical responses in water-deprived Awassi ewes treated with vitamin C. Journal of Arid Environments 2008;72: 141-149.

61. Ghanem A. The effect of vitamin C supplementation on some physiological an immunological indicators in water-deprived Awassi ewes. Master thesis. American University of Beirut; 2005. p.1-89

62. Chedid M. Physiological Responses of Feed and Water Restricted Dry Awassi Ewes to Aspirin Administration. Master thesis. American University of Beirut; 2009. p.1-64.

63. Karnib M. The Effect of Vitamin C Administration on some Physiological Parameters in Water and Feed Restricted Dry Awassi Ewes. Master Thesis, American University of Beirut; 2009. p.1-48

64. Jaber LS, Hanna N, Barbour EK, Abi Said M, Rawda N, Chedid M, Hamadeh SK. Fat mobilization in water restricted Awassi ewes supplemented with vitamin C. Journal of Arid Environment 2011;75(7) 625-628.

65. MacFarlane WV, Morris RJH, Howard B, McDonald J, Budtz-Olsen OE. Water and electrolyte changes in tropical Merino sheep exposed to dehydration during summer. Australian Journal Agricultural Research 1961;12(5) 889– 912.

66. Epstein H. The Awassi sheep with special reference to the improved dairy type. FAO Animal and Health Paper 1985;57: 34-38.

67. Acharya RM. Small ruminant production in arid and semi-arid Asia. In: Timon VM, Hanrahan JP. (eds). Small ruminant production in the developing countries: Proceedings of an Expert Consultation, 8–12 July 1985, Sofia, Bulgaria. FAO animal production and health paper 1986;58 148-169.

68. Ermias E, Yami A, Rege JEO. Fat deposition in tropical sheep as adaptive attribute to periodic feed fluctuation. Journal of Animal Breeding and Genetics 2002;119(4) 235-246.

69. Savendahl L, Underwood LE. Fasting increases serum total cholesterol, LDL cholesterol and apolipoprotein B in healthy, nonobese humans. Journal of Nutrition 1999;129: 2005-2008.

70. Umunna NN, Chineme CN, Saror DI, Ahmed A, Abed S. Response of Yankasa sheep to various lengths of water deprivation. Journal of Agricultural Science 1981;96: 619-622.

71. Abdelatif AM, Ahmed MMM. Water restriction, thermoregulation, blood constituents and endocrine responses in Sudanese desert sheep. Journal of Arid Environment 1994;26: 171–180.

72. Varady KA, Roohk DJ, Loe YC, McEvoy-Hein BK, Hellerstein M.K. Effects of modified alternate-day fasting regimens on adipocyte size, triglyceride metabolism, and plasma adiponectin levels in mice. Journal of Lipid Research 2007;48: 2212-2219.

73. Dunshea FR, Bell AW, Trigg TE. Relations between plasma nonesterified fatty acid metabolism and body tissue mobilization during chronic undernutrition in goats. British Journal of Nutrition 1988;60: 633–644.

74. Casamassima D, Pizzo R, Palazzo M, D'alessandro AG, Martemucci G. Effect of water restriction on productive performance and blood parameters in Comisana sheep reared under intensive condition. Small Ruminant Research 2008;78: 169–175.

75. Annison EF, White RR. Glucose utilization in sheep. Biochemistry Journal 1961;80: 162-169.

76. Bergman E. N. Production and utilization of metabolites by the alimentary tract as measured in portal and hepatic blood. In: MacDonald I.W., Warner A.C.I (Ed.) Digestion and Metabolism in

the Ruminants. University of New England Publishing Unit; 1975. p 292–305.

77. McDowell GH. Hormonal control of glucose homoeostasis in ruminants. In: Westerterp K.R. (ed): Proceedings of the Nutrition Society 1983;42(2) 149-167.

78. Bassett JM. Dietary and gastro-intestinal control of hormones regulating carbohydrate metabolism in ruminants. In: MacDonald I.W., Warner A.C.I (Ed.) Digestion and Metabolism in the Ruminants. University of New England Publishing Unit; 1975. p383–398.

79. Vernon RG. Effects of diet on lypolysis and its regulation. In: Westerterp K.R. (ed): Proceedings of the Nutrition Society 1992;51(3) 397-408.

80. Houseknecht KL, Portocarrero CP. Leptin and its receptors: Regulators of whole-body energy homeostasis. Domestic Animal Endocrinology 1988;15: 457475.

81. Chilliard Y, Delavaud C, Bonnet M. Leptin expression in ruminants: Nutritional and physiological regulations in relation with energy metabolism. Domestic Animal Endocrinology 2005;29(1) 2-33.

82. Schaefer A L, Jones SD M, Tong AKW, Lepage P, Murray NL. The effects of withholding feed and water on selective blood metabolites in market-weight beef steers. Canadian Journal of Animal Science 1990;70: 1151–1158.

83. Dahlborn K, Holtenius K, Olsson K. Effects of intraruminal loads of saline or water followed by voluntary drinking in the dehydrated lactating goat. Acta Physiologica Scandinavica 1988;132: 67–73.

84. Olsson K, Dahlborn K. Fluid balance during heat stress in lactating goats. Quarterly Journal of Experimental Physiology 1989;(74) 645–659.

85. Li B T, Christopherson R J, Cosgrove S.J. Effect of water restriction and environmental temperatures on metabolic rate and physiological parameters in sheep. Canadian Journal of Animal Science 2000;80: 97–104.

86. Sneddon JC. Physiological effects of hypertonic dehydration on body fluid pools in arid-adapted mammals. How do Arab-based horses

compare. Comparative Biochemistry and Physiology 1993;104(A) 201–213.

87. Ashour G, Benlamlih S. Adaptation of Mediterranean breeds to heat stress and water deprivation. In: Guessous F, Rihani N, Ilham A. (eds.), Livestock Production and Climatic Uncertainty in the Mediterranean: Proceedings of the Joint ANPA-EAAPCIHEAM-FAO Symposium. Wageningen Pers, Wageningen; 2001.

88. Caldeira A, Belo C, Santos M, Vazques A, Portugal AV. The effect of body condition score on blood metabolites and hormonal profiles in ewes. Small Ruminant Research 2007(a);68(3) 233–241.

89. Caldeira A, Belo C, Santos M, Vazques A, Portugal AV. The effect of long-term feed restriction and over-nutrition on body condition score, blood metabolites and hormonal profiles in ewes. Small Ruminant Research 2007(b);68(3) 242–255.

90. Sykes AR. An assessment of the value of plasma urea nitrogen and albumin concentrations as monitors of the protein status of sheep. British Society of Animal Production Occasional Publication 1978;1: 143-154.

91. Alamer M. Effect of deprivation and season on some biochemical constituents of blood in Awassi and Najdi sheep breeds in Saudi Arabia. Journal of Animal and Veterinary Advances 2005;48: 15–20.

92. Hamadeh SK, Hanna N, Barbour EK, Abi Said M, Rawda N, Chedid M, Jaber LS. Changes in physiological and blood parameters in water restricted Awassi ewes supplemented with different levels of Vitamin C. In: European Federation for Animal Science EAAP, 60th Annual Meeting, Barcelona 27 August 2009. Session S.26 Abstract no. 3175. http://www.eaap.org/Previous_Annual_Meetings/2009 Barcelona/Papers/26_Hamadeh.pdf

93. Cork SC, Halliwell RW. The veterinary laboratory and field manual. Nottingham University Press; 2002.

94. El-Sherif MMA, Assad F. Changes in some blood constituents in Barki ewes during pregnancy and lactation under semi-arid conditions. Small Ruminant Research 2001;40: 269-277.

95. Moorby JM, Dewhurst RJ, Evans RT, Fisher WJ. Effects of level of concentrate feeding during the second gestation of Holstein–Friesian

dairy cows. 2. Nitrogen balance and plasma metabolites. Journal of Dairy Science 2002;(85) 178–189.

96. Huntington GB, Archibeque SI. Practical aspects of urea and ammonia metabolism in ruminants. Journal of Animal Science 2000;77(E-Supplement) 1-11.

97. Moen RA, DelGuidice GD. Simulating nitrogen metabolism and urinary urea nitrogen: Creatinine ratios in ruminants. Journal of Wildlife Management 1997;61: 881–894.

98. Kataria N, Kataria AK. Compartmental water management of Marwari sheep. Veterinarski arhiv 2007; 77(6) 551-559.

99. Keenan D M, Allardyce C J. Changes of plasma creatinine levels of sheep during submaintenance feeding. Australian Veterinary Journal 1986;63: 29–30.

100. Olsson K, Benlamlih S, Hossaini-Hilali J, Dahlborn K. Regulation of fluid balance in goats and sheep from dry areas. Options Mediterranennes CIHEAM-IAMZ. 1997;34: 159- 171.

101. Li KE, Mousa HM, Hume JD. Total body water and water economy in camels, desert goats and desert sheep during water restriction and deprivation. Iugoslavica Physiologica et Pharmacologica Acta 1982; 18: 229- 236.

102. Marini JC, Klein JD, Sands JM, Van Amburgh ME. Effect of nitrogen intake on nitrogen recycling and urea transporter abundance in lambs. Journal of Animal Science 2004;82(4) 1157-1164.

103. Jacob RH, Pethick DW, Clark P, D'Souza DN, Hopkins DL, White J. Quantifying the hydration status of lambs in relation to carcass characteristics. Australian Journal of Experimental Agriculture 2006;46(4) 429–437.

104. Qinisa MM, Boomker EA, Mokoboki HK. Physiological Responses of Water-Restricted Tswana and Boer Goats. Life Science Journal 2011;8(S2) 106-111.

105. Rawda N. The effect of water restriction on some physiological and immunological indicators in dry and lactating Awassi ewes. Master thesis. American University of Beirut; 2003. p.1-123.

106. Hanna N. Physiological Changes in Water Stressed Awassi Ewes Supplemented with Vitamin C. Master Thesis. American University of Beirut; 2006. p.1-85.

107. McKinley MJ, Evered MD, Mathai ML. Renal Na excretion in dehydrated and rehydrated adrenalectomized sheep maintained with aldosterone. American Journal of Physiology: Regulatory, Integrative and Comparative Physiology 2000;279: 17-24.

108. Maltz E, Olsson K, Glick SM, Fyhrquist F, Shanikout N, Chosniak I, Shkolnik A.Homeostatic responses to water deprivation or hemorrhage in lactating and non-lactating Bedouin goats. Comparative Biochemistry and Physiology 1984;A(77A) 79-84.

109. Hossaini-Hilali J, Benlamlih S, Dahlborn K. Effects of dehydration, rehydration, and hyperhydration in the lactating and non-lactating black Moroccan goat. Comparative Biochemistry and Physiology 1994;109(A) 1017–1026.

110. Mengistu U, Dahlborn K, Olsson K. Mechanisms of water economy in lactating Ethiopian Somali goats during repeated cycles of intermittent watering. Animal 2007;1: 1009–1017.

111. Shaham D, Choshniak I, Rosenfeld J, Witenberg C, Thurau K, Shkolnik A. Modulation of plasma arginine vasopressin during rehydration in the Bedouin goat. Journal of Comparative Physiology B: Biochemical, Systemic, and Environmental Physiology 1994;164(2) 112-117.

112. Yesberg N, Henderson M, Budtz-Olsen OE. The excretion of vasopressin by normal and dehydrated sheep. Australian Journal of Experimental Biology and Medical Science 1970;48: 115–127. http://www.nature.com/icb/journal/v48/n1/abs/icb197011a.html (accessed 16 August 2012)

113. Olsson K. Fluid balance in ruminants: adaptation to external and internal challenges. Annuals of the New York Academy of Sciences. Trends in Comparative Endocrinology and Neurobiology 2005;104(1) 156-161.

114. Abdelatif AM, Elsayed SA, Hassan YM. Effect of state of hydration on body weight, blood constituents and urine excretion in Nubian goats (Capra hircus). World Journal of Agricultural Sciences 2010;6(2) 178-188.

115. Mojabi A, Abbasali Pourkabire M, Safi S, Bokaie S, Shariati T. Measurements of reference values of some biochemical parameters in serum samples of Ghezel breed sheep. Journal of the Faculty of Veterinary Medicine University of Tehran 2000;55(2) 19-27

116. Eshratkhah B, Sadaghian M, Safari Nezhad M,Sabri V, Farahmand Geyglou B. Evaluation of electrolytes normal values in blood of Moghani sheep breed. Journal of Animal and Veterinary Advances 2008; 7(4) 437-440.

117. Mengistu UK, Dahlborn K, Olsson K. Effect of intermittent watering on water balance and feed intake in Male Ethiopian Somali goats. Small Ruminant Research 2007;67: 45 – 54.

118. Tasker J.B. Fluids, electrolytes, and acid–base balance. In: Kaneko J.J., Cornelius C.E. (ed.), Clinical Biochemistry of Domestic Animals, second ed. Academic Press Inc., London; 1971. p65–75.

119. Srikandakumar A, Johnson EH, Mahgoub O. Effect of heat stress on respiratory rate, rectal temperature and blood chemistry in Omani and Australian Merino sheep. Small Ruminant Research 2003;49: 193–198.

120. El-Nouty FD, El-Naggar MI, Hassan GA, Salem MH. Effect of lactation on water requirements and metabolism in Egyptian sheep and goats. World Review of Animal Production 1991;XXVI: 40– 43.

121. Schneider E G. In water deprivation osmolality become an important determination of aldosterone secretions. News in Physiological Sciences 1990;5: 197–201.

122. Parker AJ, Hamlin GP, Coleman CJ, Fitzpatrick L. Dehydration in stressed ruminants may be the result of a cortisol-induced diuresis. Journal of Animal Science 2003;81: 512– 519.

123. Parrott RF, Thornton SN. Effects of psychological stress and dehydration on plasma cortisol, oxytocin and vasopressin in Clun Forest sheep. Journal of Physiology 1988;399(Suppl.) 91.

124. Parrot RF, Lloyd DM, Goode JA. Stress hormone response of sheep to food and water deprivation at high and low ambient temperatures. Animal Welfare 1996;5: 45– 56.

125. Fell LR, Lynch JJ, Adams DB, Hinch GN, Munro RK, Davies HI. Behavioral and physiological effects in sheep of a chronic

stressor and a parasite challenge. Australian Journal of Agricultural Research 1991;42: 1335-1346.

126. Huszenicza GY, Kulcsar M, Rudas P. Clinical endocrinology of thyroid gland functions in ruminant. Veterinarni Medicina, 2002;47(7): 199-210.

127. Latimer KS, Mahaffy EA, Prasse KW. Duncan and Prasses, Veterinary laboratory medicine, Clinical pathology. Forth edition, Iowa state press, USA; 2003

128. Thrall MA. Veterinary hematology and clinical chemistry. Lippincott Williams & Wilkins; 2004.

129. Hamadeh SK, Moussa Z, Abi Said M, Barbour E. Physiological indicators of adaptation in Awassi and Finn x Texel x Awassi sheep. Options Méditerranéennes 1994;33 : 231-236.

130. El-Barody MAA, Abdalla EB, Abd El-Hakeam AA. The changes in some blood metabolites associated with the physiological response in sheep. Livestock Production Science 2002;75(1) 45-50.

131. Novoselec J, Antunović Z, Šperanda M, Steiner Z, Šperanda T. Changes of thyroid hormones concentration in blood of sheep depending on age and reproductive status. Italian Journal Animal Science 2009;8(3) 208-210.

132. Ward MA, Neville TL, Reed JJ, Taylor JB, Hallford DM, Soto-Navarro SA, Vonnahme KA, Redmer DA, Reynolds LP, Caton JS. Effects of selenium supply and dietary restriction on maternal and fetal metabolic hormones in pregnant ewe lambs. Journal of Animal Science 2008;86: 1254–1262.

133. Nazifi S, Gheisari HR, Poorabbas H. The influences of thermal stress on serum biochemical parameters of dromedary camels and their correlation with thyroid activity. Comparative Haematology International 1999;9(1) 49-53.

134. Yokus B, Cakir DU, Kanay Z, Gulten T, Uysal E. Effects of seasonal and physiological variations on the serum chemistry, vitamins and thyroid hormone concentrations in sheep. Journal of Veterinary Medicine 2006;A (53) 271–276.

135. Nazifi S, Saeb M, Rowghani E, Kaveh K. The influences of thermal stress on serum biochemical parameters of Iranian fat-tailed

sheep and their correlation with triiodothyronine (T 3), thyroxine (T4) and cortisol concentrations. Comparative Clinical Pathology 2003;12(3) 135-139.

136. More T, Sahni KL. Recent observations on water economy and sheep production under semi-arid conditions. Indian Society of Desert Technology and University Centre of Desert Studies, Transactions 1980;5(1) 115-124.

137. Olsson K, Benlamlih L, Dahlborn K, Fyhrquist F. Effects of water deprivation and hyperhydration in pregnant and lactating goats. Acta Physiologica Scandinavica 1982;115(3) 361–367.

138. Rodriguez MN, Tebot I, Le Bas A, Nievas C, Leng L, Cirio A. Renal functions and urea handling in pregnant and lactating Corriedale ewes. Canadian Journal of Animal Science 1996;76: 469–472.

139. Sibanda LM, Ndlovu LR, Bryant MJ. Effects of feeding varying amounts of a grain/forage diet during late gestation and lactation on the performance of Matebele goats. The Journal of Agricultural Science 1997;128(4) 469-477.

140. Mittal JP, Ghosh PK. Effect of prolonged intermittent water restriction on the reproductive performance of ewes in the Indian desert. Animal Production 1986;43(2)255-260.

141. Mellado M, Valdéz R, García JE, López R, Rodríguez A. Factors affecting the reproductive performance of goats under intensive conditions in a hot arid environment. Small Ruminant Research 2006;63: 110-8.

142. Laporte-Broux B, Roussel S, Ponter AA, Perault J, Chavatte-Palmer P, Duvaux-Ponter C. Journal of Animal Science 2011;89: 2154-2163.

143. Sevi A, Albenzio M, Annicchiarco G, Caroprese M, Marino R, Taibi L. Effects of ventilation regimen on the welfare and performance of lactating ewes in summer. Journal of Animal Science 2002;8: 2362– 2372.

144. Lindsay DB, Leat, WMF. Carbohydrate and lipid metabolism. In: Hunt M.H. (ed.) The Blood of Sheep Composition and Function. Springer Verlag, Berlin, Heidelberg, New York; 1975. p. 53.

145. Fell BF, Mackie WS, Campbell RM. Quantitative changes occurring in the serum proteins during lactation in the ewe. Research in Veterinary Science 1968;9: 563–572.

146. Collier RJ. Nutritional, metabolic, and environmental aspects of lactation. In: Larson B.L. (ed.). Lactation. Iowa State University Press 1985.p. 102–107.

147. Maltz E, Shkolnik A. Milk production in the desert: lactation and water economy in the black Bedouin goat. Physiological Zoology 1980;53: 12-18.

148. Dahlborn K, Nielsen MO, Hossaini-Hilali J. Mechanisms causing decreased milk production in water deprived goats. Options Mediterranéennes CIHEAM- Zaragoza 1997;34: 199-202.

149. Dahlborn K. Effect of temporary food or water deprivation on milk secretion and milk composition in the goat. Journal of Dairy Research 1987;54(2) 153-163.

150. Alamer M. Effect of water restriction on lactation performance of Aardi goats under heat stress conditions. Small Ruminant Research 2009;84(1) 76-81.

151. Mittal SB. Dairy development and seasonal variation in the quality of milk in the arid zone of western Rajasthan. In: H. S. Mann (ed.) Arid zone research and development. Scientific Publishers, Jodhpur, India; 1980. p381-387.

152. Yagil R, Amir H, Abu-Rabiya Y, Etzion Z. Dilution of milk: a physiological adaptation of mammals to water stress? Journal of Arid Environments 1986;11: 243-247.

153. Renaudeau D, Collin A, Yahav S, de Basilio V, Gourdine JL, Collier RJ. Adaptation to hot climate and strategies to alleviate heat stress in livestock production. Animal 2012;65: 707–728.

154. Apple JK, Minton JE, Parsons KM, Unruh JA. Influence of repeated restraint and isolation stress and electrolyte administration on pituitary-adrenal secretions, electrolytes, and other blood constituents of sheep. Journal of Animal Science 1993;71: 71-77.

155. Baldock NM, Sibly RM. Effects of handling and transportation on the heart rate and behaviour of sheep. Applied Animal Behaviour Science 1999;28(1–2) 15-39.

156. Ali BH, Al-Qarawi AA, Mousa HM. Stress associated with road transportation in desert sheep and goats, and the effect of pretreatment with xylazine or sodium betaine. Research in Veterinary Science 2005;80 (3) 343-348.

157. Schaefer AL, Jones SDM, Stanley RW. The use of electrolyte solutions for reducing transport stress. Journal of Animal Science 1997;75: 258–265.

158. Al-Haidary AA. Physiological responses of naimey sheep to heat stress challenge under semi-arid environments. International Journal of Agriculture and Biology 2006;6(2) 307–309.

159. Christensen RA, Christopherson RJ, Kennedy JJ. Effect of somatostatin and chronic cold exposure on hormonal and metabolite concentration, metabolic rate, thermoregulation and gut motility in sheep. Canadian Journal of Animal Science 1990;70: 1073–1083.

160. Abdel-Samee AM, Abd-Alla OAM, EL-Adawy SAI. Nutritional treatments for alleviation of heat stress in Awassi sheep using acacia and olive pulp in subtropics. Egypt. J. Comp. Path. & Clinic. Path. 2008;21(1) 466- 477.

161. Morrison SR. Ruminant heat stress: effect on production and means of alleviation. Journal of Animal Science 1983;57(6) 1594-1600.

162. Darcan N, Güney O. Alleviation of climatic stress of dairy goats in Mediterranean climate. Small Ruminant Research 2008;74: 212–215.

163. Ayo JO, Minka NS, Mamman M. Excitability scores of goats administered ascorbic acid and transported during hot-dry conditions. Journal of Veterinary Science 2006;7(2) 127–131.

164. Minka NS, AYO JO. Physiological responses of transported goats treated with ascorbic acid during the hot-dry season. Animal Science Journal 2007;78: 164–172.

165. Hathcock JN, Azzi A, Blumberg J, Bray T, Dickinson A, Frei B, Jialal I, Johnston CS, Kelly FJ, Kraemer K, Packer L, Parthasarathy S, Sies H, Traber MG. Vitamins E and C are safe across a broad range of intakes. American Journal of Clinical Nutrition 2005;81(4) 736-745.

166. Xu Z, Kornegay ET, Sweet LA, Lindmann MD, Veit HP, Watkins BA. Effects of feeding aspirin and soybean oil to weanling pigs. Journal of Animal Science 1990;68: 1639-1647.

167. Fields M, Lewis C, Bureau I. Aspirin reduces blood cholesterol in copper deficient rats: A potential antioxidant agent? Metabolism 2001;50(5) 558-561.

168. Drew J, Arthur J, Farquharson A, Russel W, Morrice P, Duthie G. Salicylic acid modulates oxidative stress and glutathione peroxidase activity in the rat colon. Biochemical Pharmacology 2005;70(6) 888-893.

169. McDowell LR. Vitamins in animal and human nutrition. State University, Ames IA, Iowa; 2000.

170. de Rodas BZ, Maxwell CV, Davis ME, Mandali S, Broekman E, Stoecker BJ,. L-ascorbyl-2-polyphosphate as a vitamin C source for segregated and conventially weaned pigs. Journal of Animal Science 1998;76: 1636–1643.

171. Avci M, Yertürk M, Kaplan O. Effects of ascorbic acid on the performance and some blood parameters of Japanese quails reared under hot climate conditions. Turkish Journal of Veterinary and Animal Sciences 2005;29: 829-833.

172. Yousef MI. Aluminium-induced changes in hemato-biochemical parameters, lipid peroxidation and enzyme activities of male rabbits: protective role of ascorbic acid. Toxicology 2004; 47-57.

173. McKee JS, Harrison PC, Riskowski GL. Effects of supplemental ascorbic acid on the energy conversion of broiler chicks during heat stress and feed withdrawal. Poultry Science 1997;76: 1278-1286.

174. Jariwalla R. J. ,Harakech S. In: Harris, R.J. (ed.) Ascorbic acid: Biochemistry and biomedical cell biology, Plenum, New York; 1996. p. 215-231.

175. Bouda J, Jagos P, Dvorak R, Ondrova J. Vitamin E and C in the blood plasma of cows and their calves fed from buckets. Acta Veterinaria Brno 1980;49: 53-58.

176. Politis I, Hidiroglou M, Batra TR, Gilmore JR, Gorewit RC, Scherf H. Effects of vitamin E on immune function of dairy cows. American Journal of Veterinary Research 1995;56: 179.

177. Cummins KA, Brunner CJ. Effect of calf housing on plasma ascorbate and endocrine and immune function. Journal of Dairy Science 1990;74: 1582-1588.

178. Sahin K, Kucuk O, Sahin N, Sari M. Effects of vitamin C and vitamin E on lipid peroxidation status, serum hormone, metabolite, and mineral concentrations of Japanese quails reared under heat stress (34 1C). International Journal of Vitamin Nutrition Research 2002;72: 91–100.

179. Kannan G, Terrill H, Kouakou B, Gazal OS, Gelaye S, Amoah EA, Samake S. Transportation of goats: Effects on physiological stress responses and liveweight loss. Journal of Animal Science. 2000;78: 1450–1457.

180. Mahan DC, Ching S, Dabrowski K, Developmental aspects and factors influencing the synthesis and status of ascorbic acid in the pig. Annual Review of Nutrition 2004;24: 79–103.

181. Civen M, Leeb JE, Wishnow RM, Morin R. Effect of dietary ascorbic acid and vitamin E deficiency on rat adrenal cholesterol ester metabolism and corticosteroidogenesis. International Journal of Vitamin Nutrition Research 1980;50: 70–78.

182. Sivakumar AVN, Singh G, Varshney VP. Antioxidants supplementation on acid base balance during heat stress in goats. Asian-Australian Journal of Animal Science 2010;23(11) 1462 – 1468.

183. Jeffreys D. Aspirin: The story of a wonder drug. London: Bloomsbury Publishing; 2004.

184. He J, Whelton PK, Klag MJ. Aspirin and risk of hemorrhagic stroke: a meta-analysis of randomized controlled trials. The Journal of the American Medical Association 1998;280: 1930-1935.

185. Galil MA. Effect of using some anti-heat stress compounds on the performance of some local breeds of chicken under hot climatic condition. Egyptian Poultry Science Journal 2004;24(2) 417-427.

186. McDaniel CD, Balog JM, Freed M, Elkin RG, Wellenreiter RH, Hester PY. Response of layer breeders to dietary acetylsalicylic acid. 1. Effects on Hen Performance and Eggshell Quality. Poultry Science 1993;72(6) 1084-1092

187. Stilborn HL, Harris GC Jr, Bottje WG, Waldroup PW. Ascorbic acid and acetylsalicylic acid (aspirin) in the diet of broilers maintained under heat stress conditions. Poultry Science 1988;67(8) 1183-1187.

188. Traş B, Inal F, Baş AL, Altunok V, Elmas M, Yazar E. Effects of continuous supplementations of ascorbic acid, aspirin, vitamin E and selenium on some haematological parameters and serum superoxide dismutase level in broiler chickens. British Poultry Science 2000;41: 664-66.

189. Gudev D, Popova-Ralcheva S, Yanchev I, Kozelov L, Moneva P. Effects of acetylsalicylic acid on adrenal response in lambs exposed to transport and new environment. Bulgarian Journal of Agricultural Science 2002;8(4) 433-438.

190. Pollard JC, Roos V, Littlejohn RP. Effects of an oral dose of acetyl salicylate at tail docking on the behaviour of lambs aged three to six weeks. Applied Animal Behaviour Science 2000;71: 29-42.

CHAPTER 3

Metabolic Interactions between Vitamin A and Conjugated Linoleic Acid

Gianfranca Carta [1], Elisabetta Murru [1], Lina Cordeddu [2], Berenice Ortiz [3], Elena Giordano [4], Martha A. Belury [5], Loredana Quadro [6] and Sebastiano Banni [1,*]

[1]Department of Biomedical Sciences, University of Cagliari, Cittadella Universitaria, S.S. 554, km. 4500, Monserrato, Cagliari 09042, Italy

[2]Department of Bioscience and Nutrition, Karolinska Institute, Stockholm 141 83, Sweden

[3]Gerstner Sloan-Kettering Graduate School in Biomedical Sciences, Memorial Sloan-Kettering Cancer Center, New York, NY 10065, USA

[4]Laboratory of Functional Foods, Madrid Institute for Advanced Studies (IMDEA)-Food, CEI UAM + CSIC, Madrid 28049, Spain

[5]Department of Human Nutrition, Ohio State University, Columbus, OH 43210, USA

[6]Department of Food Science and Rutgers Center for Lipid Research, Rutgers University, New Brunswick, NJ 08901, USA

ABSTRACT

Lipid-soluble molecules share several aspects of their physiology due to their common adaptations to a hydrophilic environment, and may interact to regulate their action in a tissue-specific manner. Dietary conjugated linoleic acid (CLA) is a fatty acid with a conjugated diene structure that is found in low concentrations in ruminant products and available as a nutritional supplement. CLA has been shown to increase tissue levels of retinol (vitamin A alcohol) and its sole specific circulating carrier protein retinol-binding protein (RBP or RBP4). However, the precise mechanism of this action has not been elucidated yet. Here, we provide a summary of the current knowledge in this specific area of research and speculate that retinol and CLA may compete for catabolic pathways modulated by the activity of PPAR-α and RXR heterodimer. We also present preliminary data that may position PPAR-α at the crossroads between the metabolism of lipids and vitamin A.

KEYWORDS

conjugated linoleic acid (CLA); retinol; vitamin A; retinol binding protein (RBP); oleoylethanolamide (OEA); palmitoylethanolamide (PEA); peroxisome proliferator-activated receptor-alpha (PPARα)

1. INTRODUCTION

Conjugated linoleic acid (CLA) is a group of unusual fatty acids containing conjugated double bonds. The CLA present in mammalian tissues is derived from the diet or, in a smaller amount, from the intestinal flora [1]. CLA is produced by bacteria within the rumen or upon δ-9 desaturation of vaccenic acid, another fatty acid produced by ruminal bacteria [2]. Therefore, the main dietary sources of CLA are dairy products and the meat of ruminants [3].

The predominant "natural" isomer of CLA, *cis* 9, *trans* 11 (CLA c9, t11), exists in food among a plethora of other isomers that are present in much lower concentrations. After the discovery that CLA possesses anticarcinogenic activity [4], the need to produce large amounts of CLA prompted the production of synthetic CLA by alkali isomerization of linoleic acid [5]. This process resulted in the formation of several isomers, among which the CLA c9, t11 and the CLA t10, c12 were the most abundant. With such CLA preparations, several studies were carried out to investigate its anti-carcinogenic, anti-atherogenic, and anti-adiposity biological activities [6]. Some of the biological properties of CLA are specifically attributed to the t10, c12 isomer. For instance, CLA t10, c12 has the unique capability to inhibit lipid accumulation in adipose tissue, likely through downregulation of the peroxisome proliferator-activated receptor-γ (PPAR-γ) via activation of upstream signals which may induce integrated stress response (ISR), free fatty acid release and activation of NFκB and MAPKs that may directly antagonize PPAR-γ activity [7]. In addition, it has been shown that activation of AMPK by t10, c12 CLA may also be involved in adipocyte lipid reduction [8]. In contrast, CLA properties linked to the activation of PPAR-α are equally shared by the c9, t11 and t10, c12 CLA isomers [9], and it seems not directly correlated to the antiadiposity activity exerted by t10, c12 CLA [10].

PPARs are ligand-activated nuclear receptors involved in the transcriptional regulation of lipid and energy homeostasis [11]. Three isoforms are known with a high degree of sequence conservation across various species: PPAR-α, PPAR-β/δ and PPAR-γ [11]. In humans and mice, the three PPAR isoforms are encoded by separate genes on different chromosomes [12]. PPARs function as sensors for fatty acids and fatty acid derivatives, and control metabolic pathways involved in energy homeostasis [13]. The three PPAR isoforms exhibit distinct and non-interchangeable functional roles in energy metabolism [12]. PPAR-α predominantly regulates lipid catabolism. In the liver, PPAR-α regulates mitochondrial, peroxisomal, and microsomal fatty acid oxidation, as well as senses the influx of fatty acids during fasting to enhance the fatty acid burning capacity [14]. Activation of PPAR-β/δ also induces expression of genes required for fatty acid oxidation and energy dissipation in skeletal muscle and adipose tissue, which in turn lead to improved lipid profiles and reduced adiposity [15]. In the liver, PPAR-β/δ can be activated by plasma free fatty acids influxed during fasting [16]. PPAR-γ is an essential regulator of adipogenesis and fat storage in adipocytes [17].

Not all fatty acids have similar binding affinity for PPARs [18]. Therefore, modulation of PPAR activity is dependent on the intracellular total fatty acid concentrations as well as their specific profile. CLA is a strong ligand of PPARα [9] and is believed to exert some of its metabolic effects by activating this receptor [6]. In competitive binding scintillation proximity assays, CLA isomers bind to PPARα with submicromolar IC_{50} values (140 nM to 400 nM) [9]. Although one cannot directly compare the IC_{50}s obtained in different studies, due to different methodologies and/or receptor constructs utilized, CLA isomers appear to have an IC_{50} similar to that of the most avid fatty acid ligands of PPARα reported, such as that of oxygenated eicosanoids [19]. Interestingly, most of the naturally occurring fatty acids have lower affinity for PPARα (IC_{50} ~1 μM [20]).

In an attempt to evaluate the link between CLA metabolism and mammary anticarcinogenic activity, we serendipitously discovered that tissue levels of retinol, vitamin A alcohol derivative, were increased in rats fed a mixture of CLA isomers (primarily c9, t11 and t10, c12) [21]. This prompted us to investigate the possible mechanism(s) of action of enhanced retinol levels and to determine whether the altered vitamin A levels are physiologically significant.

2. VITAMIN A METABOLISM AND FUNCTIONS

Vitamin A is an essential nutrient that controls many crucial biological functions such as vision, reproduction, development, growth, and immunity [22,23]. Vitamin A exerts its functions through oxidized metabolites of retinol: retinaldehyde and retinoic acid. Retinaldehyde is involved in the visual cycle [24], and retinoic acid regulates the expression of many target genes through receptor-mediated events [25,26,27,28,29,30]. In animals, vitamin A and its derivatives are derived from animal products as preformed dietary vitamin A (retinyl esters, retinol and very small amounts of retinoic acid), or as β-carotene from vegetables and fruits [31].

Within the intestinal mucosa all retinol, regardless of its dietary origin, is enzymatically re-esterified with long-chain fatty acids. Together with other dietary lipids, the newly synthesized retinyl esters are packaged into chylomicrons and secreted into the lymphatic system [32]. Once in the general circulation, nascent chylomicrons undergo lipolysis of the triglycerides giving rise to free fatty acids and chylomicron remnants [33]. Approximately 75% of chylomicron-retinoids are cleared by the liver, the major site of vitamin A storage and metabolism [34,35]. The remaining 25% are cleared by extrahepatic tissues [36].

To meet tissue vitamin A needs the liver secretes retinol into the circulation bound to its specific transport protein, retinol-binding protein (RBP or RBP4) [37,38]. RBP is a 21 kDa protein with a single binding site for one molecule of all-*trans*-retinol. The major function of RBP is to mobilize hepatic vitamin A stores and deliver retinol to peripheral tissues [37,39]. It is mainly, but not exclusively, synthesized within hepatocytes [37,38]. RBP circulates in the blood as a 1:1 molar complex with another serum protein, transthyretin (TTR) [40]. In the fasting circulation, retinol-RBP accounts for approximately 99% of all serum retinoids. In the postprandial circulation, the concentration of chylomicrons and chylomicron remnants with retinoids can greatly exceed the concentration of plasma retinol. Blood levels of retinol-RBP in both humans and animals are maintained very constant, except in extreme cases of nutrition and in certain disease states [22,41].

3. INFLUENCE OF DIETARY CLA ON RETINOL AND RBP TISSUE LEVELS

In the first report on the effects of dietary CLA on mammalian tissue, we found increased retinol levels in plasma, mammary tissue and liver of female rats fed increasing concentration of CLA (0%–2%) [21]. More specifically both retinol and retinyl ester levels were elevated upon CLA feeding, with only liver retinol levels increasing in a CLA dose-dependent manner [21]. In order to meet tissue retinoid needs, the liver secretes retinol-RBP into the bloodstream [37]. We first became interested in the influence of CLA on RBP when a report demonstrated an increase of RBP mRNA in pig liver upon CLA feeding [42]. We therefore investigated whether the peculiar effect of dietary CLA on retinol metabolism was RBP-mediated. Additionally, in contrast to the majority of the literature data obtained by feeding laboratory animals with an equimolar mixture of CLA c9, t11 and CLA t10, c12, we sought to establish whether CLA c9, t11 and CLA t10, c12 had distinct and/or common effects on vitamin A metabolism.

We performed our study using wild-type and RBP knockout (RBP KO) mice chronically fed with separate enriched preparations of the two CLA isomers [43]. We showed that consumption of CLA c9, t11 or CLA t10, c12 induced hepatic retinyl ester accumulation in wild-type mice [43]. Only the CLA t10, c12 isomer increased hepatic retinol secretion and elevated the levels of serum retinol and its carrier RBP. In addition, we demonstrated that chronic intake of CLA t10, c12 redistributed retinoids from the hepatic stores toward the adipose tissue via RBP [43]. RBP KO mice did not exhibit increased serum retinol levels or redistribution of hepatic stores when fed the CLA t10, c12, confirming the crucial role of RBP in mediating the changes of vitamin A metabolism [43]. Within this study, we also provided evidence that dietary CLA t10, c12 could stimulate hepatic retinoid oxidation, as indicated by the upregulation of the mRNA levels of Cyp26A1 and Cyp2c39 in the liver of the wild-type mice. These two enzymes are members of the cytochrome P450 (CYP) family of enzymes and contribute to retinoid hemostasis by oxidizing retinoic acid into more polar inactive metabolites [44,45]. It is interesting to note that the expression of CYP isoforms is regulated by PPAR-α [46] of which CLA isomers are strong ligands [9]. Specifically, CLA has been shown to increase CYP4A1 expression in liver [47]. Furthermore, the chronic feeding of mice (and rats) with CLA resulted in a dramatic reduction of adipose tissue and a concomitant massive fat deposition in the liver [48]. Based on this evidence, we speculated that the stimulatory effect of dietary CLA t10, c12 on hepatic retinol secretion and probably oxidation of retinoids may be part of a compensatory mechanism of the tissue to counter the excessive accumulation of lipids, including retinoids. Remarkably, serum RBP levels have been shown to be elevated in patients with nonalcoholic fatty liver disease [49].

Short-term intake of CLA also alters vitamin A metabolism. We demonstrated this by administering a bolus dose of radiolabeled retinol in combination with either CLA c9, t11 or CLA t10, c12 to both WT and RBP KO mice [50]. Upon ingestion of either CLA isomer hepatic accumulation of retinoids took place rapidly (only three hours after gavage), and recently ingested vitamin A was rapidly re-secreted from the liver mainly in the form of retinol likely bound to RBP [50]. In addition, the use of an inhibitor of the clearance of chylomicrons suggested that CLA intake rapidly enhances intestinal absorption of dietary vitamin A, and that the vitamin A that accumulates in the liver upon CLA intake may be of intestinal origin [50].

Overall, these studies showed a dramatic effect of dietary CLA on murine retinoid metabolism. However, the molecular details of such interference await a study that would full unravel them.

How does CLA regulate the secretion of the complex retinol-RBP? How does CLA seem to promote hepatic retinoid oxidation and intestinal vitamin A absorption? These and other questions need to be answered in order to understand the real biological implications of this nutrient-nutrient interaction. In the paragraph below we speculate on how some of these interactions may take place.

4. IS PPAR-A THE KEY REGULATOR OF CLA AND RETINOID METABOLIC ROUTES?

Lipid-soluble molecules share several aspects of their physiology due to common adaptations that allow them to function in a hydrophilic environment. Absorption, transport, tissue storage, receptor-binding activities, metabolism and catabolism need specific strategies to overcome lipophilicity and differences in polarity. Molecules that share these strategies include dietary vitamin A and fatty acids. Their biological functions are exerted through binding to specific nuclear receptors. Interestingly, the fatty acid receptor [51] PPAR-α and the retinoid X receptor, RXR, must form heterodimers to regulate gene expression, implying that their action is strictly regulated by the intracellular concentration of both fatty acids and retinoids. While it is clear that fatty acids are the ligands for PPARα [52], it is still debated which are the endogenous ligands of RXR [53]. The 9-*cis* isomer of retinoic acid was initially considered the endogenous ligand of RXR, but the inability to detect this compound *in vivo* in all the tissues has raised some concerns. Interestingly, it has been demonstrated that several fatty acids can also bind this nuclear receptor at low microM [54], which may enhance the transcriptional activity of RXR-PPARs heterodimer.

RXR activation has been shown to enhance catabolic pathways regulated by PPAR-α, including the catabolism of lipid molecules such as eicosanoids via peroxisomal β-oxidation, and xenobiotics, retinoids and fatty acids by the activity of the CYP enzymes [51,55]. It is known that catabolism of CLA is predominantly mediated by peroxisomal β-oxidation, while retinoids are preferentially catabolized via the action of the CYP enzymes [45]. However, there is solid evidence indicating that different lipid molecules compete for these same two pathways. For example, it has been clearly shown that CLA and isoprostanes compete for peroxisomal β-oxidation, [56] and competition for CYP action has been well demonstrated between several drugs and alcohol [57]. We hypothesize that a competition for common catabolic pathways activated by PPAR-α may explain, at least in part, the interference between dietary CLA and tissue retinoid metabolism. Preliminary data from our laboratories seem to support this possibility. Feeding wild-type mice with WY-14,643 (0.01%), a potent synthetic PPAR-α inducer [58], dramatically decreased liver retinol and retinyl esters content (Table 1), possibly due to an increased PPAR-α-mediated CYP activity, even though we cannot rule out that the decrease may be related to other causes, such as decreased dietary uptake and/or increased mobilization to extrahepatic tissues. While a decrease of retinyl esters may be ascribed to a possible increase in retinyl ester hydrolase activity. In contrast, when mice were fed a 1:1 mixture of CLA c9, t11 and CLA t10, c12 (1.5%), known endogenous PPAR-α ligands, hepatic retinoid levels increased (Table 1), as previously shown [21]. This discrepancy could be explained by a competition for the same catabolic pathway that takes place between retinoids and CLA, but not WY-14,643. Even though it has been shown that fatty acids are metabolized by the CYP enzymes [59], to our knowledge there is no data on CLA as substrate for these enzymes. Therefore, the competition for CYP enzymes proposed above is only a potential possibility and further studies should be carried out to prove that it actually takes place. We speculate that if CLA undergoes degradation by CYP action, which is likely stimulated by CLA through PPAR-α activation [47], hepatic retinoids would indeed accumulate. To our knowledge, it is not known how WY-14,643 is metabolized within the body. However, given the extremely low concentration of this drug in the diet fed to our mice, we speculate that it is unlikely that WY-14,643 could compete with retinoids for degradation by CYP activity

Table 1. Retinol and retinyl esters concentration in liver of female mice fed for 42 weeks control diet (CTRL), or CTRL diet + 1.5% of conjugated linoleic acid (CLA) (mixture 1:1 of c9, t11 and t10, c12), or CTRL diet + 0.01% of WY-14,643.

Treatment	Retinol	Retinyl Esters
	µg/g of Liver	
CTRL	4.69 ± 0.73 [a]	273.69 ± 67.37 [a]
CLA	6.68 ± 2.03 [a]	470.64 ± 46.06 [b]
WY-14,643	1.41 ± 0.45 [b]	75.74 ± 14.23 [c]

[a,b,c] Different letters denote significant differences ($p < 0.05$) with one-way ANOVA and Tukey post-hoc analysis.

It is interesting that PPAR-α could directly regulate RBP expression, at least in certain tissues [60]. This data suggests that the ability of CLA to increase the levels of hepatic and circulating RBP could be the result of a direct action of PPAR-α on RBP levels. On the other hand, in a sort of positive feedback mechanism, the increase of RBP levels may potentially enhance retinoid availability for RXR-PPARα action, by promoting retinol transport to the periphery of the body. In this sense, RBP could be essential to amplify the PPAR-α-mediated metabolic pathways, including those involved in energy homeostasis. We have evidence that RBP could indeed play such an important role.

We have recently shown that PPAR-α activation in rodent brain slices increases the biosynthesis of the endocannabinoid congeners, oleoylethanolamide (OEA) and palmitoylethanolamide (PEA) [61]. Interestingly, both OEA and PEA also induce PPAR-α affecting energy homeostasis [62,63], probably by sustaining PPARα activity with a positive feedback. Also, PEA is a TRPV1 ligand [64]. Our preliminary data show that RBP KO mice have lower levels of OEA and PEA in the liver (Figure 1A) and adipose tissue (Figure 1B) when compared to age- and sex-matched wild-type animals. Whether CLA feeding, through induction PPAR-α activity and thus RBP expression, also enhances PEA and OEA tissue levels is not known at the moment and it would be an important question to address.

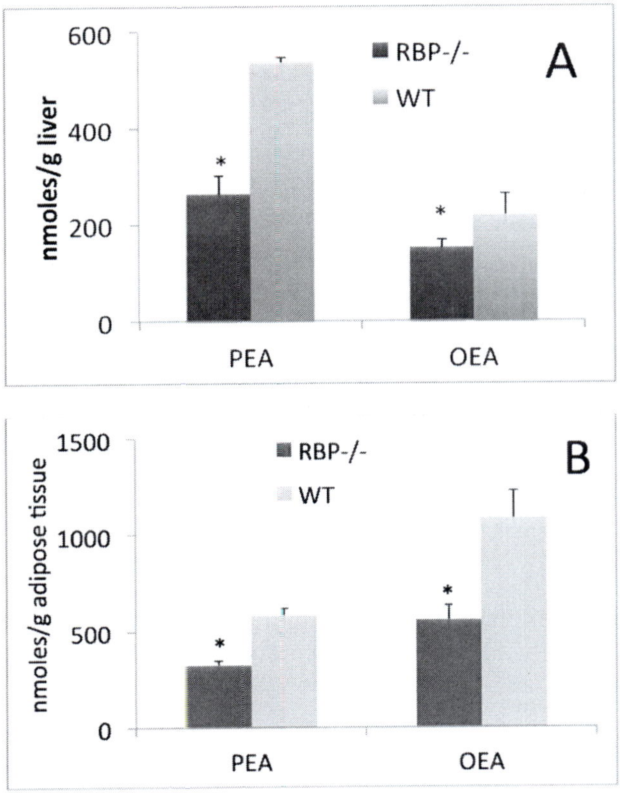

Figure 1. Palmitoylethanolamide (PEA) and oleoylethanolamide (OEA) levels in liver (**A**) and adipose tissue (**B**) of $RBP^{-/-}$ or wild type mice. * significant difference ($p < 0.05$) with *t*-student test.

5. CONCLUSIONS

The influence of dietary CLA on retinol metabolism is yet another example of nutrient-nutrient interaction that points to the difficulties in isolating and/or claiming the nutritional property of a single dietary component, without taking into account all the possible interactions with other nutrients. Once all the molecular mechanisms of the interaction between CLA and retinol are elucidated, novel dietary strategies can be explored to regulate retinol and fatty acid metabolism by modulating the heterodimer PPAR-α-RXR activities in different tissues to promote health in humans.

CONFLICTS OF INTEREST

The authors declare no conflict of interest.

REFERENCES

1. McIntosh, F.M.; Shingfield, K.J.; Devillard, E.; Russell, W.R.; Wallace, R.J. Mechanism of conjugated linoleic acid and vaccenic acid formation in human faecal suspensions and pure cultures of intestinal bacteria. Microbiology 2009, 155, 285–294.

2. Griinari, J.M.; Cori, B.A.; Lacy, S.H.; Chouinard, P.Y.; Nurmela, K.V.V.; Bauman, D.E. Conjugated linoleic acid is synthesized endogenously in lactating dairy cows by δ(9)-desaturase. J. Nutr. 2000, 130, 2285–2291.

3. Ritzenthaler, K.L.; McGuire, M.K.; Falen, R.; Shultz, T.D.; Dasgupta, N.; McGuire, M.A. Estimation of conjugated linoleic acid intake by written dietary assessment methodologies underestimates actual intake evaluated by food duplicate methodology. J. Nutr. 2001, 131, 1548–1554.

4. Pariza, M.W.; Ha, Y.L. Newly recognized anticarcinogenic fatty acids. Basic Life Sci. 1990, 52, 167–170.

5. Ma, D.W.L.; Wierzbicki, A.A.; Field, C.J.; Clandinin, M.T. Preparation of conjugated linoleic acid from safflower oil. J. Am. Oil Chem. Soc. 1999, 76, 729–730.

6. Belury, M.A. Dietary conjugated linoleic acid in health: Physiological effects and mechanisms of action. Ann. Rev. Nutr. 2002, 22, 505–531.

7. Kennedy, A.; Martinez, K.; Schmidt, S.; Mandrup, S.; LaPoint, K.; McIntosh, M. Antiobesity mechanisms of action of conjugated linoleic acid. J. Nutr. Biochem. 2010, 21, 171–179.

8. Jiang, S.; Chen, H.; Wang, Z.; Riethoven, J.J.; Xia, Y.; Miner, J.; Fromm, M. Activated AMPK and prostaglandins are involved in the response to conjugated linoleic acid and are sufficient to cause lipid reductions in adipocytes. J. Nutr. Biochem. 2011, 22, 656–664.

9. Moya-Camarena, S.Y.; Vanden Heuvel, J.P.; Blanchard, S.G.; Leesnitzer, L.A.; Belury, M.A. Conjugated linoleic acid is a potent naturally occurring ligand and activator of PPARα. J. Lipid Res. 1999, 40, 1426–1433.

10. Peters, J.M.; Park, Y.; Gonzalez, F.J.; Pariza, M.W. Influence of conjugated linoleic acid on body composition and target gene expression in peroxisome proliferator-activated receptor α-null mice. Biochim. Biophys. Acta 2001, 1533, 233–242.

11. Desvergne, B.; Wahli, W. Peroxisome proliferator-activated receptors: Nuclear control of metabolism. Endocr. Rev. 1999, 20, 649–688.

12. Bookout, A.L.; Jeong, Y.; Downes, M.; Yu, R.T.; Evans, R.M.; Mangelsdorf, D.J. Anatomical profiling of nuclear receptor expression reveals a hierarchical transcriptional network. Cell2006, 126, 789–799.

13. Chawla, A.; Repa, J.J.; Evans, R.M.; Mangelsdorf, D.J. Nuclear receptors and lipid physiology: Opening the X-files. Science 2001, 294, 1866–1870.

14. Hashimoto, T.; Cook, W.S.; Qi, C.; Yeldandi, A.V.; Reddy, J.K.; Rao, M.S. Defect in peroxisome proliferator-activated receptor α-inducible fatty acid oxidation determines the severity of hepatic steatosis in response to fasting. J. Biol. Chem. 2000, 275, 28918–28928.

15. Grimaldi, P.A. Metabolic and nonmetabolic regulatory functions of peroxisome proliferator-activated receptor β. Curr. Opin. Lipidol. 2010, 21, 186–191.

16. Sanderson, L.M.; Degenhardt, T.; Koppen, A.; Kalkhoven, E.; Desvergne, B.; Muller, M.; Kersten, S. Peroxisome proliferator-activated receptor β/δ (PPAR β/δ) but not PPARα serves as a plasma free fatty acid sensor in liver. Mol. Cell. Biol. 2009, 29, 6257–6267.

17. Rosen, E.D.; Spiegelman, B.M. Ppargamma: A nuclear regulator of metabolism, differentiation, and cell growth. J. Biol. Chem. 2001, 276, 37731–37734.

18. Forman, B.M.; Chen, J.; Evans, R.M. Hypolipidemic drugs, polyunsaturated fatty acids, and eicosanoids are ligands for peroxisome proliferator-activated receptors α and δ. Proc. Natl. Acad. Sci. USA 1997, 94, 4312–4317.

19. Kliewer, S.A.; Sundseth, S.S.; Jones, S.A.; Brown, P.J.; Wisely, G.B.; Koble, C.S.; Devchand, P.; Wahli, W.; Willson, T.M.; Lenhard, J.M.; et al. Fatty acids and eicosanoids regulate gene expression through direct interactions with peroxisome proliferator-activated receptorsα and γ. Proc. Natl. Acad. Sci. USA 1997, 94, 4318–4323.

20. Xu, H.E.; Lambert, M.H.; Montana, V.G.; Parks, D.J.; Blanchard, S.G.; Brown, P.J.; Sternbach, D.D.; Lehmann, J.M.; Wisely, G.B.; Willson, T.M.; et al. Molecular recognition of fatty acids by peroxisome proliferator-activated receptors. Mol. Cell 1999, 3, 397–403.

21. Banni, S.; Angioni, E.; Casu, V.; Melis, M.P.; Scrugli, S.; Carta, G.; Corongiu, F.P.; Ip, C. An increase in vitamin a status by the feeding of conjugated linoleic acid. Nutr. Cancer 1999, 33, 53–57.

22. Goodman, D.S. Vitamin A and retinoids in health and disease. N. Engl. J. Med. 1984, 310, 1023–1031.

23. Napoli, J.L. Biochemical pathways of retinoid transport, metabolism, and signal transduction. Clin. Immunol. Immunopathol. 1996, 80, S52–S62.

24. Saari, J.C. Retinoids in Mammalian Vision. In Retinoids; Nau, H., Blaner, W.S., Eds.; Springer Verlag Publishing: Heidelberg, Germany, 1999; Volume 139, pp. 563–588.

25. Mangelsdorf, D.J.; Thummel, C.; Beato, M.; Herrlich, P.; Schutz, G.; Umesono, K.; Blumberg, B.; Kastner, P.; Mark, M.; Chambon, P.; et al. The nuclear receptor superfamily: The second decade. Cell 1995, 83, 835–839.

26. Kurokawa, R.; Soderstrom, M.; Horlein, A.; Halachmi, S.; Brown, M.; Rosenfeld, M.G.; Glass, C.K. Polarity-specific activities of retinoic acid receptors determined by a co-repressor.Nature 1995, 377, 451–454.

27. Leblanc, B.P.; Stunnenberg, H.G. 9-cis retinoic acid signaling: Changing partners causes some excitement. Genes Dev. 1995, 9, 1811–1816.

28. Chen, J.D.; Evans, R.M. A transcriptional co-repressor that interacts with nuclear hormone receptors. Nature 1995, 377, 454–457.

29. Pfahl, M.; Chytil, F. Regulation of metabolism by retinoic acid and its nuclear receptors. Ann. Rev. Nutr. 1996, 16, 257–283.

30. Kastner, P.; Mark, M.; Chambon, P. Nonsteroid nuclear receptors: What are genetic studies telling us about their role in real life? Cell 1995, 83, 859–869.

31. Sporn, M.B.; Roberts, A.B.; Goodman, D.S. The Retinoids: Biology, Chemistry, and Medicine, 2nd ed.; Raven Press: New York, NY, USA, 1994.

32. Vogel, S.; Gamble, M.V.; Blaner, W.S. Biosynthesis, Absorption, Metabolism and Transport of Retinoids. In Retinoids—The Biochemical and Molecular Basis of Vitamin A and Retinoid Action; Nau, H., Blaner, W.S., Eds.; Springer Verlag Publishing: Heidelberg, Germany, 1999; Volume 139, pp. 31–95.

33. Goldberg, I.J. Lipoprotein lipase and lipolysis: Central roles in lipoprotein metabolism and atherogenesis. J. Lipid Res. 1996, 37, 693–707.

34. Cooper, A.D. Hepatic uptake of chylomicron remnants. J. Lipid Res. 1997, 38, 2173–2192.

35. Blaner, W.S.; Olson, J.A. Retinol and Retinoic Acid Metabolism. In The Retinoids, Biology, Chemistry and Medicine; Sporn, M.B., Roberts, A.B., Goodman, D.S., Eds.; Raven Press: New York, NY, USA, 1994; pp. 229–256.

36. Goodman, D.S.; Huang, H.S.; Shiratori, T. Tissue distribution of newly absorbed vitamin A in the rat. J. Lipid Res. 1965, 6, 390–396.

37. Quadro, L.; Hamberger, L.; Colantuoni, V.; Gottesman, M.E.; Blaner, W.S. Understanding the physiological role of retinol-binding protein in vitamin A metabolism using transgenic and knockout mouse models. Mol. Aspects Med. 2003, 24, 421–430.

38. Soprano, D.R.; Blaner, W.S. Plasma Retinol-Binding Protein. In The Retinoids, Biology, Chemistry and Medicine; Sporn, M.B., Roberts, A.B., Goodman, D.S., Eds.; Raven Press: New York, NY, USA, 1994; pp. 257–282.

39. Quadro, L.; Blaner, W.S.; Salchow, D.J.; Vogel, S.; Piantedosi, R.; Gouras, P.; Freeman, S.; Cosma, M.P.; Colantuoni, V.; Gottesman, M.E. Impaired retinal function and vitamin A availability in mice lacking retinol-binding protein. EMBO J. 1999, 18, 4633–4644.

40. Monaco, H.L.; Rizzi, M.; Coda, A. Structure of a complex of two plasma proteins: Transthyretin and retinol-binding protein. Science 1995, 268, 1039–1041.

41. Biesalski, H.K.; Frank, J.; Beck, S.C.; Heinrich, F.; Illek, B.; Reifen, R.; Gollnick, H.; Seeliger, M.W.; Wissinger, B.; Zrenner, E. Biochemical but not clinical vitamin a deficiency results from mutations in the gene for retinol-binding protein. Am. J. Clin. Nutr. 1999, 69, 931–936.

42. Dugan, M.; Rolland, D.C.; Best, D.R. The effects of feeding conjugated linoleic acid on pig liver vitamin A and retinol binding protein mRNA. Can. J. Anim. Sci. 2002, 82, 461–463.

43. Ortiz, B.; Wassef, L.; Shabrova, E.; Cordeddu, L.; Banni, S.; Quadro, L. Hepatic retinol secretion and storage are altered by dietary CLA: Common and distinct actions of CLA c9, t11 and t10, c12 isomers. J. Lipid Res. 2009, 50, 2278–2289.

44. Andreola, F.; Hayhurst, G.P.; Luo, G.; Ferguson, S.S.; Gonzalez, F.J.; Goldstein, J.A.; de Luca, L.M. Mouse liver CYP2C39 is a novel

retinoic acid 4-hydroxylase. Its down-regulation offers a molecular basis for liver retinoid accumulation and fibrosis in aryl hydrocarbon receptor-null mice. J. Biol. Chem. 2004, 279, 3434–3438.

45. White, J.A.; Guo, Y.D.; Baetz, K.; Beckett-Jones, B.; Bonasoro, J.; Hsu, K.E.; Dilworth, F.J.; Jones, G.; Petkovich, M. Identification of the retinoic acid-inducible all-trans-retinoic acid 4-hydroxylase. J. Biol. Chem. 1996, 271, 29922–29927.

46. Johnson, E.F.; Palmer, C.N.; Griffin, K.J.; Hsu, M.H. Role of the peroxisome proliferator-activated receptor in cytochrome P450 4A gene regulation. FASEB J. 1996, 10, 1241–1248.

47. Belury, M.A.; Moya-Camarena, S.Y.; Liu, K.L.; Vanden Heuvel, J.P. Dietary conjugated linoleic acid induces peroxisome-specific enzyme accumulation and ornithine decarboxylase activity in mouse liver. J. Nutr. Biochem. 1997, 8, 579–584.

48. Clement, L.; Poirier, H.; Niot, I.; Bocher, V.; Guerre-Millo, M.; Krief, S.; Staels, B.; Besnard, P. Dietary trans-10, cis-12 conjugated linoleic acid induces hyperinsulinemia and fatty liver in the mouse. J. Lipid Res. 2002, 43, 1400–1409.

49. Seo, J.A.; Kim, N.H.; Park, S.Y.; Kim, H.Y.; Ryu, O.H.; Lee, K.W.; Lee, J.; Kim, D.L.; Choi, K.M.; Baik, S.H.; et al. Serum retinol-binding protein 4 levels are elevated in non-alcoholic fatty liver disease. Clin. Endocrinol. 2008, 68, 555–560.

50. Giordano, E.; Banni, S.; Quadro, L. A single dose of c9, t11 or t10, c12 conjugated linoleic acid isomers perturbs vitamin A metabolism in mice. Nutr. Res. 2011, 31, 855–862.

51. Keller, H. Fatty acids and retinoids control lipid metabolism through activation of peroxisome proliferator-activated receptor-retinoid X receptor heterodimers. Proc. Natl. Acad. Sci. USA 1993, 90, 2160–2164.

52. Contreras, A.V.; Torres, N.; Tovar, A.R. PPAR-α as a key nutritional and environmental sensor for metabolic adaptation. Adv. Nutr. 2013, 4, 439–452.

53. Al Tanoury, Z.; Piskunov, A.; Rochette-Egly, C. Vitamin A and retinoid signaling: Genomic and nongenomic effects. J. Lipid Res. 2013, 54, 1761–1775.

54. Lengqvist, J.; Mata de Urquiza, A.; Bergman, A.C.; Willson, T.M.; Sjovall, J.; Perlmann, T.; Griffiths, W.J. Polyunsaturated fatty acids including docosahexaenoic and arachidonic acid bind to the retinoid X receptor α ligand-binding domain. Mol. Cell. Proteomics 2004, 3, 692–703.

55. Rakhshandehroo, M.; Knoch, B.; Muller, M.; Kersten, S. Peroxisome proliferator-activated receptor α target genes. PPAR Res. 2010, 2010, 1–20.

56. Iannone, A.; Petroni, A.; Murru, E.; Cordeddu, L.; Carta, G.; Melis, M.P.; Bergamini, S.; Casa, L.D.; Cappiello, L.; Carissimi, R.; et al. Impairment of 8-iso-PGF (2α) isoprostane metabolism by dietary conjugated linoleic acid (CLA). Prostaglandins Leukot. Essent. Fatty Acids 2009, 80, 279–287.

57. Morimoto, M.; Reitz, R.C.; Morin, R.J.; Nguyen, K.; Ingelman-Sundberg, M.; French, S.W. Cyp-2E1 inhibitors partially ameliorate the changes in hepatic fatty acid composition induced in rats by chronic administration of ethanol and a high fat diet. J. Nutr. 1995, 125, 2953–2964.

58. Kliewer, S.A.; Forman, B.M.; Blumberg, B.; Ong, E.S.; Borgmeyer, U.; Mangelsdorf, D.J.; Umesono, K.; Evans, R.M. Differential expression and activation of a family of murine peroxisome proliferator-activated receptors. Proc. Natl. Acad. Sci. USA 1994, 91, 7355–7359.

59. Capdevila, J.H.; Falck, J.R.; Harris, R.C. Cytochrome P450 and arachidonic acid bioactivation. Molecular and functional properties of the arachidonate monooxygenase. J. Lipid Res. 2000, 41, 163–181.

60. Rosell, M.; Hondares, E.; Iwamoto, S.; Gonzalez, F.J.; Wabitsch, M.; Staels, B.; Olmos, Y.; Monsalve, M.; Giralt, M.; Iglesias, R.; et al. Peroxisome proliferator-activated receptors-α and -γ, and camp-mediated pathways, control retinol-binding protein-4 gene expression in brown adipose tissue. Endocrinology 2012, 153, 1162–1173.

61. Melis, M.; Carta, G.; Pistis, M.; Banni, S. Physiological role of peroxisome proliferator-activated receptors type α on dopamine systems. CNS Neurol. Disord. Drug Targets 2013, 12, 70–77.

62. Fu, J.; Oveisi, F.; Gaetani, S.; Lin, E.; Piomelli, D. Oleoylethanolamide, an endogenous PPAR-α agonist, lowers body weight and hyperlipidemia in obese rats. Neuropharmacology 2005, 48, 1147–1153.

63. Lo Verme, J.; Fu, J.; Astarita, G.; La Rana, G.; Russo, R.; Calignano, A.; Piomelli, D. The nuclear receptor peroxisome proliferator-activated receptor-α mediates the anti-inflammatory actions of palmitoylethanolamide. Mol. Pharmacol. 2005, 67, 15–19.

64. Re, G.; Barbero, R.; Miolo, A.; Di Marzo, V. Palmitoylethanolamide, endocannabinoids and related cannabimimetic compounds in protection against tissue inflammation and pain: Potential use in companion animals. Vet. J. 2007, 173, 21–30.

CHAPTER 4

The Role iNDF in the Regulation of Feed Intake and the Importance of Its Assessment in Subtropical Ruminant Systems (the Role of iNDF in the Regulation of Forage Intake)

Karen J. Harper [1,*] **and David M. McNeill** [2]

[1] School of Agriculture and Food Science, the University of Queensland, Gatton Campus, Lawes, Queensland 4343, Australia

[2] School of Veterinary Science, the University of Queensland, Gatton Campus, Lawes, Queensland 4343, Australia

ABSTRACT

The intake and digestibility of forages is largely influenced by the fibre content and specifically the neutral detergent fibre (NDF). Currently, the focus in commercial diet formulation and the modelling of animal performance is on the total NDF so as to achieve higher ruminant feed intakes, higher production performance and rumen health. Rations are often formulated for a specific level of NDF in the diet assuming that the digestibility of NDF operates over a narrow range. Forage NDF, particularly in C4 forages, varies greatly in potential digestibility within the rumen. This potential digestibility is defined as the NDF fraction which disappears after a long incubation period and the remaining indigestible component of NDF (iNDF) is unavailable for microbial digestion. It is hypothesized that this dietary iNDF has an important role in contributing to rumen digesta load and voluntary intake. Formulating a diet to a specific level of NDF without reference to the iNDF could markedly affect the resulting intake, digestibility and metabolisable energy (ME) content of the diet. It is concluded that nutritional models need to be modified to accept directly determined iNDF.

KEYWORDS

iNDF; NDF; C4 forages; intake

1. INTRODUCTION

The digestibility of forage and the capacity of ruminants to consume it, are largely influenced by its content of neutral detergent fibre (NDF). Currently, the focus in commercial diet formulation and the modelling of animal performance is on the total NDF to optimise feed intakes and productivity without unduly compromising rumen health. Diets are often formulated to a specific level of NDF with an underlying assumption being that the digestibility of that NDF spans a narrow range. When feeding temperate forages or corn silage this assumption holds true, however, when considering a broader selection of tropical (C4) forages the NDF does vary in ruminal digestibility, to an important extent. This digestibility of the fibre is directly related to potentially digestible portion of NDF (pdNDF) [1,2]. Potential digestibility is defined as the NDF fraction which disappears after a long incubation period, leaving the indigestible component of NDF (iNDF) which is unavailable for microbial digestion [3,4]. Tropical forages can have the same NDF content but differ vastly in iNDF. Nutritional models predict dietary iNDF to be an important driver of rumen digesta load and therefore feed intake [5]. *In vivo*, strongly negative relationships between iNDF and feed intake have also been demonstrated when iNDF content exceeds 15% of total dietary dry matter [6]. *In vivo* data also implicate iNDF as a predictor of organic matter (OM) digestibility in forage-based diets [7]. Formulating a diet to a specific level of NDF without reference to iNDF could markedly affect energy supply through reduced feed intake, digestibility and metabolisable energy (ME) content of the diet. The digestible proportion of NDF has the greatest impact of all energy supplying nutrients on energy supply to the ruminant, in forage-based feeding systems. However, in Australia, despite the importance of iNDF, few nutritionists knowingly take iNDF into direct account when formulating diets. Moreover, iNDF values for a range of forages commonly used in diets are not readily available in feed analysis databases. Better assessment and awareness amongst nutritionists of the importance of iNDF for the range of tropical pasture species commonly used will improve the capacity of nutritionists to predict NDF digestibility and therefore to more effectively develop balanced diets.

2. DEFINING NDF

The NDF is important as a source of ME, a controller of rumen turnover rate, and is required to stimulate saliva production, which buffers the rumen thereby promoting rumen health. The NDF comprises the cell wall fraction of forages and includes a complex matrix of lignin, small amounts of protein, and various polysaccharides, particularly cellulose, hemicellulose and pectin [8]. Cell wall structure and composition vary greatly across species, plant tissues, and within a plant as it matures. The polysaccharides of plant cell walls are potentially digested by the rumen bacteria, protozoa and fungi. Classical reviews [9,10,11,12] identify the significance of the anatomical and chemical structure of leaf and stem, and the detrimental effect of lignin on the capacity of rumen microbes to digest NDF. In initial development, a plant cell wall consists of a primary wall principally made of cellulose. After the cell completes growth and elongation, the secondary thickening in the cell wall occurs with large proportions of cellulose and hemicellulose being laid down. In the rumen, cellulose and hemicellulose are slowly digested, but given enough time can be completely digested if these components are not protected by lignin [13]. Once the secondary wall deposition is completed the primary wall is lignified and the lignification then progresses to deposition of the secondary wall which occurs on the inner surface of the cell lumen. The concentration of lignin is highest in the primary wall but because the secondary wall is greater in volume and mass it contains the greater amount of lignin [14]. Also during the maturation phase there is a greater accumulation of stem mass compared with leaf material with the stems containing more tissue with secondary thickening [15] and subsequently higher concentrations of cellulose, xylan and lignin [16]. In the rumen the lignin component is the primary factor responsible for limiting the digestion of the cell walls [16] and its presence further restricts the digestion of the polysaccharide fraction to which it is cross-linked.

Non-lignified cell walls parts can be accessed by rumen bacteria either on the external exposed surface of the cell wall, internally through the lumen and via adjoining cells walls [17,18]. A high concentration of lignin in the primary walls creates a barrier to microbes, thus preventing total digestion of the plant cell [15,19]. Rumen bacteria digest cells from the interior initially attacking the secondary wall and then progressively digesting the primary wall [19]. The low accessibility of the secondary wall surface to the rumen microorganisms highlights the importance of the mastication and rumination process that causes the physical disruption of the lignified plant cells [15], as well as increasing the surface area available for microbial colonization [16]. It is estimated that within an ingested piece of a typical grass forage approximately one third of the cells leave the rumen without being digested due to the inaccessibility or lack of exposure to rumen microbes [12].

Lignin is a non-carbohydrate polymer composed of phenolic units that are extensively cross-linked to form a lignin-carbohydrate complex. The chemistry of forage lignin and the biosynthesis of various forms have been reviewed by Jung (2011) [20]. Forage lignin is a mixture of guaiacyl and syringyl monolignols with the latter increasing in proportion with maturity. The nature of the lignin-carbohydrate complex varies with the type of cell wall and the species of plant [21]. In grasses, lignin is bound to the hemicellulose fraction through ferulate cross-links which result in lower cell wall digestibility, independent of lignin concentration [22,23]. Various corn mutants, e.g., brown midrib mutants, have been discovered with reduced ferulate ester deposition resulting in improved cell wall digestibility [24]. It is unknown whether C4 forages have a greater amount of ferulate cross-links and would be worthy of examination. In legumes, it is assumed that lignin is cross-linked to the cell wall polysaccharides but the chemical linkage has not been identified [25]. More needs to be understood about the structure and chemistry of lignin within forage tissues if the often presumed predictive relationship between lignin and dietary energy supply is to be improved. Alternatively and more practically, a direct determination of iNDF may circumvent the need to use lignin as a predictor.

3. INDIGESTIBLE NDF AND ITS IMPORTANCE TO VOLUNTARY FEED INTAKE

NDF has been proposed as a reliable predictor of voluntary dry matter (DM) intake under certain conditions [26]. The relationship between

NDF content and food intake is complex and not a linear relationship [27]. The amount and quality of NDF in a diet can either enhance or limit intake. At lower NDF concentrations (7.5%–35.5%), DM intakes increased with increasing dietary NDF concentration, but DM intakes decreased sharply as NDF concentration increased over the range of 22.2%–45.8% in high-producing animals [28]. Low DM intakes at higher NDF contents is associated with rumen fill constraints [2]. The relationship between DM intake and NDF is more than just NDF content in the diet but also dependent on the potential digestibility of NDF (pdNDF) [2,3,29]. The pdNDF fraction, is defined as the difference between the NDF and iNDF. It is the iNDF component that is the rate-limiting constituent of forages at higher NDF level [6].

The iNDF is unavailable to microbial digestion in ruminants even if the total tract residence time of fibre is extended to effectively an infinite time [30]. The lack of digestibility in the iNDF fraction of forage is attributable to the cross-linking between cell wall lignin and hemicellulose [31].

A higher iNDF intake limits a ruminant's ability to consume sufficient forage to meet nutrient requirements. The intake of forage-based diets by ruminants is often controlled by rumen fill and the rate of disappearance [32]. The rate of disappearance is largely influenced by the inherent rate of digestion and passage rate [1,2]. The indigestible portion is removed from the rumen by passage only [3] and will accumulate in the rumen relative to the potentially digestible portion [1], therefore having a longer rumen retention time [33]. A longer retention time in the rumen results in a lower intake [34]. According to Ellis *et al.*, (1999) [5] the determination of iNDF should be included in all basic feedstuff analysis because it is an ideal fraction which has zero digestibility and can be used for the estimation of pdNDF. It is therefore recommended that there should be a defined proportion of iNDF in the diet. Lippke (1986) [6] suggested that maximum iNDF consumption is about 20 g/kgBW$^{0.75}$.day, however more research is required to resolve if this value is relevant for different production systems and different forages.

4. THE USE OF NDF IN DIET FORMULATION

Currently the commercial focus in diet formulation and modelling of higher performance ruminant animals is on total NDF so as to achieve higher intakes of diets containing a higher energy content. The National

Research Council [35] recommends dairy cow diets should contain approximately 30%–35% NDF in the DM. This range represents a balance between increasing voluntary intake of DM and milk production without compromising milk fat [28]. This low level of NDF can be difficult to achieve in subtropical/tropical areas due to the predominance of high NDF forages and so 40% is considered a more achievable target in subtropical grazing systems. In terms of feedlot beef production, the NDF content in the diet is often much lower (16%) [28] due to the high starch diets. Lower levels of NDF in the diet could cause rumen dysfunction therefore a minimal level of NDF is required for animal health, whilst still meeting market specifications.

Balancing diets for high-producing ruminants based on NDF content is still problematic due to the high variability in forage NDF. In our laboratory 200 summer dairy forages (temperature and subtropical) were analysed for NDF where content ranged from 20%–80%. This is not dissimilar to the range (30%–80%) published by Jung and Allen (1995) [36]. The NDF content of a forage increases as the plant matures [16] and is also dependent on the growth environment [16] including temperature, light intensity, water availability and latitude. Even accounting for the large variation in NDF contents of forages, the relationship between NDF level and animal performance often does not hold. Even when diets are composed of similar NDF concentrations the NDF in the constituent forages may have variable degrees of potential digestibility. This variability is especially apparent in subtropical areas. In the analyses of the 200 summer dairy forages the pdNDF ranged from 26% to 90% of the NDF. These large differences in the pdNDF will have a large impact on DM intake. When ruminants are fed a forage diet, increased fibre digestibility generally improves animal performance. Accurate estimation of pdNDF is a prerequisite for improving diet formulation, forage evaluation and productivity responses.

A model based on the concept of potential digestibility was developed by Waldo *et al*., (1972) [3] and is still useful for the examination of intake regulation and digestion [1,2]. Indigestible NDF is a prime determinate of the utilization of forages in many models [37]. Nutritional models such as Cornell Net Carbohydrate and Protein System (CNCPS) [38,39] and the Nordic model of dairy cow metabolism “Karoline” [40,41] also utilise an indirect calculation of iNDF to calculate OM and NDF digestibility, rumen NDF pool and microbial N flow which consequently has effects on the supply of energy and microbial protein.

The effect iNDF content has on food intake is especially apparent in grazing ruminant systems in subtropical and tropical regions. The challenge from the producer's perspective is to ensure higher intakes from forage. Increased production can be achieved by ensuring the iNDF content of the diet is low without necessarily decreasing the NDF content of the diet. This would involve identifying agronomic options to improve forage quality by lowering the iNDF content of both pasture and silages and requires a better assessment of the iNDF content rather than relying on just the NDF content.

However, despite the known importance of iNDF in both intensive and extensive ruminant production systems, particularly in subtropical and tropical areas, such data are still not present in readily available feed analysis data bases used in Australia. The actual values do not exist in well-known databases such as Feedipedia or in feed tables associated with popular nutritional models. The most likely explanation for this lack of information is due to the difficulty in assessing iNDF.

5. ASSESSMENT OF INDF

There are a number of methodologies including *in situ* or *in vitro* techniques, used in the determination of iNDF. The *in situ* techniques attain the degradability of forages by incubating samples in nylon bags which are placed in the rumen of fistulated animals. Bags are then extracted at 10–12 days. This technique is a useful predictor as it ferments feed in the actual rumen environment, and requires minimal effort to perform [42,43]. However, this technique often has poor repeatability due to a lack of standardized procedures [42,44,45].

The alternative long term ruminal *in vitro* methodology is more suitable for larger scale feed evaluations and is more cost effective. Commercial forage testing laboratories offer *in vitro* assays for NDF digestibility, however there is no standard analytical procedure, therefore results across laboratories are essentially not comparable. In addition these are generally short term fermentations (24 h or 48 h) and do not measure the actual iNDF component which require the 10 day fermentation. *In vitro* methodologies are commonly modifications of the Goering and Van Soest (1970) [46] *in vitro* procedure whereby forage is incubated in ruminal fluid, buffer and minerals, followed by a neutral detergent rinse and the residues dried and weighed. This technique is a modification of the two-stage *in vitro* DM digestion technique of Tilley and Terry (1963) [47], whereby feed is incubated in ruminal fluid and buffer, the remaining solid residue is then incubated in acid pepsin to mimic the digestion in the abomasum and the final residues are then dried and weighed. The use of rumen fluid provides the system with a large array of enzymes to effectively digest the feed samples, however, there is significant inter-assay error introduced by the rumen fluid inoculum due to the variable activity related to the host animal from which it is collected and its dietary quality. To maximise the effectiveness of the rumen fluid it is advised that the donor animals should be fed a diet similar to what is being analysed [42,44], the rumen ammonia level must be above 50 mg/L [48], fluid needs to be collected at a fixed time, contamination with oxygen and saliva needs to be avoided, the collected rumen fluid needs to be kept at 39 °C, and there needs to be a rapid transfer to the *in vitro* system [49]. The development of the ANKOM Daisy system was based on a modification of the Goering and Van Soest (1970) [50] technique This technique uses the F57 Ankom filter bags which streamline NDF analysis following digestion, by eliminating the need to filter digesta samples. With all long term fermentations, media needs to be replenished after five days.

Any inter-laboratory variations in the determination of digestion rates and iNDF must be reduced if meaningful estimates of ruminal degradation can be generated for use in predictive models. Some sources of variation are inevitable due to rumen fluid variations, but a large amount of variation can be reduced with the adoption of a common methodology such as the standardisation of forage grinding length, bag porosity, open surface area, and sample weight to bag size [42,44].

Many laboratories find long term fermentations for determining the iNDF fraction laborious and expensive therefore many nutritional models attempt to predict iNDF from the acid detergent lignin (ADL) concentration [51,52,53]. Chandler (1980) [51] estimated the iNDF fraction as the measured ADL content multiplied by 2.4 and this relationship is used in a number of laboratories and nutritional models. This concept was validated using several forage species (corn, alfalfa, grasses and wheat straw) and obtained a satisfactory regression (R^2 = 0.94) between observed and predicted iNDF [54]. This (2.4) factor is presumed to be universal across forage species and growth environments. However, this simplistic mathematical relationship does not hold across C4 forages where large variation in the iNDF/ADL relationship is observed [55]. In this study a diverse set of forages were analysed for iNDF using long term fermentation and was also analysed for ADL. The iNDF/ADL ratios ranged from 1.5 to 6.2 and averaged 3.82 which were markedly higher than the 2.4 ratio used in forages in the CNCPS model used to predict iNDF. This work is supported by other studies [30,56] where attempts to describe forage iNDF from ADL contents generally had low accuracy and precision.

It has been suggested that specific equations for forage groups, based on morphological similarities, may improve the accuracy of iNDF prediction from ADL content [44,57]. In the study by Harper *et al.*, (2014) [55] the iNDF/ADL ratio variation within most subtropical forage groups, was very large. The iNDF/ADL ratio for kikuyu pasture ranged from 1.4–4.8. Variability within single forage species was also found by Huhtanen *et al.*, (2006) [30] whereby diets for individual forage species varied between 2.8 and 5.5. Therefore, iNDF estimations using a simplistic ADL model even within a forage species, would still be unacceptably inaccurate. This raises the issue of the relevance of the calculation of iNDF from ADL in models in many subtropical and tropical areas. Using this equation on C4 forages leads to inaccuracies with regard to predicting NDF digestibility of forages and would result in errors in predicting ME content.

Lignin is not a uniform entity of cell walls and the variability of the ADL relationship with iNDF is probably attributable to the differences in cross-linking between the lignin and hemicelluloses between C4 forage species and as these forages mature [31]. In addition, the ADL relationship may be prone to environmental conditions such as temperature and light intensity [30]. C4 forages are grown across variable and harsh growing conditions and have an extreme nutritive range with maturity. With this in mind, it is not surprising that there were differences in the ADL/iNDF relationship when comparing temperate forages grown in cooler environments with C4 forages grown over a range of harsh environments.

6. IMPACT ON THE PREDICTION OF PRODUCTION

The inclusion of iNDF in nutritional models such as CNCPS, albeit indirectly, indicates its importance as a predictor of animal production. The way in which it is determined (ADL × 2.4) is likely to be unreliable in situations where C4 forages are fed. This is due to the poor relationship between iNDF and ADL found across a wide range of C4 forages. In general, an underestimation of iNDF results in the overestimation of the fractional rate of digestion (K_d) for many of the C4 forages. Modelling readily indicates how true iNDF values can change the predictions of ME supply and milk production in lactating dairy cattle [58] (Table 1). Generally the nutritional models underestimate iNDF content in most C4 forage groups and therefore over-predict potential milk production. The hypothesis that directly measured iNDF will increase model accuracy needs to be vigorously tested using C3 as well as C4 forages.

An over-prediction of ME supply and estimated milk production by CNCPS, when iNDF is calculated from ADL in contrast to actual iNDF estimates, is due to the way in which the potential ME is estimated from the NDF fraction in this model. The potentially digested NDF, therefore the ME available, is estimated from the difference between NDF and iNDF content. In turn, the iNDF content is estimated from the ADL content multiplied by 2.4 [53]. The estimation of iNDF based on the ADL × 2.4 model has further implications on other calculations used to estimate key metabolic activities within the model. This includes the prediction of the available NDF fraction (CB3), the ruminally degraded CB3 (RDCB3), the ruminally escaped CB3 (RECB3) and ruminally escaped iNDF (RECC), which are all affected by the iNDF input [39]. These fractions comprise part of the carbohydrate intake variable which is applied in the equation for calculating ME supply. The implications are, that when the CNCPS model under-estimates the iNDF content in the forages, it could over-predict milk yield in dairy cows.

7. ALTERNATIVE ASSESSMENT OF INDF

Indigestible NDF is a useful index for predicting OM digestibility and its inclusion in feed evaluation models increases accuracy [30]. Its use, however, is constrained by the difficulty in conducting long term fermentation procedures, and it is not accurately predicted from ADL content. It is therefore important to develop short-term techniques to predict iNDF. The application of near infrared reflectance spectroscopy (NIRS) is a rapid, cost effective and accurate forage analytical technique [59]. A close relationship between NIRS and iNDF of grass silage was reported by Nousiainen *et al.*, (2004) [56,60]. The NIR spectral data from a forage sample has a strong correlation with OM components such as NDF, acid detergent fibre, ADL and crude protein [61] which is directly related to organic carbon bonds. These relationships have been used as predictors of various OM components [62,63] with universal calibration equations developed from large (>100) sample sets. The prediction accuracy is dependent on a significant and strong correlation ($R^2 > 0.85$) of the spectral data and wet chemistry [64].

Table 1. Milk production (kg/day) and total metabolisable energy (ME) supply (MJ/day) estimates from the CNCPS, based on predicted iNDF2.4 or *in vitro* iNDF240 of forage groups relevant to dairy diets in Queensland, Australia.

Forage Group	iNDF2.4 (%DM)	iNDF240 (%DM)	K_d (%/h)	ME Supply iNDF2.4	Milk Yield iNDF2.4	ME Supply iNDF240	Milk Yield iNDF240	Δ Milk Yield
Barley silage	13.1	10.5	7	135	12.2	140	13.5	+1.3
Sorghum silage	12.7	22.4	5	117	6.9	99	0.6	−6.3
Tropical pasture	12.2	21.4	5	126	9.5	110	4.4	−5.1
Ryegrass fresh	7.9	10.3	9	146	12.5	141	11.2	−1.3

Inputs were: continuously grazed, 45 month old dairy cows, shrunk LW 576 kg, 120 days in milk, milk fat 4%, DMI 15 kg/day; iNDF240 was inputted by adjusting the ADL content as determined by iNDF240/2.4 [58].

Robust NIRS calibrations have been developed using largely C4 forages for the prediction for iNDF [65]. The use of FT-NIR provides rapid and reproducible results. These calibrations have acceptable prediction for iNDF for use in practical feed evaluation for a broad range of forage samples. Using the iNDF values developed from NIRS calibrations would improve the accuracy and utility of feed and diet evaluation models for ruminants compared with the current technique that uses ADL to estimate iNDF values.

8. MODEL RECOMMENDATIONS

A frustration with many of the current nutrition models is that they do not allow direct input of iNDF values. The CNCPS for example, only allows direct input of ADL values and then calculates iNDF from ADL according to iNDF = ADL × 2.4. In such models, input of actual forage iNDF values can only be achieved by modifying ADL values. For example, if the intended iNDF is 12 units, then the input value for ADL needs to be 12/2.4 = 5 units. This adjustment is awkward and time consuming. Computer models need to be modified to allow easy direct input of iNDF values which are just as easily estimated as chemical determined ADL values.

9. CONCLUSIONS

The quantity of iNDF in a forage-based diet plays a significant role in the regulation of the digestibility of diet and the feed intake of ruminants to which it is offered. This is particularly apparent for C3 *versus* C4 forages, the latter having characteristics such as targeted lignification that allow the plant to cope with harsh environmental stress. These morphological and anatomical characteristics provide support and strength to C4 plants but also provide resistance to microbial degradation [15] and therefore energy availability. The C4 forages often have high NDF concentrations and, depending on stage of maturity and environmental growing conditions, have a generally higher but variable proportion of iNDF in the total NDF. A better assessment of iNDF rather than just NDF is beneficial as a model input to accurately estimate total energy intake of ruminants fed C4 forages. Long-term fermentation experiments on forages provide an accurate assessment of the iNDF content. Alternatively, the iNDF can be estimated with high accuracy and quickly using NIRS. The iNDF of individual forage samples should be included in diet formulation models, especially for diets based on C4 forages. Nutritional models need to be modified to accept directly determined rather than mathematically derived iNDF values as inputs. This would improve the capacity of nutritionists to predict NDF digestibility and therefore balance diets more effectively, particularly in subtropical and tropical ruminant production systems.

ACKNOWLEDGMENTS

The authors would like to acknowledge and thank Dennis Poppi, David Barber and Mark Callow for their participation in the collaborative C4 milk survey. We gratefully acknowledge the funding and support provided by the Queensland Government, The University of Queensland and Dairy Australia.

AUTHOR CONTRIBUTIONS

Karen Harper conceived and wrote this review with the assistance of David McNeill.

CONFLICTS OF INTEREST

The authors declare no conflict of interest.

REFERENCES

1. Allen, M.S.; Mertens, D.R. Evaluating constraints on fiber digestion by rumen microbes. J. Nutr. 1988, 118, 261–270. [PubMed]

2. Poppi, D.P.; France, J.; McLennan, S.R. Intake, passage and digestibility. In Feeding Systems and Feed Evaluation Models; Theodorou, M.K., France, J., Eds.; CAB International: Oxfordshire, UK, 2000; pp. 35–52.

3. Waldo, D.R.; Smith, L.W.; Cox, E.L. Model of cellulose disappearance from the rumen. J. Dairy Sci. 1972, 55, 125–129.

4. Wilkins, R.J. Potential digestibility of cellulose in grasses and its relationship with chemical and anatomical parameters. J. Agric. Sci. 1972, 78, 457–464.

5. Ellis, W.C.; Poppi, D.P.; Matis, J.H.; Lippke, H.; Hill, T.M.; Rouquette, F.M. Dietary-Digestive-Metabolic Interactions Determining the Nutritive Potential of Ruminant Diets; American Society of Animal Science: Champaign, IL, USA, 1999; pp. 423–481.

6. Lippke, H. Regulation of voluntary intake of ryegrass and sorghum forages in cattle by indigestible neutral detergent fiber. J. Anim. Sci. 1986, 63, 1459–1468.

7. Nousiainen, J.; Rinne, M.; Hellamaki, M.; Huhtanen, P. Prediction of the digestibility of the primary growth of grass silages harvested at different stages of maturity from chemical composition and pepsin-cellulase solubility. Anim. Feed Sci. Technol. 2003, 103, 97–111.

8. Moore, K.J.; Hatfield, R.D. Carbohydrates and Forage Quality; Alliance of crop, soul, and environmental science society: Madison, WI, USA, 1994; pp. 229–280.

9. Wilson, J.R. Influence of plant anatomy on digestion and fibre breakdown. In Microbial and Plant Opportunities to Improve the Utilization of Lignocellulose by Ruminants; Akin, D.E., Ljungdahl, J.R., Wilkins, R.J., Harris, P.J., Eds.; Elsevier Science Publishing Company: New York, NY, USA, 1990; pp. 99–117.

10. Wilson, J.R. Plant structures: Their digestive and physical breakdown. In Recent Advances on the Nutrition of Herbivores; Ho, Y.W., Wong, H.K., Abdullah, N., Tajudding, Z.A., Eds.; Malaysian Society of Animal Production: Kuala Lumpur, Malaysia, 1991; pp. 207–216.

11. Wilson, J.R. Organization of forage plant tissues. In Forage Cell Wall Structure and Digestibility; Jung, H.G., Buxton, D.R., Hatfield, R., Ralph, J., Eds.; ASA-CSSA-SSSA: Madison, WI, USA, 1993; pp. 1–32.

12. Wilson, J.R.; Mertens, D.R. Cell-wall accessibility and cell structure limitations to microbial digestion of forage. Crop Sci. 1995, 35, 251–259.

13. Weimer, P.J. Why don't ruminal bacteria digest cellulose faster? J. Dairy Sci. 1996, 79, 1496–1502.

14. Terashima, N.; Fukushima, K.; He, L.F.; Takabe, K. Comprehensive model of the lignified plant cell wall. In Forage Cell Wall Structure and Digestibility; Jung, H.G., Buxton, D.R., Hatfield, R., Ralph, J., Eds.; ASA-CSSA-SSSA: Madison, WI, USA, 1993; pp. 247–270.

15. Wilson, J.R.; Kennedy, P.M. Plant and animal constraints to voluntary feed intake associated with fibre characteristics and particle breakdown and passage in ruminants. Aust. J. Agric. Res. 1996, 47, 199–225.

16. Jung, H.G. Forage Digestibility: The Intersection of Cell Wall Lignification and Plant Tissue Anatomy. Available online: http://dairy.ifas.ufl.edu/rns/2012/12jungrns2012.pdf (accessed on 18 April 2015).

17. Jung, H.G.; Engels, F.M. Alfalfa stem tissues: Rate and extent of cell-wall thinning during ruminal degradation. Neth. J. Agric. Sci. 2001, 49, 3–13.

18. Engels, F.M.; Jung, H.J.G. Alfalfa stem tissues: Impact of lignification and cell length on ruminal degradation of large particles. Anim. Feed Sci. Technol. 2005, 120, 309–321.

19. Engels, F.M. Some properties of cell wall layers determining ruminant digestion. In Physio-Chemical Characterization of Plant Residues for Industrial and Feed Use; Chesson, A., Orskov, E.R., Eds.; Elsevier Applied Science: London, UK, 1989; pp. 80–87.

20. Jung, H.G. Forage lignins and their effects on fiber digestibility. Agron. J. 1989, 81, 33–38.

21. Chesson, A. Lignin polysaccharide complexes of the plant-cell wall and their effect on microbial-degradation in the rumen. Anim. Feed Sci. Technol. 1988, 21, 219–228.

22. Casler, M.D.; Jung, H.J.G. Selection and evaluation of smooth bromegrass clones with divergent lignin or etherified ferulic acid concentration. Crop Sci. 1999, 39, 1866–1873.

23. Grabber, J.H. How do lignin composition, structure, and cross-linking affect degradability? A review of cell wall model studies. Crop Sci. 2005, 45, 820–831.

24. Jung, H.G.; Phillips, R.L. Putative seedling ferulate ester (sfe) maize mutant: morphology, biomass yield, and stover cell wall composition and rumen degradability. Crop Sci. 2010, 50, 403–418.

25. Jung, H.G.; Raeth-Knight, M.; Linn, J.G. Forage fiber digestibility: Measurement, variability, and impact. In Proceedings of the 65th Minnesota Nutritional Conference; University of Minnesota: Minneapolis, MN, USA, 2004. Available online: http://google.sidney-aldebaran.me/url?sa=t&rct=j&q=&esrc= s&source=web&cd=2&ved=0CCUQFjABahUKEwju6K_2-ujHAhXHSo4KHWN5Boc&url=https%3A%2F%2Fbooks.google.com%2Fbooks%2Fabout%2F65th_Minnesota_Nutrition_Conference.html%3Fid%3Di2LXMgEACAAJ&usg=AFQjCNFrf8x1VOoIvKqeSh4UQSGLANmfFA (accessed on 9 September 2015).

26. Minson, D.J. Forage in Ruminant Nutrition; Academic Press: San Diego, CA, USA, 1990; p. 483.

27. Mertens, D.R. Regulation of forage intake. In Forage Quality, Evaluation, and Utilization; American Society of Agronomy: WI, USA, 1994; pp. 450–493.

28. Arelovich, H.M.; Abney, C.S.; Vizcarra, J.A.; Galyean, M.L. Effects of dietary neutral detergent fiber on intakes of dry matter and net energy by dairy and beef cattle: Analysis of published data. Prof. Anim. Sci. 2008, 24, 375–383.

29. Wilkins, R.J. Potential digestibility of cellulose in forage and faeces. J. Agric. Sci. 1969, 73, 57–64.

30. Huhtanen, P.; Nousiainen, J.; Rinne, M. Recent developments in forage evaluation with special reference to practical applications. Agr. Food Sci. 2006, 15, 293–323.

31. Van Soest, P.J. Nutritional Ecology of the Ruminant, 2nd ed.; Comstock: Ithaca, NY, USA, 1994; p. 476.

32. Poppi, D.P. Nutritional constraints for grazing animals and the importance of selective grazing behaviour. In Grassland Productivity and Ecosystem Services; CABI: Wallingford, UK, 2011; pp. 19–26.

33. Poppi, D.P.; Minson, D.J.; Ternouth, J.H. Studies of cattle and sheep eating leaf and stem fractions of grasses. 2. Factors controlling the retention of feed in the reticulo-rumen. Aust. J. Agric. Res. 1981, 32, 109–121.

34. Poppi, D.P.; Minson, D.J.; Ternouth, J.H. Studies of cattle and sheep eating leaf and stem fractions of grasses. 1. The voluntary intake, digestibility and retention time in the reticulo-rumen. Aust. J. Agric. Res. 1981, 32, 99–108.

35. NRC. Nutrient Requirements of dairy Cattle, 7th ed.; National Academy Press: Washington, DC, USA, 2001.

36. Jung, H.G.; Allen, M.S. Characteristics of plant-cell walls affecting intake and digestibility of forages by ruminants. J. Anim. Sci. 1995, 73, 2774–2790. [PubMed]

37. Poppi, D.P. Predictions of food intake in ruminants from analyses of food composition. Aust. J. Agric. Res. 1996, 47, 489–504.

38. Fox, D.G.; Tedeschi, L.O.; Tylutki, T.P.; Russell, J.B.; van Amburgh, M.E.; Chase, L.E.; Pell, A.N.; Overton, T.R. The Cornell Net Carbohydrate and Protein System model for evaluating herd nutrition and nutrient excretion. Anim. Feed Sci. Technol. 2004, 112, 29–78.

39. Tylutki, T.P.; Fox, D.G.; Durbal, V.M.; Tedeschi, L.O.; Russell, J.B.; van Amburgh, M.E.; Overton, T.R.; Chase, L.E.; Pell, A.N. Cornell Net Carbohydrate and Protein System: A model for precision feeding of dairy cattle. Anim. Feed Sci. Technol. 2008, 143, 174–202.

40. Danfaer, A.; Huhtanen, P.; Uden, P.; Sveinbjornsson, J.; Volden, H. The Nordic Dairy Cow Model, Karoline-Description; CABI Publishing: Wallingford, UK, 2006; pp. 383–406.

41. Danfaer, A.; Huhtanen, P.; Uden, P.; Sveinbjornsson, J.; Volden, H. The Nordic Dairy Cow Model, Karoline-Evaluation; CABI Publishing: Wallingford, UK, 2006; pp. 407–415.

42. Nocek, J.E. In situ and other methods to estimate ruminal protein and energy digestibility—A review. J. Dairy Sci. 1988, 71, 2051–2069.

43. Damiran, D.; DelCurto, T.; Bohnert, D.W.; Pulsipher, G.D.; Findholt, S.L. Comparison of techniques and grinding size to estimate digestibility of forage base ruminant diets. Proc. West. Sect. Am. Soc. Anim. Sci. 2002, 53, 341–344.

44. Vanzant, E.S.; Cochran, R.C.; Titgemeyer, E.C. Standardization of in situ techniques for ruminant feedstuff evaluation. J. Anim. Sci. 1998, 76, 2717–2729. [PubMed]

45. Lund, P.; Weisbjerg, M.R.; Ahvenjarvi, S.; Huhtanen, P.; Uden, P.; Olafsson, B.; Volden, H. Nordic ringtest on INDF content and NDF degradation characteristics in three feeds. J. Anim. Feed Sci. 2004, 13, 139–142.

46. Goering, H.K.; Van Soest, P.J. Forage fiber analyses. (Apparatus, reagents, procedures, and some applications). Agriculture Handbook, United States Department of Agriculture; U.S. Agricultural Research Service: Washington, DC, USA, 1970; p. 20.

47. Tilley, J.M.A.; Terry, R.A. A two-stage technique for the in vitro digestion of forage crops.Grass and Forage Sci. 1963, 18, 104–111.

48. Satter, L.D.; Slyter, L.L. Effect of ammonia concentration on rumen microbial protein production in vitro. Br. J. Nutr. 1974, 32, 199–208.

49. Rymer, C.; Huntington, J.A.; Williams, B.A.; Givens, D.I. In vitro cumulative gas production techniques: History, methodological considerations and challenges. Anim. Feed Sci. Technol.2005, 123, 9–30.

50. Goering, H.K.; van Soest, P.J. Forage Fiber Analyses (Apparatus, Reagents, Procedures, and Some Applications); ARS-USDA: Washington, DC, USA, 1970.

51. Chandler, J.A.; Jewell, W.J.; Gossett, J.M.; Vansoest, P.J.; Robertson, J.B. Predicting methane fermentation biodegradability. Biotechnol. Bioeng. 1980, 22, 93–107.

52. Weiss, W.P.; Conrad, H.R.; Stpierre, N.R. A theoretically-based model for predicting total digestible nutrient values of forages and concentrates. Anim. Feed Sci. Technol. 1992, 39, 95–110.

53. Traxler, M.J.; Fox, D.G.; van Soest, P.J.; Pell, A.N.; Lascano, C.E.; Lanna, D.P.D.; Moore, J.E.; Lana, R.P.; Velez, M.; Flores, A. Predicting forage indigestible NDF from lignin concentration. J. Anim. Sci. 1998, 76, 1469–1480.

54. Van Soest, P.J.; Van Amburgh, M.E.; Robertson, J.B.; Knaus, W.F. Validation of the 2.4 times lignin factor for ultimate extent of NDF digestion, and curve peeling rate of fermentation curves into pools. In Proceedings of the Cornell Nutrition Conference for Feed Manufactures, East Syracuse, New York, NY, USA, 18–20 October 2005; Cornell University: Ithaca, NY, USA, 2005; pp. 139–149.

55. Harper, K.J.; Barber, D.G.; Callow, M.; McNeill, D.M.; Poppi, D.P. Assessment of the iNDF of Subtropical Pastures. Available online: http://espace.library.uq.edu.au/view/UQ:354480(accessed on 18 April 2015).

56. Nousiainen, J.; Ahvenjarvi, S.; Rinne, M.; Hellamaki, M.; Huhtanen, P. Prediction of indigestible cell wall fraction of grass silage by near infrared reflectance spectroscopy. Anim. Feed Sci. Technol. 2004, 115, 295–311.

57. Raffrenato, E.; Fievisohn, R.; Cotanch, K.W.; Grant, R.J.; Chase, L.E.; Amburgh, M.E.V. Effect of lignin linkages with other plant cell wall components on in vitro and in vivo NDF digestibility of forages and potential energy yield. In Proceedings of the Energy and Protein Metabolism and Nutrition. 3rd EAAP International Symposium on Energy and Protein Metabolism and Nutrition, Parma, Italy, 6–10 September 2010; Crovetto, G.M., Ed.; Wageningen Academic Publishers: Wageningen, The Netherlands, 2010; pp. 723–724.

58. Harper, K.J.; McNeill, D.M. Direct in vitro measurement of iNDF to improve the prediction of milk production from subtropical forages. In Proceedings of the 8th International Workshop on Modelling Nutrient Digestion and Utilization in Farm Animals, Cairns,

Australia, 15–17 September 2014; McNeill, D.M., Bell, A., Eds.; CSIRO Publishing: Victoria, Australia, 2014; p. 21.

59. Deboever, J.L.; Cottyn, B.G.; Vanacker, J.M.; Boucque, C.V. The use of NIRS to predict the chemical-composition and the energy value of compound feeds for cattle. Anim. Feed Sci. Technol. 1995, 51, 243–253.

60. Kramer, M.; Weisbjerg, M.R.; Lund, P. Estimation of indigestible NDF in feedstuffs for ruminants. Available online: http://forskningsbasen.deff.dk/Share.external?sp=Sce3d98ad-7aaa-444b-b101-2c4e7b1490ab&sp=Sau (accessed on 18 April 2015).

61. Norris, K.H.; Barnes, R.F.; Moore, J.E.; Shenk, J.S. Predicting forage quality by infrared reflectance spectroscopy. J. Anim. Sci. 1976, 43, 889–897.

62. Shenk, J.S.; Westerhaus, M.O.; Hoover, M.R. Analysis of forages by infrared reflectance. J. Dairy Sci. 1979, 62, 807–812.

63. Abrams, S.M.; Shenk, J.S.; Westerhaus, M.O.; Barton, F.E. Determination of forage quality by near-infrared reflectance spectroscopy—Efficacy of broad-based calibration equations. J. Dairy Sci. 1987, 70, 806–813.

64. Mentink, R.L.; Hoffman, P.C.; Bauman, L.M. Utility of near-infrared reflectance spectroscopy to predict nutrient composition and in vitro digestibility of total mixed rations. J. Dairy Sci.2006, 89, 2320–2326.

65. Harper, K.J.; Isherwood, P.; Barber, D.G.; Callow, M.; McNeill, D.M.; Poppi, D.P. Prediction of iNDF of subtropical forages by Near Infrared Reflectance Spectroscopy. Available online:http://espace.library.uq.edu.au/view/UQ:354481 (accessed on 18 April 2015).

CHAPTER 5

Retroviral Infections in Sheep and Goats: Small Ruminant Lentiviruses and Host Interaction

Amaia Larruskain and Begoña M. Jugo *

Genomics and Health Group, Department of Genetics, Physical Anthropology and Animal Physiology, Faculty of Science and Technology, University of the Basque Country UPV/EHU, Bilbao 48080, Spain

ABSTRACT

Small ruminant lentiviruses (SRLV) are members of the Retrovirus family comprising the closely related Visna/Maedi Virus (VMV) and the Caprine Arthritis-Encephalitis Virus (CAEV), which infect sheep and goats. Both infect cells of the monocyte/macrophage lineage and cause lifelong infections. Infection by VMV and CAEV can lead to Visna/Maedi (VM) and Caprine Arthritis-Encephalitis (CAE) respectively, slow progressive inflammatory diseases primarily affecting the lungs, nervous system, joints and mammary glands. VM and CAE are distributed worldwide and develop over a period of months or years, always leading to the death of the host, with the consequent economic and welfare implications. Currently, the control of VM and CAE relies on the control of transmission and culling of infected animals. However, there is evidence that host genetics play an important role in determining Susceptibility/Resistance to SRLV infection and disease progression, but little work has been performed in small ruminants. More research is necessary to understand the host-SRLV interaction.

KEYWORDS

small ruminants; lentivirus; Visna/Maedi; caprine arthritis encephalitis; host genetics

1. SMALL RUMINANT LENTIVIRUSES (SRLVS)

Lentiviruses are a subgroup of the Retrovirus family, which infect a wide range of species including humans and livestock species, causing slow infections that may lead to progressive degenerative diseases. Visna/Maedi Virus (VMV) and Caprine Arthritis-Encephalitis Virus (CAEV) infect sheep and goats. Phylogenetic analyses have established that they are part of a viral continuum and therefore are grouped together as Small Ruminant Lentiviruses (SRLVs) [1,2,3].

1.1. SRLV Infection and Disease

Although SRLVs can infect a variety of cells, their target cells are of the monocyte/macrophage lineage where viral expression is closely dependent on the monocyte to macrophage maturation [4,5,6,7].

SRLVs are mainly transmitted horizontally or from dam to offspring via the respiratory and lactogenic routes. Both asymptomatic and diseased animals are a source of viruses for naive animals [8,9,10, 11,12, 13,14,15,16].

Lesions induced by SRLVs are characterized by mononuclear infiltration in target organs and persistent inflammation [17,18]. SRLVs cause Visna/Maedi (VM) in sheep and Caprine Arthritis-Encephalitis (CAE) in goats, multisystemic inflammatory diseases affecting lungs, the central nervous system (CNS), joints and mammary glands. In both cases, symptoms take several months to years to develop (see Figure 1 for disease timeline). The syndromes may be present to varying degrees in the same animal or occur independently [6,10,19,20,21,22].

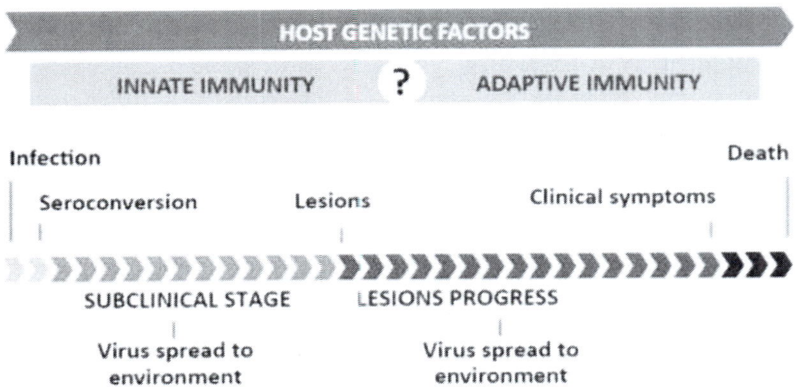

Figure 1. Small Ruminant Lentivirus (SRLV) timeline. SRLVs are transmitted via the respiratory or lactogenic route. Proviral DNA transcription and gene expression is suppressed until infected monocytes mature into macrophages. Once in the target organs, infected macrophages initiate viral replication, which induces an inflammatory cascade. Despite a vigorous immune response, the virus is not cleared and the animal will remain a carrier for life. Most animals remain asymptomatic, but a certain percentage of the animals develop the clinical disease in one or several target organs. Subclinical infection and disease progression may range from months to years. Both Visna/Maedi (VM) and Caprine Arthritis-Encephalitis (CAE) are ultimately fatal. Host genetic factors, infecting viral strain and management influence the occurrence, length and spectrum of affected target organs.

1.2. Relevance and Control of SRLVs

SRLV infection and associated diseases are present in most sheep-rearing countries worldwide, negatively affecting production and animal welfare [1,17]. The economic impact depends on the breed, management system and type of production [6,10,11,23].

SRLV infections are difficult to control due to the ability of the virus to integrate into the host's genome, induce persistent infections and elude neutralization. Vaccination attempts have occasionally caused increased viremia and more severe disease [1,6,18,24,25,26,27,28,29]. Currently, the control of VM and CAE relies on the control of transmission and test and cull schemes of infected animals and their progeny [30,31,32].

The development of disease-resistant breeding stock has been undertaken in other diseases due to lack of effectiveness of vaccines and development of resistance of pathogens to drugs and chemicals, which are becoming increasingly common [33]. However, the detection of SRLV-resistant animals remains problematic since DNA markers have to be identified and the interaction of SRLVs with the host and the pathogenesis of VM and CAE are not fully understood. It is therefore necessary to dissect the antigen and immune response if appropriate control methods are to be developed [34].

2. HOST GENETIC RESOURCES

Since the domestication of sheep and goats human-mediated breeding has generated specialized animals suitable for a wide range of purposes, and hundreds of breeds exist worldwide which often have their own genetic signature [35,36].

There has been a huge increase in the research and application of disease genetics in small ruminants due to the growing knowledge of the influence of host genetics on disease control and the increase in available genetic and genomic tools. Apart from the scientific interest, there is a growing pressure on breeders to select animals that are healthier and more resistant both to infectious and metabolic diseases, which could be feasible with the available biotechnology tools and quickly augmenting knowledge on lentivirus molecular biology/pathogenesis [32,33].

However, in general the development of small ruminant genetic and genomic resources lags behind that of other domestic species such as cattle [37,38,39].

Different markers and strategies can be used to discover disease-associated loci, including microsatellites, single nucleotide polymorphisms (SNP), copy number polymorphisms or variants,*etc.*, which in addition can be complemented with expression assays [38,40]. Current high-throughput techniques allow the rapid genotyping of markers such as SNPs. (see Table 1 for limitations of association studies).

3. HOST-VIRAL PATHOGEN INTERACTION

The interaction between retroviruses (RV) and their hosts is complex. RVs are obligate pathogens that require the host's cellular machinery to complete their cycle. In response, the hosts have developed sophisticated mechanisms to fight viral infections, including innate and adaptive immune systems [50,51]. Observation of retroviral diseases in various species has shown that even in a population infected by a particular virus strain there are usually differences in the outcome of infection and disease of individual animals [52], and studies on host immune response, gene variability and expression could reveal why.

In humans, host genetic factors are important determinants that influence susceptibility to HIV infection and subsequent progression to AIDS, which has been thoroughly studied. Inter-individual variability in susceptibility to HIV-1 infection, transmission, disease progression, and response to antiviral therapy have been attributed to host determinants and variability in multiple genes [53,54].

SRLVs share many characteristics with HIV and indeed, in small ruminants, differences in susceptibility have been observed as well. However, a characteristic that differentiates SRLVs from other lentiviruses is that they do not cause immunodeficiency. In fact, one of the unresolved issues in SRLV infections is the fact that the virus is not cleared despite the host mounting an immune response. It is believed that SRLV proviral DNA escapes detection by the immune system by persisting in monocytes, thus allowing viral dissemination to target tissues. Furthermore, SRLVs may interfere with the function of effector cells implicated in the defense against the virus, such as macrophages and dendritic cells, altering the type of immune responses that are induced [5,11,17,55,56].

Other lentivirus species, although not as extensively studied as HIV, have added information on host genetic factors that may influence lentiviral pathogenesis including Equine Infectious Anemia Virus (EIAV), bovine immunodeficiency virus (BIV), Jembrana disease virus (JDV) and feline immunodeficiency virus (FIV). Among these species, SRLVs, EIAV and BIV/JDB have a tropism for cells of the monocyte/macrophage lineage, while the others also infect lymphocytes. In these non-primate lentivirus species differential susceptibility and associated genes may be found as well [1,7,57].

Table 1. Limitations of genetic association studies in complex multifactorial diseases such as SRLV-induced pathologies. Factors leading to confounding that may be encountered as well as their consequences are shown (information compiled from [41,42,43,44,45, 46,47,48,49].

HOST		GENOTYPING	
Population stratification	Often present in livestock due to breeding practices	**Bias/Partiality**	Case samples are treated with preference
Sample size	Affects power to detect association	**Marker density**	Often few markers are analyzed per gene
Phenotype	Description must be accurate	**Marker frequency**	Frequency of marker variants affect odds of detecting association
Age	Older animals have been exposed for a longer time	**STATISTICAL ANALYSIS**	
Gene effect	Genes involved have small/moderate effects	**Power**	Depends on sample size, marker frequency and gene effect
Presence of other diseases	May facilitate SRLV pathogenesis	**Gene interaction**	Often unaccounted for
SRLVs		**Confounding factors**	Failure to account for them may lead to erroneous interpretations
Different virus strains	Different virulence and host/organ spectrum complicates research; Strain variability may affect detection of infected individuals	**Multiple corrections**	Necessary yet may lead to reject real associations
ENVIRONMENTAL FACTORS		**Consistency/ replicability**	Results must be replicated in different populations/ Can results be replicated in a different population?
Husbandry	Prolonged and crowded housing enhances infection		

4. EVIDENCE POINTING TO THE EXISTENCE OF HOST GENETICS CONTROLLING SRLV PATHOGENESIS

SRLV-induced pathogenesis is a complex process determined by multiple factors (See Table 1 for more details) such as host genetics, virus strain, management practices and presence of other diseases [10,17,32,58,59,60,61,62,63].

The above factors must be accounted for, but do not rule out genetic susceptibility. There is evidence that host genetics play an important role in determining susceptibility/resistance to SRLV infection and pathogenesis, but little work has been performed in small ruminants [64].

Research indicates that a breed (or genetic) predisposition may exist regarding infection susceptibility, provirus levels and lesion occurrence and progression, and thus the influence of the host's genetics can be studied at different levels. The VMV-infected asymptomatic Karakul sheep imported to Iceland in the 1930s never developed signs of the disease but their importation resulted in a widespread dissemination of VM to the extremely susceptible native Icelandic sheep [34]. The research carried out by Icelandic scientists showed that some flocks/family lines had a slower disease progression. In addition, progression of Maedi was delayed in Icelandic x Border Leicester cross individuals [17,32,65]. Research carried out in other countries/breeds has provided further evidence of differential individual and breed susceptibility as well [55,66,67,68,69].

4.1. Infection

Resistance to SRLV infection despite repeated exposure could be interpreted as an innate resistance to infection [64,70]. Significant breed effects on seroprevalence have been observed in comingled flocks, with some sheep and goat breeds being more likely to seroconvert, indicating a possible host genetic restriction. For instance, a higher seroprevalence has been demonstrated in Texel, Border Leicester and Finnish sheep breeds as opposed to Ile de France, Rambouillet and Suffolk [10,23,32,68,71,72]. Saanen and Toggenburg goats are more likely to seroconvert whereas Bedouin Black could be resistant to infection under natural conditions [73,74].

In Latxa dairy-sheep, seroconversion was strongly associated with lifetime maternal VMV-serological status. Offspring of consistently seronegative ewes (>4 years old) were less likely to become infected, suggesting the existence of an inheritable factor of VMV-infection resistance [31].

Herrmann-Hoesing *et al.* [64] detected breed differences in seroprevalence and provirus levels (Rambouillet had lower values in both characters in comparison with Columbia and Polipay), which could imply mechanisms preventing infection and/or aiding in virus control after infection occurs. Interestingly, a correlation has been observed between virus load and the severity of lesions in both sheep and goats [17,21,26,34,75,76] suggesting that breeds/individuals able to control virus levels may be less likely to develop VM and CAE. After infection, there is an acute viremia followed by an immune response that restricts virus replication to low levels; although the infection is not cleared, virus replication is kept at a low level [77]. Some hosts may be better at regulating the extent of viral gene expression in macrophages and perhaps disease progression [1,64]. Latency appears to be controlled by an interaction of cellular and viral transcription factors that regulate viral RNA expression [32,78,79].

4.2. Clinical Disease

Concerning disease progression, it does not progress uniformly in all individuals and rapid progressors have been observed [19,56]. Furthermore, the extent of SRLV-induced lesions and the spectrum of affected organs may depend on the host's genetics as well as the infecting virus strain [10,80]. Certain breeds seem to be more likely to develop the clinical signs, while others remain in a subclinical infection stage. Coarse wool type sheep may be more susceptible than fine wool type sheep [66]. Border Leicester sheep could be more likely than Columbia to develop SRLV-specific lesions [66,67]. Chios and Awassi sheep seem to be highly susceptible to infection but resistant to the development of the clinical disease [32,74,81]. Interestingly, Assaf sheep, a stabilized Awassi x East Friesian cross, show high SRLV seroprevalence and VM clinical disease occurrence including the neurological manifestation [82].

Further evidence is provided by experiments carried out with isogenic twin lambs, which showed that the host's genetics influence the extent and severity of SRLV-induced pulmonary lesions [70].

5. HOST GENETIC FACTORS INVOLVED IN SRLV-INDUCED PATHOGENESIS

SRLV pathogenesis is difficult to analyze due to the strains, different host species and breeds, differential disease progression and affected organ spectrums. It seems that the relationship between small ruminants and SRLVs is complex and that the pathogenesis is likely induced by a number of genes with small or moderate effects [7,56].

The cellular receptor(s) for SRLVs has not been conclusively identified and therefore that information cannot be exploited yet. Most likely it is a common cell membrane molecule as SRLVs can enter other cells apart from target cells, and thus the receptor does not dictate cell tropism [17,34,83,84]. However, classical VMV and CAEV strains appear to use different receptors [7,85,86]. The mannose receptor (MR) is a putative receptor for SRLVs. Crespo *et al.* [87] characterized the ovine MR nucleotide and protein sequence, and its role in VMV infection is currently being studied.

Both innate and adaptive (humoral and cellular) immune responses are induced by SRLVs and various works have identified immune response loci that could influence resistance/susceptibility to SRLV infection and disease, providing evidence that genetic factors might modulate the outcome. Below are some host genetic factors that have been involved in SRLV infection and disease.

5.1. Major Histocompatibility Complex (MHC)

The major histocompatibility complex (MHC) region, a polymorphic multi-gene complex located on chromosome 20 in sheep and chromosome 23 in goats [88,89], has been implicated in SRLV infection and SRLV-induced disease. The MHC Class I and II genes encode receptor glycoproteins that bind and present antigenic peptides to T cells initiating the immune response. The MHC is one of the few polymorphic systems for which it has been possible to establish a functional significance for the observed genetic variation [90]. Its polymorphism and association with infectious disease has been intensely studied in a myriad of diseases with viral etiology, although such studies are scarce in small ruminants.

In the ovine species, VMV infection increases the expression level of the MHC Class II in the lung, the CNS and the synovium [76,91,92,93,94]. In addition, Class I and II polymorphic gene variants have been linked to SRLV provirus levels and disease progression (See Table 2 for more details; [64,95,96]).

Table 2. Major histocompatability complex (MHC) genes/variants associated with VM or CAE pathogenesis.

Region	Variant	Typing method/marker	Species/Breed(s) studied	Reference
Class I	allele CLA Be7	Alloantisera	Goat/Saanen	[97]
	Allele OMHC1* 205	Microsatellite	Sheep/Latxa	[96]
Class II	Alleles DRB1*0403- and DRB1*07012 and various amino acid positions	Cloning and direct sequencing	Sheep/Rambouillet, Polipay & Columbia	[64]
	DRB1*0325	PCR-Sequence-Based Typing	Sheep/Latxa	[95]
	Allele DRB2* 275	Microsatellite	Sheep/Latxa	[96]

The highly polymorphic nature of many loci in the MHC, among which the DR beta 1 (DRB1) gene is a clear example (over 100 alleles identified to date in sheep [64]), makes it difficult to investigate its involvement in SRLV pathogenesis [90]. Genotyping methods with enough resolution are necessary to type MHC loci, and sample sizes are often inadequate to test rare variants. The effect of different MHC alleles remains to be elucidated but it is possible that polymorphisms, depending on the location, may alter the ability of MHC molecules to bind and present antigens, affect the interaction with T cells or elicit differential immune responses [98,99].

5.2. Antibody and T Cell Response

SRLV infection induces antibody response, which has been well studied due to its use in SRLV serologic diagnostics [100]. However, neutralizing antibodies are slow to be induced or have low affinity and relatively low titers. A small amount of free viruses combined with the ability to spread from cell to cell means that propagation and spread cannot be fully stopped by antibodies [7,101].

Nevertheless, some antibody responses may be better at controlling SRLV. Data indicates that the antibody response is skewed in SRLV-infected sheep and goats. Only immunoglobulin G1 (IgG1) subtype responses are detected in sheep whilst in goats IgG1 dominates the humoral response [19,102,103,104,105].

Concerning T cells, both T cell proliferative and cytotoxic responses are induced after SRLV infection [7]. However, the $CD4^+/CD8^+$ cell ratio is decreased in the bronchoalveolar fluid (BALF), synovial fluid and cerebrospinal fluid of diseased animals because of higher levels of $CD8^+$ cells, which could be correlated with the severity of lung lesions [93,94,106,107,108]. Cytotoxic T lymphocytes (CTLs) appear to be important effector cells, for they may inhibit viral replication but also contribute to lesions through cytokine production or cytotoxicity [34,76,109,110].

In ruminants, IgG1 induces a type-2 T helper cell (Th2) response, while production of IgG2 is indicative of a Th1 response [111]. Interestingly, goats with a Th2-biased $CD4^+$ T cell response are known to develop arthritis, but those with a mixed IgG1 and IgG2 response remain without clinical signs, suggesting that anti-SRLV IgG2 antibodies protect against disease induction [7,19,104,112]. In addition, CD80 levels, which favor

Th1 cell differentiation, were low in clinically affected sheep compared with asymptomatic and seronegative controls [80].

5.3. Cytokines and Receptors

Cytokines are important signaling proteins that act as important mediators of the immune system, which initiate and modulate many processes. Studies in diverse species have revealed that cytokine and cytokine receptor expression levels, polymorphisms and profiles in the host can affect the pathogenesis of virus-induced diseases.

Cytokines seem to have an important role in SRLV induced pathogenesis, including the modulation of the ongoing immune activation in target organ lesions with some cytokines favoring the appearance of lesions. Cytokines induce the recruitment and differentiation of monocytes to macrophages as well, favoring additional infection by attracting target cells and thus creating a vicious circle. Indeed, several cytokines have been found up or down regulated (See Table 3) in VM and CAE, although more data on cytokines are necessary [7,10,19,20,112,113,114,115,116,117,118,119,120].

Although most research has focused on expression, a few works have explored the influence of polymorphisms. Cytokine polymorphisms are known to affect cytokine gene transcription [121,122], or they may affect cytokine function thus influencing the outcome of infection and disease. Chemokine receptor 5 (CCR5) is an example of how polymorphisms could affect SRLV pathogenesis. CCR5 participates in leukocyte recruitment against different pathogens, including ovine lentiviruses. A deletion in this locus causes an almost four-fold reduction in transcription in animals carrying the deletion allele, as well as showing association with a decreased provirus-burden. Considering that monocytes/macrophages are the main target cells of VMV, it has been suggested that a reduced chemotactic ability of CCR5 could result in reduced influx of such cells to the site of infection, which could also slow the rate of cellular infection [120,121,123].

Alteration of cytokine profile has been suggested in both sheep, and goats and it is suspected that SRLVs induce a switch to a Type 2 (Th2 type) cytokine response. Evidence backing this is the up regulation of Type 2 cytokines such as anti-inflammatory cytokines IL4 and IL10 [20,80,102,105].

5.4. Innate Immunity and Restriction Factors

The role of innate immunity is gaining interest in the field of lentiviral infections [7]. Genes of the innate immune response are being characterized in small ruminants and their association with SRLV pathogenesis is being tested.

Since the innate immunity senses highly conserved pathogen-associated molecular patterns (PAMPs) via a number of receptors and signaling molecules, polymorphisms in innate immune genes are very likely to affect the host susceptibility to infections. Indeed, such an effect has been observed in a number of species and diseases, and thus genes such as Toll-like receptors (TLRs) and their variation have been suggested to be good candidate prospects in various diseases affecting domestic livestock [124,125].

TLRs are a family of transmembrane signaling molecules that trigger both the innate and the adaptive immune response mechanisms in response to PAMPs, including viral components [50,126].

Mikula *et al*. [127] suggested that individual susceptibility existed in the Tsigai breed that may be defined by the presence of SNPs in TLR7 and TLR8, which sense double stranded viral RNA. Mutations at the TLR7 locus may affect PAMP binding onto the TLR molecule [127,128]. In addition, these two genes have recently been found to be up regulated in diseased sheep [120]. Furthermore, different Toll-like receptors may activate different signaling pathways, with different biological consequences, and thus merit additional research [50].

In response to retroviruses and other transposable elements, which invade the host's genome, hosts have developed intracellular defenses known as restriction factors. These proteins block the viral cycle using different strategies, such as directly attacking viral structures or editing viral genetic material during reverse transcription [129,130]. In recent years there has been a notable increase in this research area.

Ovine TRIM5α seems to be able to restrict VMV DNA synthesis [131]. This intracellular factor restricts retroviral replication by interacting with the capsid and preventing uncoating. A recent study has suggested an association between a tripartite motif protein 5 alpha (TRIM5α) polymorphism that inactivates antiviral activity and accelerates AIDS disease progression [56,132]. Whether such an effect is also present in sheep and goats will still have to be investigated.

Table 3. Cytokine and cytokine receptor genes implicated in SRLV pathogenesis in sheep and goats.

Gene symbol	Gene	Species	Methods	Analyzed material	Parameter analysed	References
IL1β	*Interleukin-1beta*	Sheep	Semiquantitative RT-PCR [1]	Lung	Clinical disease	[114]
IL2/IL2R	*Interleukin-2/ Interleukin-2 receptor*	Sheep	Semiquantitative RT-PCR, qPCR [2]	Lung, PBMCs [3], lymph node leukocytes	Infection, Clinical disease	[114,118,120]
		Goat	*In situ* hybridization	Joints	Infection, Clinical disease	[113]
IL4	*Interleukin-4*	Sheep	Semiquantitative RT-PCR	Lung	Clinical disease	[114]
		Goat	Semiquantitative RT-PCR	PBMCs	Clinical disease	[20]
IL6	*Interleukin-6*	Sheep	RT-PCR	Alveolar macrophages	Infection, Clinical disease	[116]
IL8	*Interleukin-8*	Sheep	RT-qPCR, *in situ* hybridization, Northern blot, Semiquantitative RT-PCR	BALF, alveolar macrophages,	Infection, Clinical disease	[115,117,119]
		Goat	*In situ* hybridization	Macrophages	Infection	[20]
IL10	*Interleukin-10*	Sheep	Semiquantitative RT-PCR	Lung, alveolar macrophages	Infection, Clinical disease	[114,116]
IFNγ	*Interferon-gamma*	Sheep	Semiquantitative RT-PCR	Lung	Clinical disease	[114]
		Goat	Semiquantitative RT-PCR, *in situ* hybridization	PBMCs, joints	Clinical disease	[19,113]
TNFα	*Tumor Necrosis factor-alpha*	Sheep	qPCR	Lung, udder	Clinical disease	[120]
TGF-β1	*Tumor growth factor beta-1*	Sheep	Semiquantitative RT-PCR	Alveolar macrophages	Infection, Clinical disease	[116]
		Goat	*In situ* hybridization	Macrophages	Infection	[20]
MCP-1	*Monocyte chemoattractant protein 1*	Goat	*In situ* hybridization	Macrophages, joints	Infection .Clinical disease	[20,113]
GM-CSF	*Granulocyte macrophage stimulating factor*	Sheep	Semiquantitative RT-PCR	Lung, alveolar macrophages	Infection, Clinical disease	[114,116]
		Goat	*In situ* hybridization	Macrophages	Infection	[20]
CCR5	*Chemokine (C-C motif) Receptor 5*	Sheep	Cloning & sequencing, qPCR	PBMCs, lung	Infection, Clinical disease	[120,121]

[1] RT-PCR: Reverse Transcriptase PCR; [2] qPCR: Quantitative PCR; [3] PBMCs: Peripheral Blood Mononuclear Cells.

Research is needed concerning additional innate immunity genes. Apolipoprotein B mRNA-editing enzyme (APOBEC3) proteins, which act against a wide variety of retroviruses including HIV, have been studied in sheep and they could be good candidates for SRLVs [7,130,133,134,135]. A human APOBEC3G variant has been associated with rapid HIV-1 disease progression [50].

Recently, allele and haplotype variants of the transmembrane protein gene 154 (TMEM154) locus have been associated with SRLV infection, which has been proposed as a locus that could be used for genetic marker-based selection [23]. Sheep with the ancestral variant haplotypes of this gene had an increased risk of becoming infected compared with individuals with mutant forms. However, the function of TMEM14 has not been established yet.

6. CONCLUSIONS

So far, a number of genes implicated in SRLV infection and disease have been identified, but more research is necessary to understand the host-SRLV interaction. A better understanding of loci involved in SRLV infection and pathology and their polymorphism could lead to the identification and selective breeding of naturally resistant animals, thus avoiding culling and reducing viral load and disease development [23,131]. There are encouraging examples where breeding programs have successfully selected animals with enhanced resistance, for example nematodes, mastitis, foot rot and scrapie in sheep; nowadays genetic tools are being employed to aid this selection, as seen in scrapie. However, a critical question will be if there is also a correlation between disease resistance and production traits. Breeding for resistance as a sole objective may not be desirable due to the potential negative impact it may have on other desirable production traits [33,90]. In addition, the molecular characterization of host genetic factors involved in SRLV infection and disease could detect new targets, which could uncover new host pathways leading to opportunities to develop antiviral therapies.

ACKNOWLEDGMENTS

We are grateful to the editorial board of this journal for the opportunity to write this review. The research of Genomics and Health group is funded by the Basque Government Industry Dpt. (grant SA-10/00120 to B.M.J.) and UPV/EHU (grants GIU07/62 and GIU10/22 to B.M.J. and UFI 11/20). We are also grateful to the four anonymous reviewers for their suggestions and constructive criticism.

CONFLICT OF INTEREST

The authors declare no conflict of interest.

REFERENCES AND NOTES

1. Patel, J.R.; Heldens, J.G.M.; Bakonyi, T.; Rusvai, M. Important mammalian veterinary viral immunodiseases and their control. Vaccine 2012, 30, 1767–1781.
2. Leroux, C.; Cruz, J.C.M.; Mornex, J.F. SRLVs: A genetic continuum of lentiviral species in sheep and goats with cumulative evidence of cross species transmission. Curr. HIV Res.2010, 8, 94–100.
3. Leroux, C.; Vuillermoz, S.; Mornex, J.F.; Greenland, T. Genomic heterogeneity in the pol region of ovine lentiviruses obtained from bronchoalveolar cells of infected sheep from France. J. Gen. Virol. 1995, 76, 1533–1537.
4. Gorrell, M.D.; Brandon, M.R.; Sheffer, D.; Adams, R.J.; Narayan, O. Ovine lentivirus is macrophagetropic and does not replicate productively in T lymphocytes. J. Virol. 1992, 66, 2679–2688.
5. Ryan, S.; Tiley, L.; McConnell, I.; Blacklaws, B. Infection of dendritic cells by the Maedi-Visna lentivirus. J. Virol. 2000, 74, 10096–10103.

6. Christodoulopoulos, G. Maedi-Visna: Clinical review and short reference on the disease status in Mediterranean countries. Small Rumin. Res. 2006, 62, 47–53.

7. Blacklaws, B.A. Small ruminant lentiviruses: Immunopathogenesis of Visna-Maedi and caprine arthritis and encephalitis virus. Comp. Immunol. Microbiol. Infect.Dis. 2012, 35, 259–269.

8. Rowe, J.D.; East, N.E. Risk factors for transmission and methods for control of caprine arthritis-encephalitis virus infection. Vet. Clin. North Am. Food Anim. Pract. 1997, 13, 35–53.

9. Blacklaws, B.A.; Berriatúa, E.; Torsteinsdottir, S.; Watt, N.J.; De Andrés, D.; Klein, D.; Harkiss, G.D. Transmission of small ruminant lentiviruses. Vet. Microbiol. 2004, 101, 199–208.

10. Pépin, M.; Vitu, C.; Russo, P.; Mornex, J.F.; Peterhans, E. Maedi-Visna Virus infection in sheep: A review. Vet. Res. 1998, 29, 1–53.

11. Peterhans, E.; Greenland, T.; Badiola, J.; Harkiss, G.; Bertoni, G.; Amorena, B.; Eliaszewicz, M.; Juste, R.A.; Krassnig, R.; Lafont, J.P.; et al. Routes of transmission and consequences of small ruminant lentiviruses (SRLVs) infection and eradication schemes. Vet. Res. 2004, 35, 257–274.

12. Leginagoikoa, I.; Daltabuit-Test, M.; Alvarez, V.; Arranz, J.; Juste, R.A.; Amorena, B.; de Andrés, D.; Luján, L.L.; Badiola, J.J.; Berriatúa, E. Horizontal Maedi-Visna Virus (MVV) infection in adult dairy-sheep raised under varying MVV-infection pressures investigated by ELISA and PCR. Res. Vet. Sci. 2006, 80, 235–241.

13. Gufler, H.; Gasteiner, J.; Lombardo, D.; Stifter, E.; Krassnig, R. Serological study of small ruminant lentivirus in goats in Italy. Small Rumin. Res. 2007, 73, 169–173.

14. Herrmann-Hoesing, L.M.; Palmer, G.G.H.; Knowles, D.P.D. Evidence of proviral clearance following postpartum transmission of an ovine lentivirus. Virology 2007, 362, 226–234.

15. Broughton-neiswanger, L.E.; White, S.N.; Knowles, D.P.; Mousel, M.R.; Lewis, G.S.; Herndon, D.R.; Herrmann-Hoesing, L.M. Non-maternal transmission is the major mode of ovine lentivirus transmission in a ewe flock: A molecular epidemiology study. Infect. Genet. Evol. 2010, 10, 998–1007.

16. Álvarez, V.; Arranz, J.; Daltabuit-Test, M.; Leginagoikoa, I.; Juste, R.A.; Amorena, B.; de Andrés, D.; Luján, L.L.; Badiola, J.J.; Berriatúa, E. Relative contribution of colostrum from Maedi-Visna Virus (MVV) infected ewes to MVV-seroprevalence in lambs. Res. Vet. Sci. 2005,78, 237–243.

17. Thormar, H. Maedi-Visna Virus and its relationship to human immunodeficiency virus.AIDS Rev. 2005, 7, 233–245.

18. Leroux, C.; Mornex, J.F. Retroviral infections in sheep and the associated diseases. Small Rumin. Res. 2008, 76, 68–76.

19. Cheevers, W.P.; Beyer, J.C.; Knowles, D.P. Type 1 and type 2 cytokine gene expression by viral gp135 surface protein-activated T lymphocytes in caprine arthritis-encephalitis lentivirus infection. J. Virol. 1997, 71, 6259–6263.

20. Lechner, F.; Machado, J.; Bertoni, G.; Seow, H.F.; Dobbelaere, D.A.E.; Peterhans, E. Caprine arthritis encephalitis virus dysregulates the expression of cytokines in macrophages. J. Virol.1997, 71, 7488–7497.

21. Ravazzolo, A.P.; Nenci, C.; Vogt, H.R.; Waldvogel, A.; Obexer-Ruff, G.; Peterhans, E.; Bertoni, G. Viral load, organ distribution, histopathological lesions, and cytokine mRNA expression in goats infected with a molecular clone of the caprine arthritis encephalitis virus. Virology 2006, 350, 116–127.

22. Benavides, J.; García-Pariente, C.; Fuertes, M.; Ferreras, M.C.; García-Marín, J.F.; Juste, R.A. Maedi-Visna: The meningoencephalitis in naturally occurring cases. J. Comp. Pathol. 2009, 141, 1–11.

23. Heaton, M.P.M.; Clawson, M.L.M.; Chitko-Mckown, C.G.; Leymaster, K.A.; Smith, T.P.L.; Harhay, G.P.; White, S.N.; Herrmann-Hoesing, L.M.; Mousel, M.R.; Lewis, G.S.; et al. Reduced lentivirus susceptibility in sheep with TMEM154 mutations. PLoS Genet. 2012, 8, e1002467.

24. Cheevers, W.P.; Ho, I.; Beyer, J.C. Immune response to caprine arthritis/encephalitis virus surface protein induced by coimmunization with recombinant vaccinia viruses expressing the caprine arthritis/encephalitis virus envelope gene and caprine interleukin-12. Vaccine2000, 18, 2494–2503.

25. Torsteinsdóttir, S.; Carlsdóttir, H.M.; Svansson, V.; Matthíasdóttir, S.; Pétursson, G. Vaccination of sheep with Maedi-Visna Virus gag gene and protein, beneficial or harmful?Vaccine 2007, 25, 6713–6720.

26. Brodie, S.J.; Marcom, K.A.; Pearson, L.D.; Anderson, B.C.; de la Concha-Bermejillo, A.; Ellis, J.A.; DeMartini, J.C. Effects of virus load in the pathogenesis of lentivirus-induced lymphoid interstitial pneumonia. J. Infect. Dis. 1992, 166, 531–541.

27. Pétursson, G.; Matthíasdóttir, S.; Svansson, V.; Andrésdóttir, V.; Georgsson, G.; Martin, A.H.; Agnarsdóttir, G.; Gísladóttir, E.; Arnadóttir, S.; Högnadóttir, S.; et al. Mucosal vaccination with an attenuated Maedi-Visna Virus clone. Vaccine 2005, 23, 3223–3228.

28. Cheevers, W.P.; Snekvik, K.R.; Trujillo, J.D.; Kumpula-McWhirter, N.M.; Pretty On Top, K.J.; Knowles, D.P. Prime-boost vaccination with plasmid DNA encoding caprine-arthritis encephalitis lentivirus env and viral SU suppresses challenge virus and development of arthritis. Virology 2003, 306, 116–125.

29. Nenci, C.; Zahno, M.L.; Vogt, H.R.; Obexer-Ruff, G.; Doherr, M.G.; Zanoni, R.; Bertoni, G. Vaccination with a T-cell-priming Gag peptide of caprine arthritis encephalitis virus enhances virus replication transiently in vivo. J. Gen. Virol. 2007, 88, 1589–1593.

30. Blacklaws, B.; Bird, P.; McConnell, I. Early events in infection of lymphoid tissue by a lentivirus, Maedi-Visna. Trends Microbiol. 1995, 3, 434–440.

31. Berriatúa, E.; Álvarez, V.; Extramiana, B.; González, L.; Daltabuit, M.; Juste, R. Transmission and control implications of seroconversion to Maedi-Visna Virus in Basque dairy-sheep flocks. Prev. Vet. Med. 2003, 60, 265–279.

32. DeMartini, J.; de la Concha-Bermejillo, A.; Carlson, J.; Bowen, R. Diseases caused by Maedi-Visna and other ovine lentiviruses. In Breeding for Disease Resistance in Farm Animals; Axford, R.F.E., Bishop, S.C., Nicholas, F.W., Owen, J.B., Eds.; CAB International: Oxon, UK, 2000; pp. 301–324.

33. Bishop, S.; Morris, C. Genetics of disease resistance in sheep and goats. Small Rumin. Res.2007, 70, 48–59.

34. Torsteinsdottir, S.; Andresdottir, V.; Arnarson, H.; Petursson, G. Immune response to Maedi-Visna Virus. Front Biosci. 2007, 12, 1532–1543.

35. Lawson-Handley, L.; Byrne, K.; Santucci, F.; Townsend, S.; Taylor, M.; Bruford, M.W.; Hewitt, G.M. Genetic structure of European sheep breeds. Heredity 2007, 99, 620–631.

36. Kijas, J.W.; Townley, D.; Dalrymple, B.P.; Heaton, M.P.; Maddox, J.F.; Wilson, P.; Ingersoll, R.G.; Mcculloch, R.; Mcwilliam, S.; Tang, D.; et al. A Genome wide survey of SNP variation reveals the genetic structure of sheep breeds. PLoS One 2009, 4, e4668.

37. The International Sheep Genomics Consortium; Archibald, A.; Cockett, N.; Dalrymple, B.P.; Faraut, T.; Kijas, J.W.; Maddox, J.F.; Mcewan, J.C.; Oddy, V.H.; Raadsma, H.W.; et al. The sheep genome reference sequence: A work in progress. Anim. Genet. 2010, 2009, 449–453.

38. Ibeagha-Awemu, E.M.; Kgwatalala, E.M.I.P.; Ibeagha, A.E.; Zhao, X. A critical analysis of disease-associated DNA polymorphisms in the genes of cattle, goat, sheep, and pig. Mamm. Genome 2008, 19, 226–245.

39. Becker, D.; Tetens, J.; Brunner, A.; Bürstel, D.; Ganter, M.; Kijas, J.; International Sheep Genomics Consortium; Drögemüller, C. Microphthalmia in Texel sheep is associated with a missense mutation in the paired-like homeodomain 3 (PITX3) gene. PLoS One 2010, 5, e8689.

40. Purdie, A.C.; Plain, K.M.; Begg, D.J.; de Silva, K.; Whittington, R.J.; de Silva, K. Candidate gene and genome-wide association studies of Mycobacterium avium subsp. paratuberculosis infection in cattle and sheep: A review. Comp. Immunol. Microbiol. Infect. Dis. 2011, 34, 197–208.

41. Silverman, E.K.; Palmer, L.J. Case-control association studies for the genetics of complex respiratory diseases. Am. J. Respir. Cell Mol. Biol. 2000, 22, 645–648.

42. Daly, A.K.; Day, C.P. Candidate gene case-control association studies: Advantages and potential pitfalls. Br. J. Clin. Pharmacol. 2001, 52, 489–499.

43. Campbell, H.; Rudan, I. Interpretation of genetic association studies in complex disease.Pharmacogenomics J. 2002, 2, 349–360.

44. Hirschhorn, J.N.; Lohmueller, K.; Byrne, E.; Hirschhorn, K. A comprehensive review of genetic association studies. Genet. Med. 2002, 4, 45–61.

45. Cardon, L.R.; Palmer, L.J. Population stratification and spurious allelic association. Lancet2003, 361, 598–604.

46. Colhoun, H.M.; McKeigue, P.M.; Smith, G.D. Problems of reporting genetic associations with complex outcomes. Lancet 2003, 361, 865–872.

47. Cordell, H.J.; Clayton, D.G. Genetic association studies. Lancet 2005, 366, 1121–1131.

48. Balding, D.J. A tutorial on statistical methods for population association studies. Nat. Rev. Genet. 2006, 7, 781–791.

49. Narum, S.R. Beyond bonferroni: Less conservative analyses for conservation genetics.Conserv. Genet. 2006, 7, 783–787.

50. Bochud, P.Y.; Bochud, M.; Telenti, A.; Calandra, T. Innate immunogenetics: A tool for exploring new frontiers of host defence. Lancet Infect. Dis. 2007, 7, 531–542.

51. Abbas, A.K.; Lichtman, A.H. Basic Immunology, Updated Edition; Saunders Elsevier: Philadelphia, PA, USA, 2006.

52. Maclachlan, N.; Dubovi, E.J. Fenner's Veterinary Virology, 4th ed.; Elsevier: London, UK, 2011.

53. Miyazawa, M.; Tsuji-kawahara, S.; Kanari, Y. Host genetic factors that control immune responses to retrovirus infections. Vaccine 2010, 26, 2981–2996.

54. Kaur, G.; Mehra, N. Genetic determinants of HIV-1 infection and progression to AIDS: Immune response genes. Tissue Antigens 2009, 74, 373–385.

55. Brodie, S. Current concepts in the epizootiology, diagnosis, and economic importance of ovine progressive pneumonia in North America: A review. Small Rumin. Res. 1998, 27, 1–17.

56. Blacklaws, B.A.; Harkiss, G.D. Small ruminant lentiviruses and human immunodeficiency virus: Cousins that take a long view. Curr. HIV Res. 2010, 8, 26–52.

57. Stump, D.S.; VandeWoude, S. Animal models for HIV AIDS: A comparative review. Comp. Med. 2007, 57, 33–43.

58. Angelopoulou, K.; Brellou, G.D.; Vlemmas, I.; Greenland, T. A novel deletion in the LTR region of a Greek small ruminant lentivirus may be associated with low pathogenicity. Virus Res. 2006, 118, 178–184.

59. Oskarsson, T.; Hreggvidsdóttir, H.S.; Agnarsdóttir, G.; Matthíasdóttir, S.; Ogmundsdóttir, M.H.; Jónsson, S.R.; Georgsson, G.; Ingvarsson, S.; Andrésson, O.S.; Andrésdóttir, V. Duplicated sequence motif in the long terminal repeat of Maedi-Visna Virus extends cell tropism and is associated. J. Virol. 2007, 81, 4052–4057.

60. Leginagoikoa, I.; Minguijón, E.; Juste, R.A.; Barandika, J.; Amorena, B.; de Andrés, D.; Badiola, J.J.; Luján, L.; Berriatua, E.; de Andrés, D. Effects of housing on the incidence of Visna/Maedi virus infection in sheep flocks. Res. Vet. Sci. 2010, 88, 415–421.

61. González, L.; Juste, R.A.; Cuervo, L.A.; Idigoras, I.; Saez de Ocariz, C. Pathological and epidemiological aspects of the coexis- tence of Maedi-Visna and sheep pulmonary adenomatosis. Res. Vet. Sci. 1993, 54, 140–146.

62. Andrésdóttir, V.; Tang, X.; Agnarsdóttir, G.; Andrésson, O.S.; Georgsson, G.; Skraban, R.; Torsteinsdóttir, S.; Rafnar, B.; Benediktsdóttir, E.; Matthíasdóttir, S.; Arnadóttir, S.; et al. Biological and genetic differences between lung- and brain-derived isolates of Maedi-Visna Virus. Virus Genes 1998, 16, 281–293.

63. Reina, R.; Grego, E.; Bertolotti, L.; de Meneghi, D.; Rosati, S. Genome analysis of small-ruminant lentivirus genotype E: A caprine lentivirus with natural deletions of the dUTPase subunit, vpr-like accessory gene, and 70-base-pair repeat of the U3 region. J. Virol. 2009, 83, 1152–1155.

64. Herrmann-Hoesing, L.M.; White, S.N.; Mousel, M.R.; Lewis, G.S.; Knowles, D.P. Ovine progressive pneumonia provirus levels associate with breed and Ovar-DRB1.Immunogenetics 2008, 60, 749–758.

65. Palsson, P. Maedi and visna in sheep. In Slow Virus Diseases of Animals and Man; Kimberlin, R.H., Ed.; North-Holland: Amsterdam, The Netherland, 1976; pp. 17–43.

66. Straub, O.C. Maedi-Visna Virus infection in sheep. History and present knowledge. Comp. Immunol. Microbiol. Infect. Dis. 2004, 27, 1–5.

67. Cutlip, R.C.; Lehmkuhl, H.D.; Brogden, K.A.; Sacks, J.M. Breed susceptibility to ovine progressive pneumonia (Maedi/Visna) virus. Vet. Microbiol. 1986, 12, 283–288.

68. Houwers, D.J.; VIsscher, A.H.; Defise, P.R. Importance of ewe/lamb relationship and breed in the epidemiology of Maedi-Visna VIRUS infections. Res. Vet. Sci. 1989, 46, 5–8.

69. Zink, M.C.; Johnson, L.K. Pathobiology of lentivirus infections of sheep and goats. Virus Res.1994, 32, 139–154.

70. De la Concha-Bermejillo, A.; Brodie, S.J.; Magnus-Corral, S.; Bowen, R.A.; DeMartini, J.C. Pathologic and serologic responses of isogeneic twin lambs to phenotypically disinct lentiviruses. J. Acquir. Immune Defic. Syndr. Hum. Retrovirol. 1995, 8, 116–123.

71. Gates, N.L.; Winward, L.D.; Gorham, J.R.; Shen, D.T. Serologic survey of prevalence of ovine progressive pneumonia in Idaho range sheep. J. Am. Vet. Med. Assoc. 1978, 173, 1575–1577.

72. Keen, J.E.; Hungerford, L.L.; Wittum, T.E.; Kwang, J.; Littledike, E.T. Risk factors for seroprevalenceof ovine lentivirus in breeding ewe flocks in Nebraska, USA. Prev. Vet. Med.1997, 30, 81–94.

73. Rowe, J.D.; East, N.E.; Franti, C.E.; Thurmond, M.C.; Pedersen, N.C.; Theilen, G.H. Risk factors associated with the incidence of seroconversion to caprine arthritis-encephalitis virus in goats on California dairies. Am. J. Vet. Res. 1992, 53, 2396–2403.

74. Perk, K. Characteristics of ovine and caprine lentivirus infections. Leukemia 1995, 9, 98–100.

75. Zhang, Z.; Watt, N.J.; Hopkins, J.; Harkiss, G.; Woodall, C.J. Quantitative analysis of Maedi-Visna Virus DNA load in peripheral blood monocytes and alveolar macrophages. J. Virol. Methods 2000, 86, 13–20.

76. Singh, I.; McConnell, I.; Blacklaws, B. Immune response to individual Maedi-Visna Virus gag antigens. J. Virol. 2006, 80, 912–919.

77. Fluri, A.; Nenci, C.; Zahno, M.-L.; Vogt, H.; Charan, S.; Busato, A.; Pancino, G.; Peterhans, E.; Obexer-Ruff, G.; Bertoni, G. The MHC-haplotype influences primary, but not memory, immune responses to an immunodominant peptide containing T- and B-cell epitopes of the caprine arthritis encephalitis virus Gag protein. Vaccine 2006, 24, 597–606.

78. Clements, J.E.; Zink, M.C. Molecular biology and pathogenesis of animal lentivirus infections. Clin. Microbiol. Rev. 1996, 9, 100–117.

79. Staskus, K.A.; Couch, L.; Bitterman, P.; Retzel, E.F.; Zupancic, M.; List, J.; Haase, A.T. In situamplification of visna virus DNA in tissue sections reveals a reservoir of latently infected cells. Microb. Pathog. 1991, 11, 67–76.

80. Reina, R.; Glaria, I.; Benavides, J.; de Andrés, X.; Crespo, H.; Solano, C.; Pérez, V.; Luján, L.; Pérez, M.M.; Pérez de la Lastra, J.M.; et al. Association of CD80 and CD86 expression levels with disease status of Visna/Maedi virus infected sheep. Viral Immunol. 2007, 20, 609–622.

81. Gudnadóttir, M.; Demosthenous, A.; Hadjisavvas, T. Vaccination delays Maedi-Visna lentivirus infection in a naturally-infected sheep flock. BMC Vet. Res. 2013, 9, e16.

82. Benavides, J.; García-Pariente, C.; Ferreras, M.C.; Fuertes, M.; García-Marín, J.F.; Pérez, V. Diagnosis of clinical cases of the nervous form of Maedi-Visna in 4- and 6-month-old lambs.Vet. J. 2007, 174, 255–258.

83. Brodie, S.J.; Pearson, L.D.; Zink, M.C.; Bickle, H.M.; Anderson, B.C.; Marcom, K.A.; Demartinit, J.C.; DeMartini, J.C. Ovine lentivirus expression and disease. Virus replication, but not entry, is restricted to macrophages of specific tissues. Am. J. Pathol. 1995, 146, 250–263.

84. Agnarsdóttir, G.; Thorsteinsdóttir, H.; Oskarsson, T.; Matthíasdóttir, S.; Haflidadóttir, B.S.; Andrésson, O.S.; Andrésdóttir, V. The long terminal repeat is a determinant of cell tropism of Maedi-Visna Virus. J. Gen. Virol. 2000, 81, 1901–1905.

85. Hötzel, I.; Cheevers, W.P. A maedi-visna virus strain K1514 receptor gene is located in sheep chromosome 3p and the syntenic region of human chromosome 2. J. Gen. Virol. 2002, 83, 1759–1764.

86. Hötzel, I.; Cheevers, W.P. Host range of small-ruminant lentivirus cytopathic variants determined with a selectable caprine arthritis-encephalitis virus pseudotype system. J. Virol.2001, 75, 7384–7391.

87. Crespo, H.; Jauregui, P.; Glaria, I.; José, I.S.; Polledo, L.; García-Marín, J.F.; Luján, L.; de Andrés, D.; Amorena, B.; Reina, R. Mannose receptor may be involved in small ruminant lentivirus pathogenesis. Vet. Res. 2012, 43, e43.

88. Amills, M.; Ramiya, V.; Norimine, J.; Lewin, H.A. The major histocompatibility complex of ruminants. Rev. Sci. Tech. 1998, 17, 108–120.

89. Dukkipati, V.S.R.; Blair, H.T.; Garrick, D.J.; Murray, A. "Ovar-Mhc"—Ovine major histocompatibility complex: Structure and gene polymorphisms. Genet. Mol. Res. 2006, 5, 581–608.

90. Sayers, G.; Sweeney, T. Gastrointestinal nematode infection in sheep—A review of the alternatives to anthelmintics in parasite control. Anim. Health Res. Rev. 2005, 6, 159–172.

91. Lee, W.C.; Bird, P.; McConnell, I.; Watt, N.J.; Blacklaws, B. The phenotype and phagocytic activity of macrophages during Maedi-Visna Virus infection. Vet. Immunol. Immunopathol.1996, 51, 113–126.

92. Bergsteinsdóttir, K.; Arnadóttir, S.; Torsteinsdóttir, S.; Agnarsdóttir, G.; Andrésdóttir, V.; Péttursson, G.; Georgsson, G. Constitutive and visna virus induced expression of class I and II major histocompatibility complex antigens in the central nervous system of sheep and their role in the pathogenesis of visna lesions. Neuropathol. Appl. Neurobiol. 1998, 24, 224–232.

93. Harkiss, G.D.; Watt, N.J.; King, T.J.; Williams, J.; Hopkins, J. Retroviral arthritis: Phenotypic analysis of cells in the synovial fluid of sheep with inflammatory synovitis associated with visna virus infection. Clin. Immunol. Immunopathol. 1991, 60, 106–117.

94. Anderson, A.A.; Harkiss, G.D.; Watt, N.J. Quantitative analysis of immunohistological changes in the synovial membrane of sheep

infected with Maedi-Visna Virus. Clin. Immunol. Immunopathol. 1994, 72, 21–29.

95. Larruskain, A.; Minguijón, E.; García-etxebarria, K.; Moreno, B.; Arostegui, I.; Juste, R.A; Jugo, B.M. MHC class II DRB1 gene polymorphism in the pathogenesis of Maedi-Visna and pulmonary adenocarcinoma viral diseases in sheep. Immunogenetics 2010, 62, 75–83.

96. Larruskain, A.; Minguijón, E.; Arostegui, I.; Moreno, B.; Juste, R.A.; Jugo, B.M. Microsatellites in immune-relevant regions and their associations with Maedi-Visna and ovine pulmonary adenocarcinoma viral diseases. Vet. Immunol. Immunopathol. 2012, 145, 438–446.

97. Ruff, G.; Regli, J.G.; Lazary, S. Occurrence of caprine leucocyte class I and II antigens in Saanen goats affected by caprine arthritis (CAE). Eur. J. Immunogenet. 1993, 20, 285–288.

98. Batalia, M.A.; Collins, E.J.; Carolina, N. Peptide binding by class I and class II MHC molecules. Biopolymers 1997, 43, 281–302.

99. Konnai, S.; Takeshima, S.; Tajima, S.; Yin, S.A.; Okada, K.; Onuma, M.; Aida, Y. The influence of ovine MHC class II DRB1 alleles on immune response in bovine leukemia virus infection.Microbiol. Immunol. 2003, 47, 223–232.

100. De Andrés, D.; Klein, D.; Watt, N.J.; Berriatua, E.; Torsteinsdottir, S.; Blacklaws, B.A.; Harkiss, G.D. Diagnostic tests for small ruminant lentiviruses. Vet. Microbiol. 2005, 107, 49–62.

101. Singh, D.K.; Chebloune, Y.; Mselli-Lakhal, L.; Karr, B.M.; Narayan, O. Ovine lentivirus-infected macrophages mediate productive infection in cell types that are not susceptible to infection with cell-free virus. J. Gen. Virol. 1999, 80, 1437–1444.

102. Torsteinsdóttir, S.; Georgsson, G.; Gísladóttir, E.; Rafnar, B.; Pálsson, P.A.; Pétursson, G. Pathogenesis of central nervous system lesions in visna: Cell-mediated immunity and lymphocyte subsets in blood, brain and cerebrospinal fluid. J. Neuroimmunol. 1992, 41, 149–158.

103. Perry, L.L.; Wilkerson, M.J.; Hullinger, G.A.; Cheevers, W.P. Depressed CD4T lymphocyte proliferative response and enhanced

antibody response to viral antigen in chronic lentivirus-induced arthritis. J. Infect. Dis. 1995, 171, 328–334.

104. Trujillo, J.D.; Hotzel, K.J.; Snekvik, K.R.; Cheevers, W.P. Antibody response to the surface envelope of caprine arthritis-encephalitis lentivirus: Disease status is predicted by SU antibody isotype. Virology 2004, 325, 129–136.

105. Bird, P.; Reyburn, H.T.; Blacklaws, B.A.; Allen, D.; Nettleton, P.; Yirrell, D.L.; Watt, N.; Sargan, D.; McConnell, I. The restricted IgG1 antibody response to maedi visna virus is seen following infection but not following immunization with recombinant gag protein. Clin. Exp. Immunol. 1995, 102, 274–280.

106. Watt, N.J.; MacIntyre, N.; Collie, D.; Sargan, D.; McConnell, I. Phenotypic analysis of lymphocyte populations in the lungs and regional lymphoid tissue of sheep naturally infected with Maedi-Visna Virus. Clin. Exp. Immunol. 1992, 90, 204–208.

107. Luján, L.; Begara, I.; Collie, D.D.; Watt, N.J. CD8+ lymphocytes in bronchoalveolar lavage and blood: In vivo indicators of lung pathology caused by Maedi-Visna Virus. Vet. Immunol. Immunopathol. 1995, 49, 89–100.

108. Jolly, P.E.; Gangopadhyay, A.; Chen, S.; Reddy, P.G.; Weiss, H.L. Changes in the leukocyte phenotype profile of goats infected with the caprine arthritis encephalitis virus. Vet. Immunol. Immunopathol. 1997, 56, 97–106.

109. Blacklaws, B.A; Bird, P.; Allen, D.; McConnell, I. Circulating cytotoxic T lymphocyte precursors in Maedi-Visna Virus-infected sheep. J. Gen. Virol. 1994, 75, 1589–1596.

110. Lee, W.C.; McConnell, I.; Blacklaws, B.A. Cytotoxic activity against Maedi-Visna Virus-infected macrophages. J. Virol. 1994, 68, 8331–8338.

111. Estes, D.M.; Brown, W.C. Type 1 and type 2 responses in regulation of Ig isotype expression in cattle. Vet. Immunol. Immunopathol. 2002, 90, 1–10.

112. Snekvik, K.R.; Beyer, J.C.; Bertoni, G.; von Beust, B.R.; Baszler, T.V; Palmer, G.H.; McElwain, T.F.; Cheevers, W.P. Characterization of caprine interleukin-4. Vet. Immunol. Immunopathol.2001, 78, 219–229.

113. Lechner, F.; Vogt, H.R.; Seow, H.F.; Bertoni, G.; Cheevers, W.P.; von Bodungen, U.; Zurbriggen, A; Peterhans, E. Expression of cytokine mRNA in lentivirus-induced arthritis.Am. J. Pathol. 1997, 151, 1053–1065.

114. Woodall, C.J.; Maclaren, L.J.; Watt, N.J. Differential levels of mRNAs for cytokines, the interleukin-2 receptor and Class II DR/DQ genes in ovine interstitial pneumonia induced by Maedi visna virus infection. Vet. Pathol. 1997, 34, 204–211.

115. Legastelois, I.; Cottin, V.; Mornex, J.F.; Cordier, G. Alveolar macrophages from sheep naturally infected by Visna-Maedi virus contribute to IL-8 production in the lung. Vet. Immunol. Immunopathol. 1997, 59, 131–139.

116. Zhang, Z.; Harkiss, G.D.; Hopkins, J. Granulocyte macrophage colony stimulating factor is elevated in alveolar macrophages from sheep naturally infected with Maedi-Visna Virus and stimulates Maedi-Visna Virus replication in macrophages in vitro. Clin. Exp. Immunol. 2002,129, 240–246.

117. Legastelois, I.; Cordier, O.; Cozon, G.; Cador, J. Visna-Maedi virus-induced expression of interleukin-8 gene in sheep alveolar cells following experimental in vitro and in vivoinfection. Res. Virol. 1996, 147, 191–197.

118. Ellis, J.A.; DeMartini, J.C. Ovine interleukin-2: Partial purification and assay in normal sheep and sheep with ovine progressive pneumonia. Vet. Immunol. Immunopathol. 1985, 8, 15–25.

119. Legastelois, I.; Levrey, H.; Greenland, T.; Mornex, J.F.; Cordier, G. Visna-Maedi virus induces interleukin-8 in sheep alveolar macrophages through a tyrosine-kinase signaling pathway. Am. J. Respir. Cell Mol. Biol. 1998, 18, 532–537.

120. Larruskain, A.; Bernales, I.; Luján, L.; de Andrés, D.; Amorena, B.; Jugo, B.M. Expression analysis of 13 ovine immune response candidate genes in Visna/Maedi disease progression.Comp. Immunol. Microbiol. Infect. Dis. 2013, 36, 405–413.

121. White, S.N.; Mousel, M.R.; Reynolds, J.O.; Lewis, G.S.; Herrmann-Hoesing, L.M. Common promoter deletion is associated with 3.9-fold differential transcription of ovine CCR5 and reduced

proviral level of ovine progressive pneumonia virus. Anim. Genet. 2009, 40, 583–589.

122. Shirakawa, I.; Deichmann, K.A.; Izuhara, I.; Mao, I.; Adra, C.N.; Hopkin, J.M. Atopy and asthma: Genetic variants of IL-4 and IL-13 signalling. Immunol. Today 2000, 21, 60–64.

123. Locati, M.; Otero, K.; Schioppa, T.; Signorelli, P.; Perrier, P.; Baviera, S.; Sozzani, S.; Mantovani, A. The chemokine system: Tuning and shaping by regulation of receptor expression and coupling in polarized responses. Allergy 2002, 57, 972–982.

124. Medzhitov, R. Recognition of microorganisms and activation of the immune response.Nature 2007, 449, 819–826.

125. Werling, D.; Jann, O.; Offord, V.; Glass, E.; Coffey, T. Variation matters: TLR structure and species-specific pathogen recognition. Trends mmunol. 2009, 30, 124–130.

126. Takeda, K.; Kaisho, T.; Akira, S. Toll-like receptors. Annu. Rev. Immunol. 2003, 21, 335–376.

127. Mikula, I.; Bhide, M.; Pastorekova, S. Characterization of ovine TLR7 and TLR8 protein coding regions, detection of mutations and Maedi-Visna Virus infection. Vet. Immunol. Immunopathol. 2010, 138, 51–59.

128. Bell, J.K.; Mullen, G.E.; Leifer, C.A.; Mazzoni, A.; Davies, D.R.; Segal, D.M. Leucine-rich repeats and pathogen recognition in Toll-like receptors. Trends Immunol. 2003, 24, 528–533.

129. Zheng, Y.H.; Peterlin, B.B. Intracellular immunity to HIV-1: Newly defined retroviral battles inside infected cells. Retrovirology 2005, 2, e25.

130. Huthoff, H.; Towers, G.J. Restriction of retroviral replication by APOBEC3G/F and TRIM5alpha. Trends Microbiol. 2008, 16, 612–619.

131. Jáuregui, P.; Crespo, H.; Glaria, I.; Luján, L.; Contreras, A.; Rosati, S.; de Andrés, D.; Amorena, B.; Towers, G.J.; Reina, R. Ovine TRIM5αcan restrict Visna/Maedi virus. J. Virol.2012, 86, 9504–9509.

132. Van Manen, D.; Rits, M.A.; Beugeling, C.; van Dort, K.; Schuitemaker, H.; Kootstra, N.A. The effect of Trim5 polymorphisms on the clinical course of HIV-1 infection. PLoS Pathog. 2008,4, e18.

133. LaRue, R.S.; Jónsson, S.R.; Silverstein, K.A.T.; Lajoie, M.; Bertrand, D.; El-Mabrouk, N.; Hötzel, I.; Andrésdóttir, V.; Smith, T.P.L.; Harris, R.S. The artiodactyl APOBEC3 innate immune repertoire shows evidence for a multi-functional domain organization that existed in the ancestor of placental mammals. BMC Mol. Biol. 2008, 9, 10–1186.

134. LaRue, R.S.; Lengyel, J.; Jonsson, S.R.; Andresdottir, V.; Harris, R.S. Lentiviral Vif degrades the APOBEC3Z3/APOBEC3H protein of its mammalian host and is capable of cross-species activity. J. Virol. 2010, 84, 8193–8201.

135. Jónsson, S.R.; Haché, G.; Stenglein, M.D.; Fahrenkrug, S.C.; Andrésdóttir, V.; Harris, R.S. Evolutionarily conserved and non-conserved retrovirus restriction activities of artiodactyl APOBEC3F proteins. Nucleic Acids Res. 2006, 34, 5683–5694.

CHAPTER 6

Aflatoxicosis: Lessons from Toxicity and Responses to Aflatoxin B_1 in Poultry

Melissa S. Monson [1], Roger A. Coulombe [2] and Kent M. Reed [1,*]

[1] Department of Veterinary and Biomedical Sciences, College of Veterinary Medicine, University of Minnesota, St Paul, MN 55108, USA

[2] Department of Animal, Dairy and Veterinary Sciences, College of Agriculture, Utah State University, Logan, UT 84322, USA

ABSTRACT

This review is a comprehensive introduction to the effects of poultry exposure to the toxic and carcinogenic mycotoxin aflatoxin B_1 (AFB_1). The relationship between AFB_1 sensitivity and metabolism, major direct and indirect effects of AFB_1, recent studies of gene expression and transcriptome responses to exposure, and mitigation strategies to reduce toxicity are discussed. Exposure to AFB_1 primarily occurs by consumption of contaminated corn, grain or other feed components. Low levels of residual AFB_1 in poultry feeds can cause reduction in growth, feed conversion, egg production, and compromised immune functions, resulting in significant economic costs to producers. Thus, AFB_1 acts as a "force multiplier" synergizing the adverse effects of microbial pathogens and other agents, and factors detrimental to poultry health. Domestic turkeys (*Meleagris gallopavo*) are one of the most sensitive animals known to AFB_1 due, in large part, to a combination of efficient hepatic bioactivation by cytochromes P450 1A5 and 3A37, and deficient hepatic glutathione-*S*-transferase (GST)-mediated detoxification. Because of their sensitivity, turkeys are a good model to investigate chemopreventive treatments and feed additives for their ability to reduce AFB_1 toxicity. Transcriptome analysis (RNA-seq) of turkey poults (liver and spleen) has identified AFB_1-induced gene expression changes in pathways of apoptosis, carcinogenesis, lipid regulation, antimicrobial activity, cytotoxicity and antigen presentation. Current research focuses on further identifying the molecular mechanisms underlying AFB_1 toxicity with the goal of reducing aflatoxicosis and improving poultry health.

KEYWORDS

turkey; aflatoxin B_1; hepatotoxicity; immunosuppression; feed additives; transcriptome; RNA-seq

1. INTRODUCTION

Dietary exposure to aflatoxins can have severe toxic and carcinogenic effects in humans and animals. The production and metabolism of aflatoxin, symptoms and biomarkers of exposure, and methods to reduce aflatoxicosis have been extensively investigated over the past 50 years [1,2,3,4,5,6,7,8,9]. Most studies have focused on humans, laboratory model species, or agricultural animals, and have identified conserved and species-specific aspects of aflatoxicosis [2,10,11,12,13,14,15]. This review specifically examines the responses of poultry, a particularly sensitive group, to aflatoxin B_1. Aflatoxin metabolism, toxicity, and expression responses in poultry are discussed, along with potential mitigation strategies.

2. ETIOLOGY OF AFLATOXICOSIS

The acute toxicity of dietary aflatoxin was discovered in 1960 when a then unknown disease, termed Turkey "X" Disease, caused the deaths of over 100,000 turkeys (*Meleagris gallopavo*) and other poultry in England [16,17]. Upon examination, the causative agent was identified as imported Brazilian peanut-meal contaminated with aflatoxins [16,18]. Aflatoxins belong to a heterologous group of fungal secondary metabolites called mycotoxins that adversely affect human and animal health. Structurally derivatives of difurocoumarin, aflatoxins are most commonly produced by strains of *Aspergillus flavus*, *A. parasiticus*, and *A. nominus*, although many other *Aspergilli* (including *Emericella* teleomorphs) have aflatoxigenic capabilities [7,19,20,21]. Named according to their blue or green fluorescence under UV light, there are four primary aflatoxins: aflatoxin B_1 (AFB_1), B_2 (AFB_2), G_1 (AFG_1), and G_2 (AFG_2) (Figure 1) [7,21]. Of these, AFB_1 is the most hepatotoxic, the most mutagenic, and the most prevalent worldwide [2,3,21,22].

Livestock, including poultry, are exposed to AFB_1 and other aflatoxins by consuming contaminated feed. Many agricultural feed commodities (corn, cottonseed, peanuts and sorghum) and other foods (figs, tree nuts and spices) are at especially high risk [21,23]. Stress from drought or insect damage can reduce crop resistance to *Aspergilli* and lead to aflatoxin contamination prior to harvest [21,24]. Warm and humid conditions during maturation, harvest, transport or storage, promote *Aspergillus* colonization and subsequent aflatoxin production [21,24]. Temperatures near 30 °C and water activity of 0.99 provide ideal conditions for AFB_1 biosynthesis, although substrate, time, CO_2 levels, and other environmental factors are also important [25,26,27,28,29,30]. Along with primary contamination of crops, aflatoxins can transfer to milk, meat and eggs of livestock and poultry fed the toxins [23,31,32,33,34,35,36,37,38]. Therefore, AFB_1 is a human food safety risk in both plant and animal products.

Animal susceptibility to the acute effects of aflatoxicosis varies widely (Table 1). Domestic turkeys and ducks (*Anas platyrhynchos*) are highly sensitive to both the acute and chronic toxicity of AFB_1 [21,39,40]. Chickens (*Gallus gallus*) are more resistant to acute aflatoxicosis than other poultry species, except during embryonic development (Table 1). Even when exposure does not cause mortality or morbidity, aflatoxicosis contributes directly and indirectly to losses for the poultry industry. While precise numbers are not available, it has been estimated that aflatoxins inflict a loss of at least $143 million each year due to AFB_1-induced hepatotoxicity, reduced performance and secondary infections [23,41].

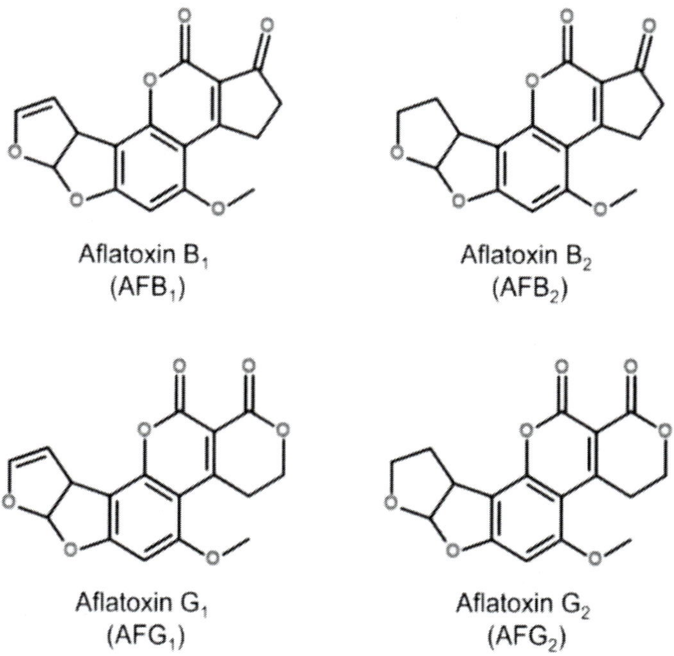

Figure 1. Molecular structures of the four primary aflatoxins.

3. AFB$_1$ METABOLISM AND SENSITIVITY

3.1. Metabolism

Bioactivation is required for AFB$_1$ to be toxic and this processing predominantly occurs in hepatocytes [1,4,7,21]. AFB$_1$ is initially absorbed in the small intestine, especially the duodenum [42]. While bioactivating enzymes with low affinity for AFB$_1$ are present in the small intestine [43], the majority of the toxin is metabolized in the liver, where AFB$_1$ is converted by hepatic cytochromes P450 (*CYP*) enzymes into the reactive and electrophilic *exo*-AFB$_1$-8,9-epoxide (AFBO) (Figure 2) [1,3,4,21,22,40]. An *endo* stereoisomer of the AFBO epoxide can also be produced, but is far less toxic and not relevant to AFB$_1$ toxicity [1,4].

Table 1. Comparative acute toxicity of a single oral dose of aflatoxin B_1 (AFB_1).

Species	Age	Oral LD_{50} (mg/kg Body weight) [1]
Baboon	A	2.0–2.2
Cat	A	0.6
Chicken	E	0.3–5.0
Chicken	Y	6.5–18.0
Dog	A	0.5–1.0
Duck	E	0.5–1.0
Duck	N	0.3–0.6
Guinea Pig	Y	1.4–2.0
Hamster	Y	10.2–12.8
Macaque (Cynomolgus)	A	2.2
Macaque (Rhesus)	A	7.8–8.0
Mouse	N	1.5
Mouse	Y	7.3–9.0
Rabbit	Y	0.3–0.5
Rat	N	0.6–1.0
Rat	Y	5.5–7.4
Rat	A	6.3–18.0
Sheep	A	2.0
Swine	Y	0.6
Trout	Y	0.5
Turkey	Y	1.4–3.2

Lethal dose in 50% (LD_{50}), adult (A), embryo (E), neonate (N), young (Y); [1] compiled from [7,10–12,14,22].

AFBO is a highly unstable intermediate ($t_{0.5} \sim 0.5$ s) and quickly reacts to form adducts with DNA, RNA, and proteins [1,4,7], which are then responsible for AFB_1 toxicity. In most mammals, AFBO is primarily detoxified by glutathione *S*-transferase (GST) enzymes that add glutathione (GSH) to form an 8,9-dihydro-8-(*S*-glutathionyl)-9-hydroxy-AFB_1 (AFB_1-SG) adduct (Figure 2) [3,4,7,21,40]. Although AFBO is the most toxic, other metabolites of AFB_1 include aflatoxicol (AFL), aflatoxin M_1 (AFM_1), aflatoxin P_1 (AFP_1), and aflatoxin Q_1 (AFQ_1,) [3,4,7].

3.2. Sensitivity: Mice to Humans

Sensitivity to AFB_1 is determined not only by the rate of P450-mediated bioactivation, but more importantly by the subsequent detoxification of the reactive AFBO intermediate [1,4]. In mice (*Mus musculus*), P450 enzymes such as P450 2A5, 3A11 and 3A13, effectively activate AFB_1 and can produce large amounts of AFBO [4,44,45,46,47,48,49,50]. However, the murine alpha-class GST (GSTA) enzyme mGSTA3 has high affinity for AFBO, making mice extremely resistant to aflatoxicosis [4,45,51,52,53,54,55]. Interestingly, the orthologous hepatic GST enzymes from rats (*Rattus norvegicus*) can have more than 50 times lower activity towards AFBO [45,54,56]. Effective activation by rat P450 3A2 and 2C11 and minimal conjugation by the rat orthologue of mGSTA3 (rGSTA3, Yc1) make the rat far more susceptible to aflatoxicosis [4,44,45,51,55,57,58,59]. Another rat AFBO-conjugating enzyme rGSTA5 (Yc2) has high affinity for AFBO, but expression is limited to neonatal tissue, some female adults and after antioxidant-induction [59,60,61,62]. Therefore even within rodents, the metabolism and sensitivity to AFB_1 is species-specific.

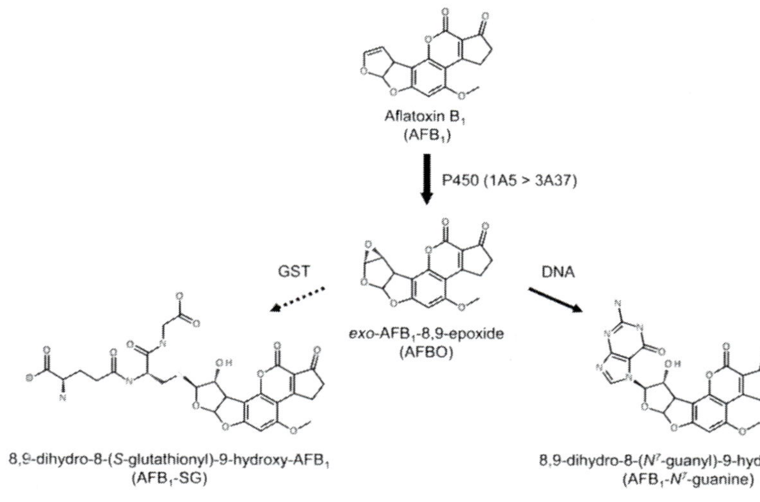

Figure 2. Metabolites and enzymes involved in aflatoxin $B_1(AFB_1)$ metabolism in the turkey liver. Arrow width shows reaction efficiency (wide > narrow), while a dashed line indicates that a reaction does not occur in all birds. Cytochrome P450 (*CYP*) enzymes effectively interact with AFB_1. Domestic turkey glutathione *S*-transferase (GST) enzymes cannot conjugate *exo*-AFB_1-8,9-epoxide (AFBO), although wild and heritage turkey GST enzymes can have activity.

In humans (*Homo sapiens*), multiple hepatic P450 enzymes can activate AFB_1, but P450 1A2 and 3A4 are the primary producers of AFBO [4,43,63,64,65,66,67]. Human P450 3A4 is active at high AFB_1 concentrations, while P450 1A2 has high affinity at low biologically-relevant concentrations [54,66]. Overall, human P450 enzymes produce far less AFBO than rodents, even at high concentrations of AFB_1 (1/4 that of rats, 1/8 that of mice) [48]. Although lower levels of AFBO reduce toxicity, the epoxide is less effectively detoxified in humans. Hepatic GST enzymes in humans can have over 3000 fold less activity against AFBO than murine GST [56]. Human enzymes hGSTA1 and hGSTA2 have little AFBO-conjugating activity [57,59,68]. Instead, some AFB_1-SG adducts are produced by the mu-class GST (GSTM) enzyme hGSTM1 [59,68,69,70,71]. Although hGSTM1 forms more AFB_1-SG adducts from the *endo* stereoisomer, it is able to detoxify mutagenic*exo*-AFBO [68].

3.3. Sensitivity: Poultry

Poultry are sensitive to even low levels of AFB_1, and among species of agricultural importance, the order of sensitivity is ducks > turkeys > Japanese quail (*Coturnix japonica*) > chickens [7,21,72,73,74,75]. Therefore, lower concentrations of AFB_1 are lethal to turkeys and ducks and more adversely affect production and health in these species (Table 2). In domestic turkey, efficient production of AFBO contributes to sensitivity. We have conducted considerable research to determine the P450 enzymes responsible for AFB_1 bioactivation and metabolism in turkey livers. This work revealed two turkey P450 enzymes, encoded by *CYP1A5* and *CYP3A37*, predominantly responsible for converting AFB_1 into AFBO *in vitro* and *in vivo* [76,77,78,79]. P450 1A5 has high-affinity (high V_{max}, K_{cat}; low K_m) and catalyzes the production of both *exo*-AFBO and the detoxified metabolite AFM_1 according to traditional Michaelis-Menten kinetics [76,77]. P450 3A37 is the lower affinity catalyst, exhibiting apparent subunit allostery conforming to Hill enzyme kinetics and producing *exo*-AFBO and AFQ_1 [76,77].

We used polyclonal anti-peptide antibodies in a series of immunoinhibition experiments to determine the relative importance of these P450 enzymes in AFB_1 bioactivation in turkey livers and liver microsomes. Turkey P450 1A5 is the predominant catalyst (>98%) at low (<50 μM) pharmacologically-relevant AFB_1 concentrations commonly seen in the livers of exposed turkeys, while P450 3A37 predominates at much higher AFB_1 concentrations not likely to be achieved in tissues under *in vivo* conditions [78]. Another P450 enzyme, likely orthologous to mammalian P450 2A6, also converts AFB_1 to AFBO in turkeys, chickens, ducks and quail [80,81,82]. However, the specific gene encoding this protein has not been identified in any poultry species [80,81,82,83]. These same studies implicated a P450 1A1 orthologue in AFBO production, which is likely encoded by *CYP1A5* in turkeys [78,80,81,82].

Table 2. Minimum AFB_1 concentrations with major effects in poultry.

Species	Minimum Dietary Contamination Level (ppb) to Cause		
	100% Lethality	Gross Hepatic Lesions	Impaired Production
Chicken	NR (>4000)	800 [1]	800 [1]
Duck	1000 [2]	500 [2]	500 [2]
Goose	4000 [2]	500 [2]	700 [3]
Pheasant	4000 [2]	500 [2]	1000 [2]
Quail (Bobwhite)	ND	ND	700 [3]
Turkey	800 [1]	400 [1]	400 [1]

Compiled from studies that examined aflatoxicosis in multiple poultry species; parts per billion (ppb); not determined (ND); not reached (NR); [1] [39]; [2] [75]; [3] [72].

The activity of P450 enzymes in the turkey, and therefore AFBO production, is inversely related to age, with extreme sensitivity to AFB_1 in young poults [39,77,84,85]. A similar age-dependent sensitivity occurs in chickens [86,87,88]. Hepatic microsomal P450 enzymes from turkeys and quail produce 2 to 4 times more AFBO than ducks or chickens [74]. A later study confirmed the highest AFBO production in turkey, with intermediate levels in duck and quail, and the lowest in chicken [83]. AFBO production correlates with the sensitivity of poultry, except for the duck. In addition, the levels of AFL produced by liver cytosol are highest in turkey, followed by ducks, chickens and quail [74]. The sensitivity of ducks may be more linked to AFL production and toxicity or to oxidative stress [74,83,89].

Like the cytochrome P450 enzymes, hepatic tGSTA enzymes contribute to the severity of aflatoxicosis in the turkey. Six *tGSTA* genes have been amplified, cloned and heterologously expressed turkey liver (*GSTA1.1*, *GSTA1.2*, *GSTA1.3*, *GSTA2*, *GSTA3*, and *GSTA4*) [90,91]. Hepatic cytosolic GST enzymes from domestic turkey have essentially no AFBO-conjugating activity [40,91]. However, tGSTA enzymes can conjugate AFBO to GSH *in vitro* using recombinant proteins in an *E. coli* expression system, suggesting that gene silencing mechanisms or post-transcriptional modifications are likely responsible for their lack of function *in vivo* [91]. The high sensitivity of the domestic turkey to AFB_1 appears to be due to an unfortunate combination of efficient P450 enzymes and dysfunctional GST enzymes that allow accumulation of AFB_1 adducts in the liver.

The effects of AFB_1 exposure on North American wild turkeys are similar to, but less severe than those seen in domestic poultry [92]. This difference in response may be the result of genetic changes that occurred during domestic selection, but could also be due to the separate genetic background of the domestic turkey. Although members of the same species, domestic turkeys were originally derived from the Mexican subspecies (*Meleagris gallopavo gallopavo*) of wild turkey native to Central America; the Eastern subspecies from North America (*Meleagris gallopavo silvestris*) was likely involved in later crosses [93]. Wild turkeys were originally exported to Europe from Mexico in the 1500's and then reintroduced to North America in the 1600's. Reintroduced turkeys were selectively bred, first forming the heritage breeds, and then developed into the modern commercial breeds. Current commercial production predominately utilizes the Broad Breasted White. Selection for production traits in poultry is known to increase metabolic disorders [94] and decrease immune functions [95,96,97,98]. Likewise, detoxification capabilities could be reduced as an unintended side-effect of selection in domestic birds.

Eastern wild turkeys are more resistant to aflatoxicosis than their domestic relatives [92]. AFBO detoxification is likely a main contributor to the differences seen among turkeys. Consistent with this hypothesis, hepatic cytosolic GST enzymes from both the Eastern and Rio Grande (*Meleagris gallopavo intermedia*) subspecies of wild turkey have activity against AFBO [99], unlike their domestic counterparts. Wild turkey GSTA enzymes heterologously expressed *in vitro* also had AFBO-conjugating activity [99]. Like wild strains, the Royal Palm turkey, a heritage breed, retains hepatic and *in vitro* activity [99]. tGSTM enzymes were also amplified, cloned and heterologously expressed from the livers of wild and domestic turkeys, but had no measurable affinity toward AFBO, and therefore are not thought to be involved in AFB_1 detoxification [100].

4. EFFECTS OF AFB_1 EXPOSURE

4.1. AFB_1 Adducts

Formation of AFB_1 adducts is detrimental to cellular processes. AFBO can react with DNA or RNA to form *trans*-8,9-dihydro-8-(N7-guanyl)-9-hydroxy-AFB_1 (AFB_1-N7-guanine) adducts (Figure 2) [1,3,4,7,21]. Some AFB_1-N7-guanine adducts convert into stable AFB_1-formamido-pyrimidine (AFB_1-FAPY) adducts [1,3,7,101]. These AFB_1-DNA and AFB_1-RNA adducts can inhibit transcription and translation, induce DNA mutations during DNA repair and replication, and even initiate apoptosis or carcinogenesis [1,3,4,21,44,102]. AFBO can hydrolyze into AFB_1-8,9-dihydrodiol (AFB_1-diol) and through a series of reactions generate adducts with lysine residues in proteins [4,89,103]. AFB_1-lysine adducts can cause toxicity by impairing protein stability and function [4,89,103]. Although not as well studied, AFB_1 adducts have similar effects on poultry. In chicken primary hepatocytes, the interaction between AFBO and DNA, RNA and proteins has been verified and shown to strongly inhibit synthesis of these macromolecules [104].

4.2. Mutagenicity

Binding of AFBO to DNA introduces G-T transversion mutations in hepatic DNA [4,65,105]. The high incidence of G-T transversions results from AFB_1-FAPY adducts predisposing bypass DNA synthesis machinery to make G-T changes [101]. A G-T transversion in codon 249 of the p53 tumor suppressor has been identified in many human liver cancers and may mechanistically contribute to cancer formation [7,65,71,101,105,106,107]. Chronic AFB_1 exposure, especially in combination with hepatitis-B infections, severely increases the risk of hepatocellular carcinoma in humans [1,3,4,5,6,9,21,105,108]. Dietary AFB_1 is known to have hepatocarcinogenic effects in other mammals, especially the sensitive rat [4,109,110,111]. Based on this evidence from humans and other mammals, AFB_1 is classified as a group I carcinogen by the International Agency for Research on Cancer [4,105]. AFBO is highly mutagenic in poultry, although adenoma and hepatocellular carcinoma have only been reported in ducks [82,112,113,114]. The potential synergistic effect of hepatitis-B virus and AFB_1 has not been consistently reproduced in these studies [112,113,114].

4.3. Production Losses

Beyond its mutagenic effects, AFB_1 negatively affects production values, resulting in economic losses for the poultry industry. Dietary exposure to AFB_1 and other aflatoxins leads to lower weight gain and absolute body weights in both chickens and turkeys [34,39,84,88,115,116,117,118,119, 120,121,122,123,124]. Reduced feed intake [34,115,118,122,123, 124,125] and decreased efficiency of nutrient usage [34,39,84,88, 115,122,125] both contribute to this impaired growth during aflatoxicosis. AFB_1 lowers feed conversion causing poultry to require more feed to produce muscle (broilers and turkeys) [39,84,88,115,122] and eggs (layers) [125]. Although less severe, AFB_1 also reduces feed intake and weight gain in wild turkeys [92]. Similarly, feed consumption was decreased in AFB_1-exposed quail, but body weight and feed conversion were unaffected [126]. In ducks fed AFB_1, both feed intake and weight gain were reduced but without affecting feed efficiency [127].

Exposure to aflatoxins lowers reproductive performance in poultry. In layers fed AFB_1, age to maturity is increased [34] and egg production is reduced [34,121,125,128,129,130,131]. Egg quality parameters such as total weight, shape, albumin or yolk percentage, and shell thickness in chickens and quail can be adversely affected by AFB_1, although the effects were variable among studies [34,125,126,128,131,132,133]. The declines in poultry production traits are often indirect effects of AFB_1 reducing the metabolic potential of the liver. For example, impaired hepatic protein production likely contributes to AFB_1-induced changes within eggs, as the liver is the chief site of synthesis of proteins and lipids incorporated into the egg yolk.

4.4. Hepatotoxicity

Critical to protein synthesis, enzymatic metabolism and detoxification processes, the liver is the primary site of AFB_1 activation and therefore toxicity [1,4,7,21]. Aflatoxicosis in poultry is characterized by an enlarged, pale, and friable liver [10,34,38,39,84,116,117, 120,121, 134,135] Although relative liver weight can initially decrease [117], longer exposure to dietary AFB_1 raises the relative weight of the liver and causes pale or yellowed pigmentation [10,84,88,115,116,117, 120, 121,122,123,124,135,136]. At the cellular level, increased vacuolation of AFB_1-exposed hepatocytes allows high levels of lipids to accumulate

[10,34,39,84,116,120,121,126]. Steatosis is therefore responsible for the changes in liver color and size during aflatoxicosis.

Both acute and chronic AFB_1 consumption by poultry cause other hepatic lesions. Common histopathological signs of AFB_1-induced liver damage include focal necrotic hepatocytes or hemorrhages [10,34,39, 84,116,121,137]. Acute damage initiates inflammatory responses and leads to leukocyte infiltration and proliferation in the liver [10,34,84, 112,116,138]. Short-term exposure to higher dose can cause morbidity and mortality from extensive liver damage [3,21]. In poultry, chronic AFB_1 consumption is mutagenic and leads to remodeling of liver tissues. Hyperplasia of bile duct epithelial cells or oval cells develops first, followed by periportal fibrosis and nodular tissue regeneration [10,34,39,84,112,116,121,126,134,139].

AFB_1 adducts with biomolecules cause damage to hepatocytes that impairs metabolic functions of the liver during AFB_1 exposure. This is exemplified by AFB_1-reduced total serum protein levels, as the liver is responsible for production of most circulating proteins [88,115,117, 118,119,123,127,129,140,141]. Aflatoxicosis negatively affects albumin, globulin, cholesterol, and triglyceride levels in serum [88,115,117, 119,127,129,136,140,141,142]. Multiple blood coagulation factors are produced in the liver; the activity of these clotting factors and the serum levels of fibrinogen are diminished by AFB_1 in both chickens and turkeys [142,143,144,145,146]. Hepatic protein concentrations also decrease in AFB_1-fed chickens [88]. Protein content likely declines because AFB_1-DNA adducts inhibit transcription or translation and AFB_1-lysine adducts result in protein degradation or excretion. Reduced synthesis of enzymes in the liver would have systemic effects on poultry metabolism. For example, the decreased hepatic fatty acid synthesis observed in AFB_1-exposed chickens [147] could be responsible for lower production of serum cholesterol and triglycerides [129,140].

4.5. Immunotoxicity

The avian immune system relies on the bursa of Fabricius, thymus and spleen to produce mature or active leukocytes. Even at low dietary concentrations, AFB_1 can damage these immune tissues and suppress innate and adaptive immune responses [3,7,148]. AFB_1 consumption during growth can lead to immune tissue atrophy, reducing relative weights of the bursa, spleen and thymus [120,122,134,139, 149,150, 151,152]. Increases in relative spleen weight have also been observed during aflatoxicosis [117,120,153,154,155,156]. In young chickens, tissue changes are concomitant to the development of histopathological lesions. AFB_1 exposure causes visible congestion in the spleen and thymus, while nuclear debris accumulates in the thymus and bursa [139,149,150,156]. In AFB_1-exposed chicks, vacuoles increase in the lymphoid follicles of the bursa and the white pulp of the spleen, especially the T-cell rich periarteriolar lymphoid sheaths [139,150,156].

At the cellular level, innate and adaptive cell-mediated immune functions are impaired by AFB_1exposure in poultry [148]. *In vivo* exposure to AFB_1 and other aflatoxins has been shown to decrease phagocytic activity in chicken leukocytes, including heterophils [157,158], macrophages [159,160,161], and monocytes [162]. Reduced phagocytosis was demonstrated *in vitro* for peritoneal macrophages isolated from both chickens and turkeys, although microsomally activated AFB_1 was required [163,164]. In contrast, the phagocytic activity of thrombocytes was not affected in chickens fed an aflatoxin-contaminated diet [165]. Dietary AFB_1 inhibits T lymphocyte activation in both chickens and turkeys as evidenced by delayed hypersensitive skin tests and graft-versus-host response tests [39,84,116,122,159,166,167].

AFB_1 can induce circulating lymphocytopenia [34,159,166,168] and cause lymphoid depletion in the bursa, spleen and thymus [34,134,139,149,150,159,166]. This likely results from increased apoptosis of splenocytes, thymocytes and bursal B-cells as seen in young chickens during aflatoxicosis [139,149,156,169,170]. In the chicken, both $CD4^+$ and $CD8^+$ T lymphocytes in the spleen, thymus, peripheral blood, and even the ileum can be affected by AFB_1 exposure [149,150,156,168,171,172]. Oxidative stress and DNA damage from AFB_1 are likely responsible for initiating apoptotic processes in lymphocytes [169,170]. Together these losses reduce the adaptive immune potential of poultry fed AFB_1.

Although some studies only observed effects on cell-mediated immunity [39,84,116], others found that AFB_1 can similarly diminish innate and adaptive humoral immune capabilities. Total serum complement is decreased by feeding AFB_1 to chickens, ducks and turkeys [115,127,173,174,175]. Antibody titers are often reduced, whether measured as total serum levels of IgA, IgG and IgM [139,141,167], as production of specific antibodies in response to sheep red blood cells [122,152,159] or as exposure response to infectious bronchitis virus, infectious bursal disease virus, Newcastle disease virus or *Pasteurella multocida* [128,176,177]. AFB_1 can also impair the effectiveness of vaccination for these poultry diseases [128,158,176,177,178].

As a consequence of AFB_1-driven immunosuppression, exposed poultry have lower resistance to secondary infections [148]. Poultry with aflatoxicosis are more susceptible to the bacterial diseases, fowl cholera [151,176,179,180,181] and salmonellosis [182,183,184]. Exposure to AFB_1 can increase the severity of the protozoan disease, cecal coccidiosis [185,186,187,188] and the fungal infection, crop mycosis [189]. Similarly, AFB_1 consumption decreases resistance to viral pathogens, including infectious bronchitis virus [128,177], infectious bursal disease virus [128,176,177,190,191], Marek's disease virus [185,192], and Newcastle disease virus [128,158,177,178,193]. These secondary infections dramatically increase the economic losses attributed to AFB_1 exposure. Thus, AFB_1 is a potent immunotoxin and acts as a synergistic "force multiplier" that can enhance the incidence and impacts of avian diseases.

4.6. Intestinal Toxicity

Nutrients are primarily absorbed by epithelial cells in the small intestine and these cells facilitate uptake of AFB_1. Most AFB_1 is transferred directly into the blood stream; however, as in mammals [43], some may be metabolized in intestinal tissues. Therefore in poultry, both local and systemic effects of AFB_1 exposure likely occur in the small intestine, but these are not well characterized. Dietary exposure to AFB_1 can lower the unit weight (length/weight) of the duodenum and jejunum [194,195] and affect tissue morphology. In chickens, AFB_1 has been shown to raise crypt depth in the jejunum [196], decrease villus height in the duodenum [197], and lower the ratio between villus height/crypt depth in all three sections of the small intestine [197]. However, these histopathological effects may not be pervasive [198].

The severity of aflatoxicosis in poultry can be affected by nutrition [7]. Deficiencies in some dietary vitamins raised AFB_1 sensitivity in chickens, although feeding vitamins in excess was not protective [199,200]. Dietary supplementation with tryptophan also increased hepatotoxicity of AFB_1 [201]. Conversely, diets high in fat or protein may be beneficial to chickens and turkeys fed AFB_1 [202,203,204].

Direct investigations of AFB_1-effects on absorption or retention of individual nutrients have had variable results [34,194,196,205, 206,207,208]. Exposure to AFB_1 decreases the apparent metabolizable energy poultry can obtain from their diet [34,195,196,205,206,207]. Therefore, increased dietary nutrients are needed to compensate for impaired uptake. It is currently unclear how much reduced nutrient uptake in the intestine contributes to AFB_1 effects on growth and feed efficiency in poultry.

4.7. Embryotoxicity

Exposure during development recapitulates many of signs of aflatoxicosis seen in hatched chicks and poults. Although most studies are carried out by *in ovo* AFB_1 injections, embryonic exposure to the toxin is a known risk to poultry. AFB_1 and its metabolites can be transferred from the laying hen into the albumin and yolk of the egg [31,34,37,128,131,133,209]. AFM_1 is a common metabolite detected in eggs and, while not as carcinogenic as AFB_1, is acutely toxic [4,37,133]. Contamination of unfertilized shell eggs is therefore a food safety risk when used for human consumption.

Transfer of aflatoxins into embryonated eggs is also a concern for poultry producers. In experimental settings, *in ovo* exposure of chickens or turkeys to AFB_1 caused DNA damage in the embryonic liver and increased embryo mortality [131,161,210,211,212,213,214,215]. When introduced into the maternal diet to simulate the natural route of embryonic exposure, AFB_1caused reduced hatchability [130,131,161]. Consistent with studies in hatched poultry, turkeys may be more sensitive to embryotoxic effects than chickens [215]. Since embryonic liver has an active protein from the *CYP1A* family [216], hepatotoxicity in turkey embryos is likely mediated by the same P450 1A5 as in poults.

Embryonic AFB_1 exposure can lead to morphological defects [212], such as abnormal area opaca cells [210,214], skeletal defects in the tibia growth plate [213], and inhibition of bursal follicle development [210,214]. These mutagenic effects can reduce embryo viability and adversely affect hatched progeny. *In ovo* AFB_1-driven immunosuppression has the potential to increase the incidence of infectious disease in young poultry and negatively affect their health and productivity. Whether *in ovo* injection or maternal feeding, chickens exposed to AFB_1 during embryogenesis have compromised cellular and humoral immune functions post hatch [160,161,211,217,218].

5. GENE EXPRESSION AND AFB_1

5.1. P450 and GST Enzymes

Gene expression can improve our understanding of responses to AFB_1 and provide targets to modulate the mechanisms of toxicity. Both cellular responses to toxicity and AFB_1 inhibition of transcription and translation will affect gene expression [102]. For example, AFB_1 down-regulates*p53* expression in human cells [67] and in tissue from hepatocellular carcinomas in rats [219], likely due to mutations introduced by AFB_1-DNA adducts. Expression of AFB_1-metabolizing genes can also be affected by AFB_1. Exposure in the rat liver to AFB_1 can increase expression of P450 (*CYP3A*,*CYP4F1*), and *rGSTM2* genes (Yb2) [220]. Another study in rats identified effects on multiple *P450*and *GST* genes, with greatest up-regulation in *rGSTA5* and pi-class GST (GSTP) *rGSTP1* [221]. In chickens, dietary AFB_1 up-regulated hepatic expression of *CYP1A1* and *CYP2H1*, and down-regulated expression of epoxide hydrolase (*EH*) and *GSTA*, although the specific *GSTA* gene target was not identified [123,124]. Other *P450* family members were significantly down-regulated [124]. Interestingly, a recent study in broilers observed the opposite changes in the liver, down-regulation of *CYP1A1* and up-regulation of *EH* and *GST* [115].

5.2. Cytokines and the MHC

AFB_1 is known to initiate hepatic inflammation, and correspondingly, was shown to affect expression of pro-inflammatory cytokines in both mammals and poultry [124,171,222,223,224,225,226]. Splenic expression of interleukins 1 beta (*IL1*β), 6 (*IL6*), 10 (*IL10*), interferon gamma (*IFN*-γ) and tumor necrosis factor alpha (*TNF*-α) in pigs (*Sus scrofa*) were increased by AFB_1 exposure [225]. Expression of *IFN*-γ and *TNF*-α increased in rats during aflatoxicosis, and protein levels of IL-1, IL-2, and IL-6 were also modulated [224,226]. In chickens, dietary AFB_1 increased hepatic expression of *IL6* [115,124]. However, in another experiment [123], expression of the IL6 receptor (*IL6R*) and IL10 receptor beta (*IL10RB*) were reduced in the liver. Exposure to AFB_1 reduced ileac expression of *IL2*, interleukin 4 (*IL4*), *IL6*, *IL10*, interleukin 17 (*IL17*), and *IFN*-γ in chickens [171,172]. Lipopolysaccharide-induced TNF factor (*LITAF*) expression also decreased in the intestine [171,172]. This gene is used as a marker for

TNF activity since *TNF-α* has not been identified and may not be present in birds. Lastly, serum protein levels of IL-2 and IFN-γ were lower in chickens exposed to AFB_1 [168]. Down-regulation of these cytokines and their receptors in poultry may result in decreased T lymphocyte activation and proliferation.

Both up- and down-regulation of major histocompatibility complex (MHC) class I genes in response to AFB_1 exposure was observed in pigs [227], rats [221], and rainbow trout (*Oncorhynchus mykiss*) [228]. The MHC is a highly polymorphic genomic region that contains genes encoding proteins essential to innate and adaptive immune functions. For example, MHC class I and class II molecules are necessary for antigen presentation to T lymphocytes and are common to the galliform MHC [229,230,231,232,233,234,235,236,237]. In turkeys and chickens, the MHC is composed of 2 genetically unlinked regions, the *B*-locus (*MHC-B*) and *Y*-locus (*MHC-Y*), co-located on a single microchromosome (GGA16 or MGA18, respectively) [229,230,231,232,233,234, 235,236, 237]. Although some genes are well characterized, the functions and expression patterns of many poultry MHC genes are still unknown. In the turkey, expression of multiple MHC genes significantly increased in response to AFB_1 exposure [238].

5.3. Moving towards Transcriptomics

Few studies have examined gene expression changes across the entire transcriptome. Although the specific genes affected by AFB_1 vary between species, exposure to the toxin has an up-regulatory effect on expression of damage responses, or when those fail, pathways of carcinogenesis. In rats, AFB_1 enhanced hepatic expression of genes involved in xenobiotic detoxification, cell cycle regulation, oxidative stress, DNA damage repair, tumor development and amino acid metabolism [221]. For example, E2F transcription factor 1 (*E2F1*) and many genes downstream of *E2F1* were up-regulated by AFB_1 exposure; E2F1 regulates DNA replication and apoptosis, and could contribute to carcinogenesis [221]. Aflatoxicosis in pigs affects expression of many of the same pathways, with greatest effects on metabolism, cell cycle, DNA damage responses and apoptotic processes in one study [239] and on protein degradation, metabolism, apoptosis, and immune responses in another [227]. Hepatocellular carcinomas from AFB_1-exposed rainbow trout showed changes in cell cycle, metabolism, immune and acute phase response genes when compared to adjacent non-cancerous liver tissue

[228]. Two of these studies utilized high throughput RNA sequencing (RNA-seq), rather than microarrays, to characterize transcriptome changes after AFB_1 exposure [221,239].

5.4. Poultry Transcriptomics and AFB_1

AFB_1 effects on the transcriptome have been investigated in chickens using microarray [123]. Exposure to AFB_1 affected hepatic expression of genes associated with fatty acid metabolism, development, detoxification, coagulation, immunity and cell proliferation [123]. Among these, the greatest proportions of differentially expressed genes were involved in cell proliferation and metabolism. For example, aflatoxicosis had the greatest effect on expression of the cell signaling inhibitor Dickkopf homolog 3 (*DKK3*) (down-regulated) and the metabolic glycogen synthase 1 (*GYS1*) (up-regulated).

More recently, RNA-seq has been applied to the domestic turkey transcriptome to elucidate hepatic and splenic responses to dietary AFB_1 [227,238,240]. Similar to expression changes in other species, transcriptome analysis in the turkey found that more genes were up-regulated than down-regulated by AFB_1 [227,238,240]. In one study of the liver, more than 80 genes belonging to cancer or focal adhesion pathways were up-regulated during aflatoxicosis, whereas smaller numbers of genes involved in cell signaling, cytoskeleton, and cell cycle were also up-regulated [227]. Conversely, the greatest numbers of down-regulated genes were involved in complement and coagulation.

Our investigation of AFB_1-hepatotoxicity in the turkey by RNA-seq found 313 transcripts significantly affected by AFB_1 exposure, with up-regulation of genes involved in apoptosis, cancer, and cell cycle regulation, and down-regulation of lipid metabolism [240]. Greatest up-regulation was observed for keratin 20 (*KRT20*), cell-death activator CIDE-3 (*CIDEC*), and E3 ubiquitin-protein ligase Mdm2 (*MDM2*). Alpha-2-macroglobulin (*A2M*) was the most down-regulated gene. Exposure to AFB_1 also turned on expression of HEPACAM family member 2 (*HEPACAM2*), a protein important in modulating cell adhesion and migration, and *S*-adenosylmethionine synthase isoform type-2 (*MAT2A*), important in methylation pathways.

Utilizing spleen samples from the same AFB_1-challenge trial, immunotoxicity in domestic turkey was also examined through RNA-seq [238]. Exposure to AFB_1 induced significant expression changes in 391 *de novo* assembled transcripts; the greatest up-regulation was seen in E3 ubiquitin-protein ligase CBL-B (*CBLB*) and ubiquitin specific peptidase 40 (*USP40*). Of the significantly altered transcripts, 27.6% encoded proteins with known immune functions representing both innate and adaptive responses. Antimicrobial genes, including beta-defensin 1 (*THP1*) and 2 (*THP2*), were down-regulated, while cytotoxic and antigen presentation genes, such as granzyme A (*GZMA*), perforin 1 (*PRF1*), MHC class IA and class IIB, were up-regulated in AFB_1-exposed tissue.

These studies were also designed to evaluate the ability of a *Lactobacillus*-based oral probiotic to reduce AFB_1-effects on the liver and spleen [238,240]. In the same challenge trial, domestic turkeys were exposed to probiotics alone or in combination with AFB_1. Addition of probiotics during AFB_1exposure mitigated AFB_1-induced expression changes in genes such as serine/arginine repetitive matrix protein 1 (*SRRM1*), 28S ribosomal RNA (*RNA28S*), and ISG12-2 protein-like (*ISG12-2*) (Figure 3A) [240]. In the spleen, probiotics had greater ameliorating properties, significantly reducing AFB_1-effects on multiple immune genes, including *THP2*, *GZMA*, and *PRF1* (Figure 3B) [238]. However, probiotics were unable to reverse most AFB_1-induced expression changes and even had synergistic effects with AFB_1. For example, combined treatment increased differential expression in apolipoprotein A-IV (*APOA4*) in the liver and interestingly both *RNA28S* and *ISG12-2* in the spleen (Figure 3). Therefore, oral probiotics modulated expression in both tissues, but did not restore normal transcriptome profiles.

6. STRATEGIES TO REDUCE AFB_1 TOXICITY

6.1. Chemical Detoxification

Since prevention of AFB_1 contamination is often impractical, methods to chemically detoxify this mycotoxin have been intensely investigated [7]. Several candidate chemicals have been examined for their ability to detoxify AFB_1 in crops such as grain, rice, corn, and cottonseed. These include ammonium hydroxide [241,242,243,244,245,246], calcium hydroxide [243,247], hydrogen peroxide [243,248], sodium hydroxide [243,249], and sodium hypochlorite [243,250] all of which reduce AFB_1 concentrations through hydrolysis and produce a degraded form with reduced or no toxicity. However, most of these chemicals are of themselves, hazardous and safe use is often expensive and may decrease the nutrient value of feed components. In addition, it is not possible to fully preclude low levels of AFB_1 in feed production, especially in crops that are heavily contaminated, severely reducing returns for producers.

6.2. Feed Additives

Given that eliminating all potential for exposure to AFB_1 is not feasible, feed additives have been examined for their ability to protect poultry from aflatoxicosis [7,21]. Some additives, such as selenium supplementation, attempt to boost detoxification, metabolic, or immune functions to counteract the effects of AFB_1 [139,149,150,168, 170,171, 183,251]. However, most additives have been investigated for their potential to reduce AFB_1 uptake by the intestine [21]. Many natural absorbents have been shown to decrease the effects of AFB_1 in poultry, including super-activated charcoal [153], zeolites like hydrated sodium calcium aluminosilicate [115,143,198,252,253,254,255,256,257], clinoptilolite [134,138], and sodium bentonite [158,258]. Antioxidants like butylated hydroxytolouene (BHT) [137,259,260,261,262,263] and turmeric [124,264] can also mitigate the severity of aflatoxicosis.

6.3. Probiotics

Many Gram-positive bacteria, including *Streptococcus*, *Enterococcus*, *Lactococcus*, and*Berevibacillus*, can bind AFB_1 *in vitro* [265,266,267,268,269]. However, most research has focused on probiotic strains of *Lactobacillus*, *Bifidobacterium*, and *Propionibacterium* [42,266,267,270,271,272,273,274,275,276,277,278,279,280,281,282,283 ,284,285,286,287,288]. Interactions between AFB_1 and *Lactobacillus rhamnosus* GG (LGG), *L. rhamnosus* LC-705 (LC-705), *Propionibacterium freudenrieichii* strain *shermanii* JS (PJS) or mixtures of these strains have been shown to be especially effective [42,270, 271,272,273,274,276,277,278,279,280,284,285,286,288]. A mixture of these strains was also utilized in our transcriptomic analyzes [238,240]. As gastrointestinal commensals or cultures used in cheese-making, yogurt and other dairy products, the safety of these lactic acid bacterial strains is well-established and easily applicable as potential chemopreventatives.

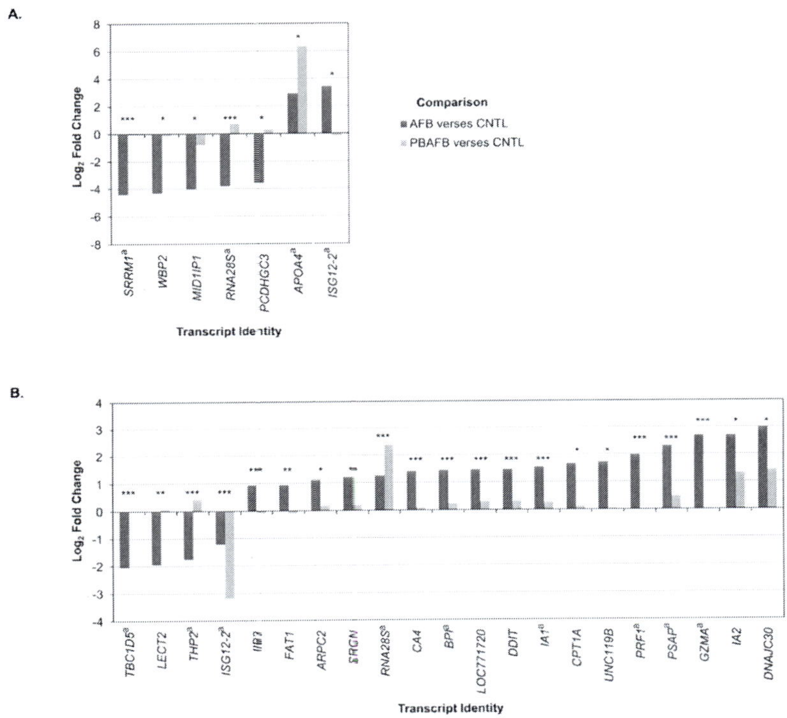

Figure 3. Oral probiotics mitigate gene expression changes induced by aflatoxin B_1 (AFB_1). (**A**) Liver. (**B**) Spleen. Graphs show annotated transcripts from each tissue with significant differential expression in the AFB_1-treated group (AFB) *versus*the control group (CNTL) and in the probiotic + AFB_1 group (PBAFB) verses AFB. Bars illustrate $\log_2$ fold change in AFB verses CNTL (**blue**) or PBAFB verses CNTL (**green**). Significance of the probiotics (PBAFB verses AFB) is represented by the number of asterisks (*: 0.05 > *p*-value > 0.01, **: 0.01 > *p*-value > 0.001, ***: *p*-value < 0.001). Genes with multiple significant transcripts are indicated by [a]; only the most significant transcript in PBAFB verses AFB is shown. Apolipoprotein A-IV (*APOA4*), actin-related protein 2/3 complex subunit 2-like (*ARPC2*), bactericidal permeability-increasing protein (*BPI*), carbonic anhydrase IV (*CA4*), carnitine O-palmitoyltransferase 1, liver isoform (*CPT1A*), DNA-damage-inducible transcript 4 (*DDIT*), DnaJ homolog subfamily C member 30 (*DNAJC30*), FAT tumor suppressor homolog 1 (*FAT1*), granzyme A (*GZMA*), MHC class I antigen alpha chain 1 (*IA1*), MHC class I antigen alpha chain 2 (*IA2*), MHC class II antigen beta chain 3 (*IIB3*), ISG12-2 protein-like (*ISG12-2*), leukocyte cell-derived chemotaxin-2 (*LECT2*), uncharacterized LOC771720 (*LOC771720*), MID1 interacting protein 1 (*MID1IP1*), protocadherin gamma subfamily C3 (*PCDHGC3*), perforin 1 (*PRF1*), presaposin (*PSAP*), 28S ribosomal RNA gene (*RNA28S*), serglycin (*SRGN*), serine/arginine repetitive matrix protein 1 (*SRRM1*), TBC1 domain family member 5 (*TBC1D5*), beta-defensin 2 (*THP2*), unc-119 homolog B (*UNC119B*), and WW domain-binding protein 2 (*WBP2*).

Strains LGG and LC-705 can sequester up to 80% of AFB_1 introduced into growth media [272,280]. AFB_1 interacts with the thick peptidoglycan layer characteristic of the Gram-positive bacterial cell wall [282,285]. When bound to AFB_1 *in vitro*, LGG and a mixture of LC-705 and PJS interacted less with intestinal mucus [276]. Incubation of LGG with AFB_1 reduced LGG adhesion to a Caco-2 intestinal cell monolayer [284] and decreased transport of AFB_1 across the monolayer [278]. These *in vitro* models suggest the probiotic and the toxin would be excreted together *in vivo*and thereby decrease the effective dose of AFB_1.

Reduced AFB_1 toxicity has been demonstrated in both mice [270] and in rats [277] after addition of dietary LGG. In humans, probiotics enhanced excretion of AFB_1 [289,290]. Furthermore, addition of LGG, LC-705 or PJS *ex vivo* through injection of AFB_1 into the lumen of the chicken duodenum significantly reduced AFB_1 uptake (by 74%, 63% and 37%, respectively) [274]. A mixture of LC-705 and PJS caused a 40% reduction in AFB_1 absorption into chicken duodenal tissue in a repeat experiment [42]. Therefore, a probiotic mixture including LGG or LC-705 could be an effective preventative for aflatoxicosis if added to poultry feeds.

6.4. Selection for Resistance

Another option to minimize the adverse effects of AFB_1 is to increase the resistance of domestic poultry [7]. Selection for AFB_1 resistant lines of chicken [291,292,293,294,295] and quail [296,297,298,299] has been investigated; however, selection studies have not been performed in the far more sensitive domestic turkey. Improvement in AFB_1 resistance is highly dependent on starting population [299] and selection is most effective during AFB_1 challenge since correlations of phenotypic measures like weight gain or blood parameters are most informative during exposure [291,292,298]. The requirement for concurrent aflatoxicosis makes phenotypic selection difficult to implement in a commercial setting. However, understanding the molecular mechanism of aflatoxicosis in domestic turkey and identifying the genetic differences underlying decreased sensitivity in wild turkeys could allow targeted genetic selection for resistant alleles without constant AFB_1-exposure. Due to the similarity of symptoms of aflatoxicosis across species, potential genetic targets may be translatable to other poultry species.

7. CONCLUSIONS: SUGGESTED AREAS FOR FURTHER RESEARCH

Future gene expression analyzes can provide insight into the mechanisms of aflatoxicosis and methods to reduce its effects. Our characterization of AFB_1-induced changes in the turkey liver and spleen transcriptomes identified genes responding to aflatoxicosis and host responses to toxicity. Since both proliferation and apoptosis occur during aflatoxicosis, the expression of cell cycle genes in the liver needs to be quantified alongside measures of apoptotic and mitotic cells. Similarly, gene expression in immune tissues such as the spleen should be measured concurrent with lymphocyte numbers, activation or apoptotic state to better clarify gene functions. Investigation of individual cell types could also detect cell-specific gene modulation. For example, the effects of AFB_1 on expression of immune genes could be measured in heterophils or T lymphocyte subsets (e.g., $CD4^+$ verses $CD8^+$ T cells). Gene expression in cells of the bile duct, known targets of AFB_1 mutagenesis, could be used to characterize hyperplasia. Expression patterns, as determined from mRNA, do not always directly correlate with protein levels or stability; therefore, proteomics could confirm effects on cell cycle regulators or immune mediators and provide a measure for AFB_1 inhibition of protein synthesis.

Analyses of systemic responses to AFB_1 in other tissues, such as the bursa, thymus, kidney or small intestine are needed to fully elucidate AFB_1 effects. Intestinal epithelial cells are directly exposed to AFB_1 during absorption and the potential prevention of AFB_1 uptake by feed additives also occurs within the small intestine. Therefore, investigation of intestinal transcriptome responses to AFB_1 is a priority. The utility of a *Lactobacillus*-based probiotics as preventative for aflatoxicosis requires further examination *in vivo*. Higher concentrations or different compositions of dietary probiotics should be examined for their ability to protect poultry from the development of hepatic lesions and in restoring gene expression profiles. Furthermore, no definitive conclusions can be made regarding the effects of probiotics without characterizing the intestinal microbiota. Shifts in bacterial population structure of the microbiome could be investigated by 16S NGS sequencing for direct comparison to the host effects shown by RNA-seq.

Comparative analysis of transcriptome responses of poultry to AFB_1 could help resolve differences in their sensitivity. RNA-seq studies investigating both domestic and wild turkey using an *in ovo* exposure model [300] and dietary challenge of poults [301] are currently underway. Preliminary data from embryonic exposures illustrates conserved effects on cell cycle regulators and variation in metabolic and anti-oxidant enzymes [300]. Genes and pathways identified in these studies will provide targets for selection efforts to improve resistance to aflatoxicosis in domestic poultry.

ACKNOWLEDGMENTS

The authors gratefully acknowledge support from the United States Department of Agriculture, Agriculture and Food Research Initiative Grants 2005-01326, 2007-35205-17880, 2009-35205-05302, and 2013-01043 from the USDA National Institute of Food and Agriculture Animal Genome Program. Support from the Minnesota and Utah Agricultural Experiments Stations is also acknowledged.

AUTHOR CONTRIBUTIONS

The work presented is a part of a Ph.D. thesis; the information was collected and summarized by Melissa Monson who took the lead on writing the manuscript. Kent Reed and Roger Coulombe supervised the work and edited the manuscript.

CONFLICTS OF INTEREST

The authors declare no conflict of interest.

REFERENCES

1. Bedard, L.L.; Massey, T.E. Aflatoxin B1-induced DNA damage and its repair. Cancer Lett.2006, 241, 174–183.

2. Bennett, J.W.; Klich, M. Mycotoxins. Clin. Microbiol. Rev. 2003, 16, 497–516.

3. Coulombe, R.A., Jr. Biological action of mycotoxins. J. Dairy Sci. 1993, 76, 880–891.

4. Eaton, D.L.; Gallagher, E.P. Mechanisms of aflatoxin carcinogenesis. Annu. Rev. Pharmacol. Toxicol. 1994, 34, 135–172.

5. Groopman, J.D.; Kensler, T.W. Role of metabolism and viruses in aflatoxin-induced liver cancer. Toxicol. Appl. Pharmacol. 2005, 206, 131–137.

6. Kensler, T.W.; Roebuck, B.D.; Wogan, G.N.; Groopman, J.D. Aflatoxin: A 50-Year odyssey of mechanistic and translational toxicology. Toxicol. Sci. 2011, 120, S28–S48.

7. Leeson, S.; Diaz, G.J.; Summers, J.D. Aflatoxins. In Poultry Metabolic Disorders and Mycotoxins; University Books: Guelph, Canada, 1995.

8. Yabe, K.; Nakajima, H. Enzyme reactions and genes in aflatoxin biosynthesis. Appl. Microbiol. Biotechnol. 2004, 64, 745–755.

9. Wogan, G.N.; Kensler, T.W.; Groopman, J.D. Present and future directions of translational research on aflatoxin and hepatocellular carcinoma. A review. Food Addit. Contam. 2012, 29, 249–257.

10. Newberne, P.M.; Butler, W.H. Acute and chronic effects of aflatoxin on the liver of domestic and laboratory animals: A review. Cancer Res. 1969, 29, 236–250. [PubMed]

11. Patterson, D.S.P. Metabolism as a factor in determining the toxic action of the aflatoxins in different animal species. Food Cosmet. Toxicol. 1973, 11, 287–294.

12. Pier, A.C. Aflatoxicosis and Immunosuppression in Mammalian Animals. In Aflatoxin in Maize: A Proceedings of the Workshop; Zuber, M.S., Lillehoj, E.B., Renfro, B.L., Eds.; CIMMYT: El Batan, Mexico, 1986.

13. Robens, J.F.; Richard, J.L. Aflatoxins in animal and human health. Rev. Environ. Contam. Toxicol. 1992, 127, 69–94. [PubMed]

14. Wogan, G.N. Chemical nature and biological effects of the aflatoxins. Bacteriol. Rev. 1966, 30, 460–470. [PubMed]

15. Wogan, G.N. Aflatoxins as risk factors for hepatocellular carcinoma in humans. Cancer Res.1992, 52, 2114s–2118s. [PubMed]

16. Blount, W.P. Turkey "X" disease. Turkeys 1961, 9, 77.

17. Wannop, C.C. The histopathology of turkey "X" disease in Great Britain. Avian Dis. 1961, 5, 371–381.

18. Nesbitt, B.F.; O'Kelly, J.; Sargeant, K.; Sheridan, A. Aspergillus flavus and turkey X disease. Toxic metabolites of Aspergillus flavus. Nature 1962, 195, 1062–1063.

19. Cary, J.W.; Klich, M.A.; Beltz, S.B. Characterization of aflatoxin-producing fungi outside ofAspergillus section Flavi. Mycologia 2005, 97, 425–432.

20. Frisvad, J.C.; Skouboe, P.; Samson, R.A. Taxonomic comparison of three different groups of aflatoxin producers and a new efficient producer of aflatoxin B1, sterigmatocystin and 3-O-methylsterigmatocystin, Aspergillus rambellii sp. nov. Syst. Appl. Microbiol. 2005, 28, 442–453.

21. Rawal, S.; Kim, J.E.; Coulombe, R. Aflatoxin B1 in poultry: Toxicology, metabolism and prevention. Res. Vet. Sci. 2010, 89, 325–331.

22. Cullen, J.M.; Newberne, P.M. Acute Hepatotoxicity of Aflatoxins. In The Toxicology of Aflatoxins: Human Health, Veterinary, and Agricultural Significance; Eaton, D.L., Groopman, J.D., Eds.; Academic Press: London, UK, 1993.

23. CAST. Mycotoxins: Risks in Plant, Animal and Human Systems; No. 139; Council for Agricultural Science and Technology: Ames, IA, USA, 2003.

24. Cotty, P.J.; Jaime-Garcia, R. Influences of climate on aflatoxin producing fungi and aflatoxin contamination. Int. J. Food Microbiol. 2007, 119, 109–115.

25. Gqaleni, N.; Smith, J.E.; Lacey, J.; Gettinby, G. Effects of temperature, water activity, and incubation time on production of aflatoxins and cyclopiazonic acid by an isolate ofAspergillus flavus

in surface agar culture. Appl. Environ. Microbiol. 1997, 63, 1048–1053. [PubMed]

26. Medina, A.; Rodriguez, A.; Magan, N. Effect of climate change on Aspergillus flavus and aflatoxin B1 production. Front. Microbiol. 2014, 5, 348.

27. Mousa, W.; Ghazali, F.M.; Jinap, S.; Ghazali, H.M.; Radu, S. Modeling growth rate and assessing aflatoxins production by Aspergillus flavus as a function of water activity and temperature on polished and brown rice. J. Food Sci. 2013, 78, M56–M63.

28. Schindler, A.F.; Palmer, J.G.; Eisenberg, W.V. Aflatoxin production by Aspergillus flavus as related to various temperatures. Appl. Microbiol. 1967, 15, 1006–1009. [PubMed]

29. Schroeder, H.W.; Hein, H. Aflatoxins: Production of the toxins in vitro in relation to temperature. Appl. Microbiol. 1967, 15, 441–445. [PubMed]

30. Trenk, H.L.; Hartman, P.A. Effects of moisture content and temperature on aflatoxin production in corn. Appl. Microbiol. 1970, 19, 781–784. [PubMed]

31. Aly, S.A.; Anwer, W. Effect of naturally contaminated feed with aflatoxins on performance of laying hens and the carryover of aflatoxin B1 residues in table egg. Pakistan J. Nutr. 2009,8, 181–186.

32. Chen, C.; Pearson, A.M.; Coleman, T.H.; Gray, J.I.; Pestka, J.J.; Aust, S.D. Tissue deposition and clearance of aflatoxins from broiler chickens fed a contaminated diet. Food Chem. Toxicol.1984, 22, 447–451.

33. Galvano, F.; Galofaro, V.; Galvano, G. Occurrence and stability of aflatoxin M1 in milk and milk products: A worldwide review. J. Food Prot. 1996, 59, 1079–1090.

34. Pandey, I.; Chauhan, S.S. Studies on production performance and toxin residues in tissues and eggs of layer chickens fed on diets with various concentrations of aflatoxin AFB1. Br. Poult. Sci. 2007, 48, 713–723.

35. Purchase, I.F.H. Aflatoxin residues in food of animal origin. Food Cosmet. Toxicol. 1972, 10, 531–544.

36. Richard, J.L.; Stubblefield, R.D.; Lyon, R.L.; Peden, W.M.; Thurston, J.R.; Rimler, R.B. Distribution and clearance of aflatoxins B1 and M1 in turkeys fed diets containing 50 or 150 ppb aflatoxin from naturally contaminated corn. Avian Dis. 1986, 30, 788–793.

37. Wolzak, A.; Pearson, A.M.; Coleman, T.H.; Pestka, J.J.; Gray, J.I. Aflatoxin deposition and clearance in the eggs of laying hens. Food Chem. Toxicol. 1985, 23, 1057–1061.

38. Wolzak, A.; Pearson, A.M.; Coleman, T.H.; Pestka, J.J.; Gray, J.I.; Chen, C. Aflatoxin carryover and clearance from tissues of laying hens. Food Chem. Toxicol. 1986, 24, 37–41.

39. Giambrone, J.J.; Diener, U.L.; Davis, N.D.; Panangala, V.S.; Hoerr, F.J. Effects of aflatoxin on young turkeys and broiler chickens. Poult. Sci. 1985, 64, 1678–1684.

40. Klein, P.J.; Buckner, R.; Kelly, J.; Coulombe, R.A., Jr. Biochemical basis for the extreme sensitivity of turkeys to aflatoxin B1. Toxicol. Appl. Pharmacol. 2000, 165, 45–52.

41. CAST. Mycotoxins: Economic and Health Risks; No. 116; Council for Agricultural Science and Technology: Ames, IA, USA, 1989.

42. Gratz, S.; Mykkänen, H.; el-Nezami, H. Aflatoxin B1 binding by a mixture of Lactobacillusand Propionibacterium.: In vitro versus ex vivo. J. Food Prot. 2005, 68, 2470–2474. [PubMed]

43. Guengerich, F.P.; Johnson, W.W.; Ueng, Y.F.; Yamazaki, H.; Shimada, T. Involvement of cytochrome P450, glutathione S-transferase, and epoxide hydrolase in the metabolism of aflatoxin B1 and relevance to risk of human liver cancer. Environ. Health Perspect. 1996, 104, 557–562.

44. Godoy, H.M.; Neal, G.E. Some studies of the effects of aflatoxin B1 in vivo and in vitro on nucleic acid synthesis in rat and mouse. Chem. Biol. Interact. 1976, 13, 257–277.

45. Monroe, D.H.; Eaton, D.L. Comparative effects of butylated hydroxyanisole on hepatic in vivo DNA binding and in vitro biotransformation of aflatoxin B1 in the rat and mouse.Toxicol. Appl. Pharmacol. 1987, 90, 401–409.

46. Pelkonen, P.; Lang, M.A.; Wild, C.P.; Negishi, M.; Juvonen, R.O. Activation of aflatoxin B1 by mouse CYP2A enzymes and

cytotoxicity in recombinant yeast cells. Eur. J. Pharmacol. 1994,292, 67–73.

47. Pelkonen, P.; Lang, M.A.; Negishi, M.; Wild, C.P.; Juvonen, R.O. Interaction of aflatoxin B1with cytochrome P450 2A5 and its mutants: Correlation with metabolic activation and toxicity. Chem. Res. Toxicol. 1997, 10, 85–90.

48. Ramsdell, H.S.; Eaton, D.L Species susceptibility to aflatoxin B1 carcinogenesis: Comparative kinetics of microsomal biotransformation. Cancer Res. 1990, 50, 615–620. [PubMed]

49. Yanagimoto, T.; Itoh, S.; Sawada, M.; Hashimoto, H.; Kamataki, T. Molecular cloning and functional expression of a mouse cytochrome P-450 (Cyp3a-13): Examination of Cyp3a-13 enzyme to activate aflatoxin B1 (AFB1). Biochim. Biophys. Acta 1994, 1201, 405–410.

50. Yanagimoto, T.; Itoh, S.; Sawada, M.; Kamataki, T. Mouse cytochrome P450 (Cyp3a11): Predominant expression in liver and capacity to activate aflatoxin B1. Arch. Biochem. Biophys.1997, 340, 215–218.

51. Buetler, T.M.; Slone, D.; Eaton, D.L. Comparison of the aflatoxin B1–8,9-epoxide conjugating activities of two bacterially expressed alpha class glutathione S-transferase isozymes from mouse and rat. Biochem. Biophys. Res. Commun. 1992, 188, 597–603.

52. Hayes, J.D.; Judah, D.J.; Neal, G.E.; Nguyen, T. Molecular cloning and heterologous expression of a cDNA encoding a mouse glutathione S-transferase Yc subunit possessing high catalytic activity for aflatoxin B1-8,9-epoxide. Biochem. J. 1992, 285, 173–180.

53. Ilic, Z.; Crawford, D.; Vakharia, D.; Egner, P.A.; Sell, S. Glutathione-S-transferase A3 knockout mice are sensitive to acute cytotoxic and genotoxic effects of aflatoxin B1. Toxicol. Appl. Pharmacol. 2010, 242, 241–246.

54. Monroe, D.H.; Eaton, D.L. Effects of modulation of hepatic glutathione on biotransformation and covalent binding of aflatoxin B1 to DNA in the mouse. Toxicol. Appl. Pharmacol. 1988, 94, 118–127.

55. Quinn, B.A.; Crane, T.L.; Kocal, T.E.; Best, S.J.; Cameron, R.G.; Rushmore, T.H.; Farber, E.; Hayes, M.A. Protective activity of different hepatic cytosolic glutathione S-transferases against DNA-binding metabolites of aflatoxin B1. Toxicol. Appl. Pharmacol. 1990, 105, 351–363.

56. Slone, D.H.; Gallagher, E.P.; Ramsdell, H.S.; Rettie, A.E.; Stapleton, P.L.; Berlad, L.G.; Eaton, D.L. Human variability in hepatic glutathione S-transferase-mediated conjugation of aflatoxin B1-epoxide and other substrates. Pharmacogenetics 1995, 5, 224–233.

57. Buetler, T.M.; Bammler, T.K.; Hayes, J.D.; Eaton, D.L. Oltipraz-mediated changes in aflatoxin B1 biotransformation in rat liver: Implications for human chemointervention. Cancer Res.1996, 56, 2306–2313. [PubMed]

58. Imaoka, S.; Ikemoto, S.; Shimada, T.; Funae, Y. Mutagenic activation of aflatoxin B1 by pulmonary, renal, and hepatic cytochrome P450s from rats. Mutat. Res. 1992, 269, 231–236.

59. Johnson, W.W.; Ueng, Y.F.; Widersten, M.; Mannervik, B.; Hayes, J.D.; Sherratt, P.J.; Ketterer, B.; Guengerich, F.P. Conjugation of highly reactive aflatoxin B1 exo-8,9-epoxide catalyzed by rat and human glutathione transferases: Estimation of kinetic parameters. Biochemistry 1997,36, 3056–3060.

60. Hayes, J.D.; Judah, D.J.; McLellan, L.I.; Kerr, L.A.; Peacock, S.D.; Neal, G.E. Ethoxyquin-induced resistance to aflatoxin B1 in the rat is associated with the expression of a novel alpha-class glutathione S-transferase subunit, Yc2, which possesses high catalytic activity for aflatoxin B1–8,9-epoxide. Biochem. J. 1991, 279, 385–398.

61. Hayes, J.D.; Nguyen, T.; Judah, D.J.; Petersson, D.G.; Neal, G.E. Cloning of cDNAs from fetal rat liver encoding glutathione S-transferase Yc polypeptides. The Yc2 subunit is expressed in adult rat liver resistant to the hepatocarcinogen aflatoxin B1. J. Biol. Chem. 1994, 269, 20707–20717. [PubMed]

62. Hayes, J.D.; Pulford, D.J.; Ellis, E.M.; McLeod, R.; James, R.F.; Seidegård, J.; Mosialou, E.; Jernström, B.; Neal, G.E. Regulation of rat glutathione S-transferase A5 by cancer chemopreventive agents: Mechanisms of inducible resistance to aflatoxin B1. Chem. Biol. Interact. 1998, 111–112, 51–67.

63. Forrester, L.M.; Neal, G.E.; Judah, D.J.; Glancey, M.J.; Wolf, C.R. Evidence for involvement of multiple forms of cytochrome P-450 in aflatoxin B1 metabolism in human liver. Proc. Natl. Acad. Sci. USA 1990, 87, 8306–8310.

64. Gallagher, E.P.; Wienkers, L.C.; Stapleton, P.L.; Kunze, K.L.; Eaton, D.L. Role of human microsomal and human complementary DNA-expressed cytochromes P4501A2 and P4503A4 in the bioactivation of aflatoxin B1. Cancer Res. 1994, 54, 101–108. [PubMed]

65. Macé, K.; Aguilar, F.; Wang, J.S.; Vautravers, P.; Gómez-Lechón, M.; Gonzalez, F.J.; Groopman, J.; Harris, C.C.; Pfeifer, A.M. Aflatoxin B1-induced DNA adduct formation and p53 mutations in CYP450-expressing human liver cell lines. Carcinogenesis 1997, 18, 1291–1297.

66. Ramsdell, H.S.; Parkinson, A.; Eddy, A.C.; Eaton, D.L. Bioactivation of aflatoxin B1 by human liver microsomes: Role of cytochrome P450 IIIA enzymes. Toxicol. Appl. Pharmacol.1991, 108, 436–447.

67. Van Vleet, T.R.; Watterson, T.L.; Klein, P.J.; Coulombe, R.A., Jr. Aflatoxin B1 alters the expression of p53 in cytochrome P450-expressing human lung cells. Toxicol. Sci. 2006, 89, 399–407.

68. Raney, K.D.; Meyer, D.J.; Ketterer, B.; Harris, T.M.; Guengerich, F.P. Glutathione conjugation of aflatoxin B1 exo- and endo-epoxides by rat and human glutathione S-transferases. Chem. Res. Toxicol. 1992, 5, 470–478.

69. Gross-Steinmeyer, K.; Stapleton, P.L.; Tracy, J.H.; Bammler, T.K.; Strom, S.C.; Eaton, D.L. Sulforaphane- and phenethyl isothiocyanate-induced inhibition of aflatoxin B1-mediated genotoxicity in human hepatocytes: Role of GSTM1 genotype and CYP3A4 gene expression.Toxicol. Sci. 2010, 116, 422–432.

70. Liu, Y.H.; Taylor, J.; Linko, P.; Lucier, G.W.; Thompson, C.L. Glutathione S-transferase mu in human lymphocyte and liver: Role in modulating formation of carcinogen-derived DNA adducts. Carcinogenesis 1991, 12, 2269–2275.

71. McGlynn, K.A.; Rosvold, E.A.; Lustbader, E.D.; Hu, Y.; Clapper, M.L.; Zhou, T.; Wild, C.P.; Xia, X.L.; Baffoe-Bonnie, A.; Ofori-Adjei, D.; et al. Susceptibility to hepatocellular carcinoma is

associated with genetic variation in the enzymatic detoxification of aflatoxin B1. Proc. Natl. Acad. Sci. USA 1995, 92, 2384–2387.

72. Arafa, A.S.; Bloomer, R.J.; Wilson, H.R.; Simpson, C.F.; Harms, R.H. Susceptibility of various poultry species to dietary aflatoxin. Br. Poult. Sci. 1981, 22, 431–436.

73. Gumbmann, M.R.; Williams, S.N.; Booth, A.N.; Vohra, P.; Ernst, R.A.; Bethard, M. Aflatoxin susceptibility in various breeds of poultry. Proc. Soc. Exp. Biol. Med. 1970, 134, 683–688.

74. Lozano, M.C.; Diaz, G.J. Microsomal and cytosolic biotransformation of aflatoxin B1 in four poultry species. Br. Poult. Sci. 2006, 47, 734–741.

75. Muller, R.D.; Carlson, C.W.; Semeniuk, G.; Harshfield, G.S. The response of chicks, ducklings, goslings, pheasants and poults to graded levels of aflatoxins. Poult. Sci. 1970, 49, 1346–1350.

76. Rawal, S.; Mendoza, K.M.; Reed, K.M.; Coulombe, R.A., Jr. Structure, genetic mapping, and function of the cytochrome P450 3A37 gene in the turkey (Meleagris gallopavo). Cytogenet. Genome Res. 2009, 125, 67–73.

77. Rawal, S.; Yip, S.S.; Coulombe, R.A., Jr. Cloning, expression and functional characterization of cytochrome P450 3A37 from turkey liver with high aflatoxin B1 epoxidation activity.Chem. Res. Toxicol. 2010, 23, 1322–1329.

78. Rawal, S.; Coulombe, R.A., Jr. Metabolism of aflatoxin B1 in turkey liver microsomes: The relative roles of cytochromes P450 1A5 and 3A37. Toxicol. Appl. Pharmacol. 2011, 254, 349–354.

79. Yip, S.S.; Coulombe, R.A. Molecular cloning and expression of a novel cytochrome P450 from turkey liver with aflatoxin B1 oxidizing activity. Chem. Res. Toxicol. 2006, 19, 30–37.

80. Diaz, G.J.; Murcia, H.W.; Cepeda, S.M. Bioactivation of aflatoxin B1 by turkey liver microsomes: Responsible cytochrome P450 enzymes. Br. Poult. Sci. 2010, 51, 828–837.

81. Diaz, G.J.; Murcia, H.W.; Cepeda, S.M. Cytochrome P450 enzymes involved in the metabolism of aflatoxin B1 in chickens and quail. Poult. Sci. 2010, 89, 2461–2469.

82. Diaz, G.J.; Murcia, H.W.; Cepeda, S.M.; Boermans, H.J. The role of selected cytochrome P450 enzymes on the bioactivation of aflatoxin B1 by duck liver microsomes. Avian Pathol. 2010,39, 279–285.

83. Murcia, H.W.; Díaz, G.J.; Cepeda, S.M. Enzymatic activity in turkey, duck, quail and chicken liver microsomes against four human cytochrome P450 prototype substrates and aflatoxin B1. J. Xenobiotics 2011, 1, 17–21.

84. Giambrone, J.J.; Diener, U.L.; Davis, N.D.; Panangala, V.S.; Hoerr, F.J. Effects of purified aflatoxin on turkeys. Poult. Sci. 1985, 64, 859–865.

85. Klein, P.J.; van Vleet, T.R.; Hall, J.O.; Coulombe, R.A., Jr. Biochemical factors underlying the age-related sensitivity of turkeys to aflatoxin B1. Comp. Biochem. Physiol. C Toxicol. Pharmacol.2002, 132, 193–201.

86. Hussain, Z.; Khan, M.Z.; Khan, A.; Javed, I.; Saleemi, M.K.; Mahmood, S.; Asi, M.R. Residues of aflatoxin B1 in broiler meat: Effect of age and dietary aflatoxin B1 levels. Food Chem. Toxicol. 2010, 48, 3304–3307.

87. Lanza, G.M.; Washburn, K.W.; Wyatt, R.D. Variation with age in response of broilers to aflatoxin. Poult. Sci. 1980, 59, 282–288.

88. Quezada, T.; Cuéllar, H.; Jaramillo-Juárez, F.; Valdivia, A.G.; Reyes, J.L. Effects of aflatoxin B1 on the liver and kidney of broiler chickens during development. Comp. Biochem. Physiol. C Toxicol. Pharmacol. 2000, 125, 265–272.

89. Barraud, L.; Douki, T.; Guerret, S.; Chevallier, M.; Jamard, C.; Trepo, C.; Wild, C.P.; Cadet, J.; Cova, L. The role of duck hepatitis B virus and aflatoxin B1 in the induction of oxidative stress in the liver. Cancer Detect. Prev. 2001, 25, 192–201. [PubMed]

90. Kim, J.E.; Bauer, M.M.; Mendoza, K.M.; Reed, K.M.; Coulombe, R.A., Jr. Comparative genomics identifies new alpha class genes within the avian glutathione S-transferase gene cluster. Gene 2010, 452, 45–53.

91. Kim, J.E.; Bunderson, B.R.; Croasdell, A.; Coulombe, R.A., Jr. Functional characterization of alpha-class glutathione S-transferases

from the turkey (Meleagris gallopavo). Toxicol. Sci. 2011,124, 45–53.

92. Quist, C.F.; Bounous, D.I.; Kilburn, J.V.; Nettles, V.F.; Wyatt, R.D. The effect of dietary aflatoxin on wild turkey poults. J. Wildl. Dis. 2000, 36, 436–444.

93. Reed, K.M. Turkey. In Genome Mapping and Genomics in Domestic Animals; Cockett, N.E., Kole, C., Eds.; Springer-Verlag: Berlin, Germany, 2009.

94. Julian, R.J. Production and growth related disorders and other metabolic diseases of poultry—A review. Vet. J. 2005, 169, 350–369.

95. Bayyari, G.R.; Huff, W.E.; Rath, N.C.; Balog, J.M.; Newberry, L.A.; Villines, J.D.; Skeeles, J.K.; Anthony, N.B.; Nestor, K.E. Effect of the genetic selection of turkeys for increased body weight and egg production on immune and physiological responses. Poult. Sci. 1997, 76, 289–296.

96. Li, Z.; Nestor, K.E.; Saif, Y.M.; Luhtala, M. Flow cytometric analysis of T lymphocyte subpopulations in large-bodied turkey lines and a randombred control population. Poult. Sci. 2000, 79, 219–223.

97. Li, Z.; Nestor, K.E.; Saif, Y.M.; Anderson, J.W.; Patterson, R.A. Effect of selection for increased body weight in turkeys on lymphoid organ weights, phagocytosis, and antibody responses to fowl cholera and Newcastle disease-inactivated vaccines. Poult. Sci. 2001, 80, 689–694.

98. Van der Most, P.J.; de Jong, B.; Parmentier, H.K.; Verhulst, S. Trade-off between growth and immune function: A meta-analysis of selection experiments. Funct. Ecol. 2011, 25, 74–80.

99. Kim, J.E.; Bunderson, B.R.; Croasdell, A.; Reed, K.M.; Coulombe, R.A., Jr. Alpha-class glutathione S-transferases in wild turkeys (Meleagris gallopavo): Characterization and role in resistance to the carcinogenic mycotoxin aflatoxin B1. PLoS ONE 2013, 8, e60662.

100. Bunderson, B.R.; Kim, J.E.; Croasdell, A.; Mendoza, K.M.; Reed, K.M.; Coulombe, R.A., Jr. Heterologous expression and functional characterization of avian mu-class glutathione S-

transferases. Comp. Biochem. Physiol. C Toxicol. Pharmacol. 2013, 158, 109–116.

101. Smela, M.E.; Hamm, M.L.; Henderson, P.T.; Harris, C.M.; Harris, T.M.; Essigmann, J.M. The aflatoxin B1 formamidopyrimidine adduct plays a major role in causing the types of mutations observed in human hepatocellular carcinoma. Proc. Natl. Acad. Sci. USA 2002, 99, 6655–6660.

102. Corrier, D.E. Mycotoxicosis: Mechanisms of immunosuppression. Vet. Immunol. Immunopathol. 1991, 30, 73–87.

103. Guengerich, F.P.; Arneson, K.O.; Williams, K.M.; Deng, Z.; Harris, T.M. Reaction of aflatoxin B1 oxidation products with lysine. Chem. Res. Toxicol. 2002, 15, 780–792.

104. Iwaki, M.; Kitagawa, T.; Akamatsu, Y.; Aibara, K. Cytotoxic effects of aflatoxin B1 and its association with cellular components in chicken embryo primary cultured cells. Biochim. Biophys. Acta 1990, 1035, 146–153.

105. IARC. Aflatoxins. In A Review of Human Carcinogens. Part. F: Chemical Agents and Related Occupations; International Agency for Research on Cancer Working Group on the Evaluation of Carcinogenic Risks to Humans: Lyon, France, 2012.

106. Aguilar, F.; Hussain, S.P.; Cerutti, P. Aflatoxin B1 induces the transversion of G-->T in codon 249 of the p53 tumor suppressor gene in human hepatocytes. Proc. Natl. Acad. Sci. USA 1993,90, 8586–8590.

107. Shen, H.M.; Ong, C.N. Mutations of the p53 tumor suppressor gene and ras oncogenes in aflatoxin hepatocarcinogenesis. Mutat. Res. 1996, 366, 23–44.

108. Wild, C.P.; Montesano, R. A model of interaction: Aflatoxins and hepatitis viruses in liver cancer aetiology and prevention. Cancer Lett. 2009, 286, 22–28.

109. Butler, W.H.; Greenblatt, M.; Lijinsky, W. Carcinogenesis in rats by aflatoxins B1, G1, and B2.Cancer Res. 1969, 29, 2206–2211. [PubMed]

110. Wogan, G.N.; Newberne, P.M. Dose-response characteristics of aflatoxin B1 carcinogenesis in the rat. Cancer Res. 1967, 27, 2370–2376. [PubMed]

111. Wogan, G.N.; Paglialunga, S.; Newberne, P.M. Carcinogenic effects of low dietary levels of aflatoxin B1 in rats. Food Cosmet. Toxicol. 1974, 12, 681–685.

112. Cova, L.; Wild, C.P.; Mehrotra, R.; Turusov, V.; Shirai, T.; Lambert, V.; Jacquet, C.; Tomatis, L.; Trépo, C.; Montesano, R. Contribution of aflatoxin B1 and hepatitis B virus infection in the induction of liver tumors in ducks. Cancer Res. 1990, 50, 2156–2163. [PubMed]

113. Cullen, J.M.; Marion, P.L.; Sherman, G.J.; Hong, X.; Newbold, J.E. Hepatic neoplasms in aflatoxin B1-treated, congenital duck hepatitis B virus-infected, and virus-free Pekin ducks.Cancer Res. 1990, 50, 4072–4080. [PubMed]

114. Uchida, T.; Suzuki, K.; Esumi, M.; Arii, M.; Shikata, T. Influence of aflatoxin B1 intoxication on duck livers with duck hepatitis B virus infection. Cancer Res. 1988, 48, 1559–1565. [PubMed]

115. Chen, X.; Horn, N.; Applegate, T.J. Efficiency of hydrated sodium calcium aluminosilicate to ameliorate the adverse effects of graded levels of aflatoxin B1 in broiler chicks. Poult. Sci.2014, 93, 2037–2047.

116. Giambrone, J.J.; Diener, U.L.; Davis, N.D.; Panangala, V.S.; Hoerr, F.J. Effects of purified aflatoxin on broiler chickens. Poult. Sci. 1985, 64, 852–858.

117. Huff, W.E.; Kubena, L.F.; Harvey, R.B.; Corrier, D.E.; Mollenhauer, H.H. Progression of aflatoxicosis in broiler chickens. Poult. Sci. 1986, 65, 1891–1899.

118. Rauber, R.H.; Dilkin, P.; Giacomini, L.Z.; Araújo de Almeida, C.A.; Mallmann, C.A. Performance of turkey poults fed different doses of aflatoxins in the diet. Poult. Sci. 2007, 86, 1620–1624.

119. Richard, J.L.; Pier, A.C.; Cysewski, S.J.; Graham, C.K. Effect of aflatoxin and aspergillosis on turkey poults. Avian Dis. 1973, 17, 111–121.

120. Smith, J.W.; Hamilton, P.B. Aflatoxicosis in the broiler chicken. Poult. Sci. 1970, 49, 207–215.

121. Sims, W.M., Jr.; Kelley, D.C.; Sanford, P.E. A study of aflatoxicosis in laying hens. Poult. Sci.1970, 49, 1082–1084.

122. Verma, J.; Johri, T.S.; Swain, B.K.; Ameena, S. Effect of graded levels of aflatoxin, ochratoxin and their combinations on the performance and immune response of broilers. Br. Poult. Sci.2004, 45, 512–518.

123. Yarru, L.P.; Settivari, R.S.; Antoniou, E.; Ledoux, D.R.; Rottinghaus, G.E. Toxicological and gene expression analysis of the impact of aflatoxin B1 on hepatic function of male broiler chicks. Poult. Sci. 2009, 88, 360–371.

124. Yarru, L.P.; Settivari, R.S.; Gowda, N.K.; Antoniou, E.; Ledoux, D.R.; Rottinghaus, G.E. Effects of turmeric (Curcuma longa) on the expression of hepatic genes associated with biotransformation, antioxidant, and immune systems in broiler chicks fed aflatoxin. Poult. Sci. 2009, 88, 2620–2627.

125. Lee, J.T.; Jessen, K.A.; Beltran, R.; Starkl, V.; Schatzmayr, G.; Borutova, R.; Caldwell, D.J. Mycotoxin-contaminated diets and deactivating compound in laying hens: 1. Effects on performance characteristics and relative organ weight. Poult. Sci. 2012, 91, 2089–2095.

126. Oliveira, C.A.; Rosmaninho, J.F.; Butkeraitis, P.; Corrêa, B.; Reis, T.A.; Guerra, J.L.; Albuquerque, R.; Moro, M.E. Effect of low levels of dietary aflatoxin B1 on laying Japanese quail. Poult. Sci. 2002, 81, 976–980.

127. Chen, X.; Horn, N.; Cotter, P.F.; Applegate, T.J. Growth, serum biochemistry, complement activity, and liver gene expression responses of Pekin ducklings to graded levels of cultured aflatoxin B1. Poult. Sci. 2014, 93, 2028–2036.

128. Azzam, A.H.; Gabal, M.A. Aflatoxin and immunity in layer hens. Avian Pathol. 1998, 27, 570–577.

129. Garlich, J.D.; Tung, H.T.; Hamilton, P.B. The effects of short term feeding of aflatoxin on egg production and some plasma constituents of the laying hen. Poult. Sci. 1973, 52, 2206–2211.

130. Howarth, B., Jr.; Wyatt, R.D. Effect of dietary aflatoxin on fertility, hatchability, and progeny performance of broiler breeder hens. Appl. Environ. Microbiol. 1976, 31, 680–684. [PubMed]

131. Khan, W.A.; Khan, M.Z.; Khan, A.; Hassan, Z.U.; Rafique, S.; Saleemi, M.K.; Ahad, A. Dietary vitamin E in White Leghorn layer breeder hens: A strategy to combat aflatoxin B1-induced damage. Avian Pathol. 2014, 43, 389–395.

132. Lee, J.T.; Jessen, K.A.; Beltran, R.; Starkl, V.; Schatzmayr, G.; Borutova, R.; Caldwell, D.J. Effects of mycotoxin-contaminated diets and deactivating compound in laying hens: 2. Effects on white shell egg quality and characteristics. Poult. Sci. 2012, 91, 2096–2104.

133. Oliveira, C.A.F.; Rosmaninho, J.F.; Castro, A.L.; Butkeraitis, P.; Alves Reis, T.; Corrêa, B. Aflatoxin residues in eggs of laying Japanese quail after long-term administration of rations containing low levels of aflatoxin B1. Food Addit. Contam. 2003, 20, 648–653.

134. Ortatatli, M.; Oğuz, H. Ameliorative effects of dietary clinoptilolite on pathological changes in broiler chickens during aflatoxicosis. Res. Vet. Sci. 2001, 71, 59–66.

135. Merkley, J.W.; Maxwell, R.J.; Phillips, J.G.; Huff, W.E. Hepatic fatty acid profiles in aflatoxin-exposed broiler chickens. Poult. Sci. 1987, 66, 59–67.

136. Siloto, E.V.; Oliveira, E.F.; Sartori, J.R.; Fascina, V.B.; Martins, B.A.; Ledoux, D.R.; Rottinghaus, G.E.; Sartori, D.R. Lipid metabolism of commercial layers fed diets containing aflatoxin, fumonisin, and a binder. Poult. Sci. 2013, 92, 2077–2083.

137. Klein, P.J.; van Vleet, T.R.; Hall, J.O.; Coulombe, R.A., Jr. Dietary butylated hydroxytoluene protects against aflatoxicosis in turkeys. Toxicol. Appl. Pharmacol. 2002, 182, 11–19.

138. Ortatatli, M.; Oğuz, H.; Hatipoğlu, F.; Karaman, M. Evaluation of pathological changes in broilers during chronic aflatoxin (50 and 100 ppb) and clinoptilolite exposure. Res. Vet. Sci.2005, 78, 61–68.

139. Chen, K.; Fang, J.; Peng, X.; Cui, H.; Chen, J.; Wang, F.; Chen, Z.; Zuo, Z.; Deng, J.; Lai, W.; Zhou, Y. Effect of selenium supplementation on aflatoxin B1-induced histopathological lesions

and apoptosis in bursa of Fabricius in broilers. Food Chem. Toxicol. 2014, 74, 91–97.

140. Kubena, L.F.; Harvey, R.B.; Bailey, R.H.; Buckley, S.A.; Rottinghaus, G.E. Effects of a hydrated sodium calcium aluminosilicate (T-Bind) on mycotoxicosis in young broiler chickens. Poult. Sci. 1998, 77, 1502–1509.

141. Tung, H.T.; Wyatt, R.D.; Thaxton, P.; Hamilton, P.B. Concentrations of serum proteins during aflatoxicosis. Toxicol. Appl. Pharmacol. 1975, 34, 320–326.

142. Fernández, A.; Verde, M.T.; Gomez, J.; Gascon, M.; Ramos, J.J. Changes in the prothrombin time, haematology and serum proteins during experimental aflatoxicosis in hens and broiler chickens. Res. Vet. Sci. 1995, 58, 119–122.

143. Doerr, J.A.; Huff, W.E.; Tung, H.T.; Wyatt, R.D.; Hamilton, P.B. A survey of T-2 toxin, ochratoxin, and aflatoxin for their effects on the coagulation of blood in young broiler chickens. Poult. Sci. 1974, 53, 1728–1734.

144. Doerr, J.A.; Wyatt, R.D.; Hamilton, P.B. Impairment of coagulation function during aflatoxicosis in young chickens. Toxicol. Appl. Pharmacol. 1976, 35, 437–446.

145. Doerr, J.A.; Hamilton, P.B. Aflatoxicosis and intrinsic coagulation function in broiler chickens. Poult. Sci. 1981, 60, 1406–1411.

146. Witlock, D.R.; Wyatt, R.D. Effect of dietary aflatoxin on hemostasis of young turkey poults.Poult. Sci. 1981, 60, 528–531.

147. Donaldson, W.E.; Tung, H.T.; Hamilton, P.B. Depression of fatty acid synthesis in chick liver (Gallus domesticus) by aflatoxin. Comp. Biochem. Physiol. B 1972, 41, 843–847.

148. Hoerr, F.J. Clinical aspects of immunosuppression in poultry. Avian Dis. 2010, 54, 2–15.

149. Chen, K.; Shu, G.; Peng, X.; Fang, J.; Cui, H.; Chen, J.; Wang, F.; Chen, Z.; Zuo, Z.; Deng, J.; Geng, Y.; Lai, W. Protective role of sodium selenite on histopathological lesions decreased T-cell subsets and increased apoptosis of thymus in broilers intoxicated with aflatoxin B1. Food Chem. Toxicol. 2013, 59, 446–454.

150. Chen, K.; Peng, X.; Rang, J.; Cui, H.; Zuo, Z.; Deng, J.; Chen, Z.; Geng, Y.; Lai, W.; Tang, L.; Yang, Q. Effects of dietary selenium on histopathological changes and T cells of spleen in broilers exposed to aflatoxin B1. Int. J. Environ. Res. Public Health 2014, 11, 1904–1913.

151. Pier, A.C.; Heddleston, K.L.; Cysewski, S.J.; Patteron, J.M. Effect of aflatoxin on immunity in turkeys. II. Reversal of impaired resistance to bacterial infection by passive transfer of plasma. Avian Dis. 1972, 16, 381–387.

152. Thaxton, J.P.; Tung, H.T.; Hamilton, P.B. Immunosuppression in chickens by aflatoxin. Poult. Sci. 1974, 53, 721–725.

153. Edrington, T.S.; Kubena, L.F.; Harvey, R.B.; Rottinghaus, G.E. Influence of a superactivated charcoal on the toxic effects of aflatoxin or T-2 toxin in growing broilers. Poult. Sci. 1997, 76, 1205–1211.

154. Huff, W.E.; Kubena, L.F.; Harvey, R.B.; Hagler, W.M., Jr.; Swanson, S.P.; Phillips, T.D.; Creger, C.R. Individual and combined effects of aflatoxin and deoxynivalenol (DON, vomitoxin) in broiler chickens. Poult. Sci. 1986, 65, 1291–1298.

155. Huff, W.E.; Harvey, R.B.; Kubena, L.F.; Rottinghaus, G.E. Toxic synergism between aflatoxin and T-2 toxin in broiler chickens. Poult. Sci. 1988, 67, 1418–1423.

156. Peng, X.; Zhang, K.; Bai, S.; Ding, X.; Zeng, Q.; Yang, J.; Fang, J.; Chen, K. Histological lesions, cell cycle arrest, apoptosis and T cell subsets changes of spleen in chicken fed aflatoxin-contaminated corn. Int. J. Environ. Res. Public Health 2014, 11, 8567–8580.

157. Chang, C.F.; Hamilton, P.B. Impaired phagocytosis by heterophils from chickens during aflatoxicosis. Toxicol. Appl. Pharmacol. 1979, 48, 459–466.

158. Ibrahim, I.K.; Shareef, A.M.; Al-Joubory, K.M. Ameliorative effects of sodium bentonite on phagocytosis and Newcastle disease antibody formation in broiler chickens during aflatoxicosis. Res. Vet. Sci. 2000, 69, 119–122.

159. Ghosh, R.C.; Chauhan, H.V.; Jha, G.J. Suppression of cell-mediated immunity by purified aflatoxin B1 in broiler chicks. Vet. Immunol. Immunopathol. 1991, 28, 165–172.

160. Neldon-Ortiz, D.L.; Qureshi, M.A. Effects of AFB1 embryonic exposure on chicken mononuclear phagocytic cell functions. Dev. Comp. Immunol. 1992, 16, 187–196.

161. Qureshi, M.A.; Brake, J.; Hamilton, P.B.; Hagler, W.M., Jr.; Nesheim, S. Dietary exposure of broiler breeders to aflatoxin results in immune dysfunction in progeny chicks. Poult. Sci.1998, 77, 812–819.

162. Chang, C.F.; Hamilton, P.B. Impairment of phagocytosis in chicken monocytes during aflatoxicosis. Poult. Sci. 1979, 58, 562–566.

163. Neldon-Ortiz, D.L.; Qureshi, M.A. Direct and microsomal activated aflatoxin B1 exposure and its effects on turkey peritoneal macrophage functions in vitro. Toxicol. Appl. Pharmacol.1991, 109, 432–442.

164. Neldon-Ortiz, D.L.; Qureshi, M.A. The effects of direct and microsomal activated aflatoxin B1 on chicken peritoneal macrophages in vitro. Vet. Immunol. Immunopathol. 1992, 31, 61–76.

165. Chang, C.F.; Hamilton, P.B. Refractory phagocytosis by chicken thrombocytes during aflatoxicosis. Poult. Sci. 1979, 58, 559–561.

166. Ghosh, R.C.; Chauhan, H.V.; Roy, S. Immunosuppression in broilers under experimental aflatoxicosis. Br. Vet. J. 1990, 146, 457–462.

167. Giambrone, J.J.; Ewert, D.L.; Wyatt, R.D.; Eidson, C.S. Effect of aflatoxin on the humoral and cell-mediated immune systems of the chicken. Am. J. Vet. Res. 1978, 39, 305–308. [PubMed]

168. Chen, K.; Yuan, S.; Chen, J.; Peng, X.; Wang, F.; Cui, H.; Fang, J. Effects of sodium selenite on the decreased percentage of T cell subsets, contents of serum IL-2 and IFN-γ induced by aflatoxin B1 in broilers. Res. Vet. Sci. 2013, 95, 143–145.

169. Chen, J.; Chen, K.; Yuan, S.; Peng, X.; Fang, J.; Wang, F.; Cui, H.; Chen, Z.; Yuan, J.; Geng, Y. Effects of aflatoxin B1 on oxidative

stress markers and apoptosis in spleens in broilers.Toxicol. Ind. Health 2013.

170. Wang, F.; Shu, G.; Peng, X.; Fang, J.; Chen, K.; Cui, H.; Chen, Z.; Zuo, Z.; Deng, J.; Gene, Y.; Lai, W. Protective effects of sodium selenite against aflatoxin-B1 induced oxidative stress and apoptosis in broiler spleen. Int. J. Environ. Res. Public Health 2013, 10, 2834–2844.

171. He, Y.; Fang, J.; Peng, X.; Cui, H.; Zuo, Z.; Deng, J.; Chen, Z.; Lai, W.; Shu, G.; Tang, L. Effects of sodium selenite on aflatoxin B1-induced decrease of ileac T cell and the mRNA contents of IL-2, IL-6, and TNF-α in broilers. Biol. Trace Elem. Res. 2014, 159, 167–173.

172. Jiang, M.; Peng, X.; Fang, J.; Cui, H.; Yu, Z.; Chen, Z. Effects of aflatoxin B1 on T-cell subsets and mRNA expression of cytokines in the intestine of broilers. Int. J. Mol. Sci. 2015, 16, 6945–6959.

173. Campbell, M.L., Jr.; May, J.D.; Huff, W.E.; Doerr, J.A. Evaluation of immunity of young broiler chickens during simultaneous aflatoxicosis and ochratoxicosis. Poult. Sci. 1983, 62, 2138–2144.

174. Pier, A.C.; Heddleston, K.L.; Boney, W.A.; Lukert, P.K. The effect of aflatoxin on immunity.Proc. XIX Con. Mund. Med. Vet. Zootech. 1971, 1, 216–219.

175. Stewart, R.G.; Skeeles, J.K.; Wyatt, R.D.; Brown, J.; Page, R.K.; Russell, I.D.; Lukert, P.D. The effect of aflatoxin on complement activity in broiler chickens. Poult. Sci. 1985, 64, 616–619.

176. Azzam, A.H.; Gabal, M.A. Interaction of aflatoxin in the feed and immunization against selected infectious diseases. I. Infectious bursal disease. Avian Pathol. 1997, 26, 317–325.

177. Gabal, M.A.; Azzam, A.H. Interaction of aflatoxin in the feed and immunization against selected infectious diseases in poultry. II. Effect on one-day-old layer chicks simultaneously vaccinated against Newcastle disease, infectious bronchitis and infectious bursal disease.Avian Pathol. 1998, 27, 290–295.

178. Boulton, S.L.; Dick, J.W.; Hughes, B.L. Effects of dietary aflatoxin and ammonia-inactivated aflatoxin on Newcastle disease antibody titers in layer-breeders. Avian Dis. 1982, 26, 1–6.

179. Hegazy, S.M.; Azzam, A.; Gabal, M.A. Interaction of naturally occurring aflatoxins in poultry feed and immunization against fowl cholera. Poult. Sci. 1991, 70, 2425–2428.

180. Pier, A.C.; Heddleston, K.L. The effect of aflatoxin on immunity in turkeys. I. Impairment of actively acquired resistance to bacterial challenge. Avian Dis. 1970, 14, 797–809.

181. Pier, A.C. Effects of aflatoxin on immunity. J. Am. Vet. Med. Assoc. 1973, 163, 1268–1269. [PubMed]

182. Boonchuvit, B.; Hamilton, P.B. Interaction of aflatoxin and paratyphoid infection in broiler chickens. Poult. Sci. 1975, 54, 1567–1573.

183. Hegazy, S.M.; Adachi, Y. Comparison of the effects of dietary selenium, zinc, and selenium and zinc supplementation on growth and immune response between chick groups that were inoculated with Salmonella and aflatoxin or Salmonella. Poult. Sci. 2000, 79, 331–335.

184. Wyatt, R.D.; Hamilton, P.B. Interaction between aflatoxicosis and a natural infection of chickens with Salmonella. Appl. Microbiol. 1975, 30, 870–872. [PubMed]

185. Edds, G.T.; Nair, K.P.; Simpson, C.F. Effect of aflatoxin B1 on resistance in poultry against cecal coccidiosis and Marek's disease. Am. J. Vet. Res. 1973, 34, 819–826. [PubMed]

186. Edds, G.T.; Simpson, C.F. Cecal coccidiosis in poultry as affected by prior exposure to aflatoxin B1. Am. J. Vet. Res. 1976, 37, 65–68. [PubMed]

187. Witlock, D.R.; Wyatt, R.D.; Anderson, W.I. Relationship between Eimeria adenoeides infection and aflatoxicosis in turkey poults. Poult. Sci. 1982, 61, 1293–1297.

188. Wyatt, R.D.; Ruff, M.D.; Page, R.K. Interaction of aflatoxin with Eimeria tenella infection and monensin in young broiler chickens. Avian Dis. 1975, 19, 730–740.

189. Hamilton, P.B.; Harris, J.R. Interaction of aflatoxicosis with Candida albicans infections and other stresses in chickens. Poult. Sci. 1971, 50, 906–912.

190. Chang, C.F.; Hamilton, P.B. Increased severity and new symptoms of infectious bursal disease during aflatoxicosis in broiler chickens. Poult. Sci. 1982, 61, 1061–1068.

191. Giambrone, J.J.; Partadiredja, M.; Eidson, C.S.; Kleven, S.H.; Wyatt, R.D. Interaction of aflatoxin with infectious bursal disease virus infection in young chickens. Avian Dis. 1978,22, 431–439.

192. Batra, P.; Pruthi, A.K.; Sadana, J.R. Effect of aflatoxin B1 on the efficacy of turkey herpesvirus vaccine against Marek's disease. Res. Vet. Sci. 1991, 51, 115–119.

193. Tessari, E.N.; Oliveira, C.A.; Cardoso, A.L.; Ledoux, D.R.; Rottinghaus, G.E. Effects of aflatoxin B1 and fumonisin B1 on body weight, antibody titres and histology of broiler chicks. Br. Poult. Sci. 2006, 47, 357–364.

194. Yunus, A.W.; Ghareeb, K.; Abd-El-Fattah, A.A.; Twaruzek, M.; Böhm, J. Gross intestinal adaptations in relation to broiler performance during chronic aflatoxin exposure. Poult. Sci.2011, 90, 1683–1689.

195. Yunus, A.W.; Razzazi-Fazeli, E.; Bohm, J. Aflatoxin B1 in affecting broiler's performance, immunity, and gastrointestinal tract: A review of history and contemporary issues. Toxins2011, 3, 566–590.

196. Applegate, T.J.; Schatzmayr, G.; Prickel, K.; Troche, C.; Jiang, Z. Effect of aflatoxin culture on intestinal function and nutrient loss in laying hens. Poult. Sci. 2009, 88, 1235–1241.

197. Yang, J.; Bai, F.; Zhang, K.; Lv, X.; Bai, S.; Zhao, L.; Peng, X.; Ding, X.; Li, Y.; Zhang, J. Effects of feeding corn naturally contaminated with AFB1 and AFB2 on performance and aflatoxin residues in broilers. Czech. J. Anim. Sci. 2012, 57, 506–515.

198. Ledoux, D.R.; Rottinghaus, G.E.; Bermudez, A.J.; Alonso-Debolt, M. Efficacy of a hydrated sodium calcium aluminosilicate to ameliorate the toxic effects of aflatoxin in broiler chicks.Poult. Sci. 1999, 78, 204–210.

199. Hamilton, P.B.; Garlich, J.D. Failure of vitamin supplementation to alter the fatty liver syndrome caused by aflatoxin. Poult. Sci. 1972, 51, 688–692.

200. Hamilton, P.B.; Tung, H.T.; Wyatt, R.D.; Donaldson, W.E. Interaction of dietary aflatoxin with some vitamin deficiencies. Poult. Sci. 1974, 53, 871–877.

201. Rogers, S.R.; Pesti, G.M.; Wyatt, R.D. Effect of tryptophan supplementation on aflatoxicosis in laying hens. Poult. Sci. 1991, 70, 307–312.

202. Hamilton, P.B.; Tung, H.T.; Harris, J.R.; Gainer, J.H.; Donaldson, W.E. The effect of dietary fat on aflatoxicosis in turkeys. Poult. Sci. 1972, 51, 165–170.

203. Richardson, K.E.; Nelson, L.A.; Hamilton, P.B. Effect of dietary fat level on dose response relationships during aflatoxicosis in young chickens. Poult. Sci. 1987, 66, 1470–1478.

204. Smith, J.W.; Hill, C.H.; Hamilton, P.B. The effect of dietary modifications on aflatoxicosis in the broiler chicken. Poult. Sci. 1971, 50, 768–774.

205. Kermanshahi, H.; Akbari, M.R.; Maleki, M.; Behgar, M. Effect of prolonged low level inclusion of aflatoxin B1 into diet on performance, nutrient digestibility, histopathology and blood enzymes of broiler chickens. J. Anim. Vet. Adv. 2007, 6, 686–692.

206. Verma, J.; Swain, B.K.; Johri, T.S. Effect of various levels of aflatoxin and ochratoxin A and combinations thereof on protein and energy utilisation in broilers. J. Sci. Food Agric. 2002, 82, 1412–1417.

207. Verma, J.; Johri, T.S.; Swain, B.K. Effect of aflatoxin, ochratoxin and their combination on protein and energy utilisation in white leghorn laying hens. J. Sci. Food Agric. 2007, 87, 760–764.

208. Ruff, M.D.; Wyatt, R.D. Intestinal absorption of L-methionine and glucose in chickens with aflatoxicosis. Toxicol. Appl. Pharmacol. 1976, 37, 257–262.

209. Oliveira, C.A.; Kobashigawa, E.; Reis, T.A.; Mestieri, L.; Albuquerque, R.; Corrêa, B. Aflatoxin B1 residues in eggs of laying

hens fed a diet containing different levels of the mycotoxin. Food Addit. Contam. 2000, 17, 459–462.

210. Celik, I.; Oguz, H.; Demet, O.; Boydak, M.; Donmez, H.H.; Sur, E.; Nizamlioglu, F. Embryotoxicity assay of aflatoxin produced by Aspergillus parasiticus NRRL 2999. Br. Poul. Sci. 2000, 41, 401–409.

211. Dietert, R.R.; Qureshi, M.A.; Nanna, U.C.; Bloom, S.E. Embryonic exposure to aflatoxin-B1: Mutagenicity and influence on development and immunity. Environ. Mutagen. 1985, 7, 715–725.

212. Edrington, T.S.; Harvey, R.B.; Kubena, L.F. Toxic effects of aflatoxin B1 and ochratoxin A, alone and in combination, on chicken embryos. Bull. Environ. Contam. Toxicol. 1995, 54, 331–336.

213. Oznurlu, Y.; Celik, I.; Sur, E.; Ozaydın, T.; Oğuz, H.; Altunbaş, K. Determination of the effects of aflatoxin B1 given in ovo on the proximal tibial growth plate of broiler chickens: Histological, histometric and immunohistochemical findings. Avian Pathol. 2012, 41, 469–477.

214. Sur, E.; Celik, I. Effects of aflatoxin B1 on the development of the bursa of Fabricius and blood lymphocyte acid phosphatase of the chicken. Br. Poult. Sci. 2003, 44, 558–566.

215. Williams, J.G.; Deschl, U.; William, G.M. DNA damage in fetal liver cells of turkey and chicken eggs dosed with aflatoxin B1. Arch. Toxicol. 2011, 85, 1167–1172.

216. Perrone, C.E.; Ahr, H.J.; Duan, J.D.; Jeffrey, A.M.; Schmidt, U.; Williams, G.M.; Enzmann, H.H. Embryonic turkey liver: Activities of biotransformation enzymes and activation of DNA-reactive carcinogens. Arch. Toxicol. 2004, 78, 589–598.

217. Sur, E.; Celik, I.; Oznurlu, Y.; Aydin, M.F.; Oguz, H.; Kurtoglu, V.; Ozaydin, T. Enzyme histochemical and serological investigations on the immune system from chickens treated in ovo with aflatoxin B1 (AFB1). Revue Méd. Vét. 2011, 162, 443–448.

218. Ul-Hassan, Z.; Khan, M.Z.; Khan, A.; Javed, I. Immunological status of the progeny of breeder hens kept on ochratoxin A (OTA)- and aflatoxin B1 (AFB1)-contaminated feeds. J. Immunotoxicol. 2012, 9, 381–391.

219. Lilleberg, S.L.; Cabonce, M.A.; Raju, N.R.; Wagner, L.M.; Kier, L.D. Alterations in the structural gene and the expression of p53 in rat liver tumors induced by aflatoxin B1. Mol. Carcinog. 1992, 6, 159–172.

220. Harris, A.J.; Shaddock, J.G.; Manjanatha, M.G.; Lisenbey, J.A.; Casciano, D.A. Identification of differentially expressed genes in aflatoxin B1-treated cultured primary rat hepatocytes and Fischer 344 rats. Carcinogenesis 1998, 19, 1451–1458.

221. Merrick, B.A.; Phadke, D.P.; Auerbach, S.S.; Mav, D.; Stiegelmeyer, S.M.; Shah, R.R.; Tice, R.R. RNA-Seq profiling reveals novel hepatic gene expression pattern in aflatoxin B1 treated rats. PLoS ONE 2013, 8, e61768.

222. Dugyala, R.R.; Sharma, R.P. The effect of aflatoxin B1 on cytokine mRNA and corresponding protein levels in peritoneal macrophages and splenic lymphocytes. Int. J. Immunopharmacol. 1996, 18, 599–608.

223. Han, S.H.; Jeon, Y.J.; Yea, S.S.; Yang, K.H. Suppression of the interleukin-2 gene expression by aflatoxin B1 is mediated through the down-regulation of the NF-AT and AP-1 transcription factors. Toxicol. Lett. 1999, 108, 1–10.

224. Hinton, D.M.; Myers, M.J.; Raybourne, R.A.; Francke-Carroll, S.; Sotomayor, R.E.; Shaddock, J.; Warbritton, A.; Chou, M.W. Immunotoxicity of aflatoxin B1 in rats: Effects on lymphocytes and the inflammatory response in a chronic intermittent dosing study. Toxicol. Sci. 2003, 73, 362–377.

225. Meissonnier, G.M.; Pinton, P.; Laffitte, J.; Cossalter, A.M.; Gong, Y.Y.; Wild, C.P.; Bertin, G.; Galtier, P.; Oswald, I.P. Immunotoxicity of aflatoxin B1: Impairment of the cell-mediated response to vaccine antigen and modulation of cytokine expression. Toxicol. Appl. Pharmacol.2008, 231, 142–149.

226. Qian, G.; Tang, L.; Guo, X.; Wang, F.; Massey, M.E.; Su, J.; Guo, T.L.; Williams, J.H.; Phillips, T.D.; Wang, J.S. Aflatoxin B1 modulates the expression of phenotypic markers and cytokines by splenic lymphocytes of male F344 rats. J. Appl. Toxicol. 2014, 34, 241–249.

227. Murarolli, R.A. Effects of Aflatoxin B1 (AFB1) on Hepatic Gene Expression in Pigs and Turkeys. Doctoral Dissertation, University of Missouri, Columbia, MO, USA, 2013.

228. Tilton, S.C.; Gerwick, L.G.; Hendricks, J.D.; Rosato, C.S.; Corley-Smith, G.; Givan, S.A.; Bailey, G.S.; Bayne, C.J.; Williams, D.E. Use of a rainbow trout oligonucleotide microarray to determine transcriptional patterns in aflatoxin B1-induced hepatocellular carcinoma compared to adjacent liver. Toxicol. Sci. 2005, 88, 319–330.

229. Afanassieff, M.; Goto, R.M.; Ha, J.; Sherman, M.A.; Zhong, L.; Auffray, C.; Coudert, F.; Zoorob, R.; Miller, M.M. At least one class I gene in restriction fragment pattern-Y (Rfp.-Y.), the second MHC gene cluster in the chicken, is transcribed, polymorphic, and shows divergent specialization in antigen binding region. J. Immunol. 2001, 166, 3324–3333.

230. Bauer, M.M.; Reed, K.M. Extended sequence of the turkey MHC B-locus and sequence variation in the highly polymorphic B-G loci. Immunogenetics 2011, 63, 209–221.

231. Chaves, L.D.; Krueth, S.B.; Bauer, M.M.; Reed, K.M. Sequence of a turkey BAC clone identifies MHC class III orthologs and supports ancient origins of immunological gene clusters. Cytogenet. Genome Res. 2011, 132, 55–63.

232. Chaves, L.D.; Krueth, S.B.; Reed, K.M. Characterization of the turkey MHC chromosome through genetic and physical mapping. Cytogenet. Genome Res. 2007, 117, 213–220.

233. Chaves, L.D.; Krueth, S.B.; Reed, K.M. Defining the turkey MHC: Sequence and genes of the B locus. J. Immunol. 2009, 183, 6530–6537.

234. Delany, M.E.; Robinson, C.M.; Goto, R.M.; Miller, M.M. Architecture and organization of chicken microchromosome 16: Order of the NOR, MHC-Y, and MHC-B subregions. J. Hered.2009, 100, 507–514.

235. Kaufman, J.; Milne, S.; Göbel, T.W.F.; Walker, B.A.; Jacob, J.P.; Auffray, C.; Zoorob, R.; Beck, S. The chicken B locus is a minimal essential major histocompatibility complex. Nature 1999,401, 923–925.

236. Miller, M.M.; Goto, R.; Bernot, A.; Zoorob, R.; Auffray, C.; Bumstead, N.; Briles, W.E. TwoMhc class I and two Mhc class II genes map to the chicken Rfp.-Y. system outside the Bcomplex. Proc. Natl. Acad. Sci. USA 1994, 91, 4397–4401.

237. Reed, K.M.; Bauer, M.M.; Monson, M.S.; Benoit, B.; Chaves, L.D.; O'Hare, T.H.; Delany, M.E. Defining the turkey MHC: Identification of expressed class I- and class IIB-like genes independent of the MHC-B. Immunogenetics 2011, 63, 753–771.

238. Monson, M.S.; Settlage, R.E.; Mendoza, K.M.; Rawal, S.; El-Nezami, H.S.; Coulombe, R.A.; Reed, K.M. Modulation of the spleen transcriptome in domestic turkey (Meleagris gallopavo) in response to aflatoxin B1 and probiotics. Immunogenetics 2015, 67, 163–178.

239. Rustemeyer, S.M.; Lamberson, W.R.; Ledoux, D.R.; Wells, K.; Austin, K.J.; Cammack, K.M. Effects of dietary aflatoxin on the hepatic expression of apoptosis genes in growing barrows.J. Anim. Sci. 2005, 89, 916–925.

240. Monson, M.S.; Settlage, R.E.; McMahon, K.W.; Mendoza, K.M.; Rawal, S.; el-Nezami, H.S.; Coulombe, R.A.; Reed, K.M. Response of the hepatic transcriptome to aflatoxin B1 in domestic turkey (Meleagris gallopavo). PLoS ONE 2014, 9, e100930.

241. Bothast, R.J.; Nofsinger, G.W.; Lagoda, A.A.; Black, L.T. Integrated process for ammonia inactivation of aflatoxin-contaminated corn and ethanol fermentation. Appl. Environ. Microbiol. 1982, 43, 961–963. [PubMed]

242. Brekke, O.L.; Sinnhuber, R.O.; Peplinski, A.J.; Wales, J.H.; Putnam, G.B.; Lee, D.J.; Ciegler, A. Aflatoxin in corn: Ammonia inactivation and bioassay with rainbow trout. Appl. Environ. Microbiol. 1977, 34, 34–37. [PubMed]

243. Jalili, M.; Jinap, S.; Son, R. The effect of chemical treatment on reduction of aflatoxins and ochratoxin a in black and white pepper during washing. Food Addit. Contam. Part A Chem. Anal. Control. Expo. Risk Assess. 2011, 28, 485–493.

244. Mann, G.E.; Gardner, H.K.; Booth, A.N.; Gumbmann, M.R. Aflatoxin inactivation. Chemical and biological properties of

ammonia and methylamine treated cottonseed meal. J. Agric. Food Chem. 1971, 19, 1155–1158. [PubMed]

245. Vesonder, R.F.; Beckwith, A.C.; Ciegler, A.; Dimler, R.J. Ammonium hydroxide treatment of aflatoxin B1. Some chemical characteristics and biological effects. J. Agric. Food Chem. 1975,23, 242–243.

246. Weng, C.Y.; Martinez, A.J.; Park, D.L. Efficacy and permanency of ammonia treatment in reducing aflatoxin levels in corn. Food Addit. Contam. 1994, 11, 649–658.

247. Codifer, L.P., Jr.; Mann, G.E.; Dollear, F.G. Aflatoxin inactivation: Treatment of peanut meal with formaldehyde and calcium hydroxide. J. Am. Oil Chem. Soc. 1976, 53, 204–206.

248. Patel, U.D.; Govindarajan, P.; Dave, P.J. Inactivation of aflatoxin B1 by using the synergistic effect of hydrogen peroxide and gamma radiation. Appl. Environ. Microbiol. 1989, 55, 465–467. [PubMed]

249. Moerck, K.E.; McElfresh, P.; Wohlman, A.; Hilton, B.W. Aflatoxin destruction in corn using sodium bisulfite, sodium hydroxide and aqueous ammonia. J. Food Prot. 1980, 43, 571–574.

250. Yang, C.Y. Comparative studies on the detoxification of aflatoxins by sodium hypochlorite and commercial bleaches. Appl. Microbiol. 1972, 24, 885–890. [PubMed]

251. Gregory, J.F., 3rd; Edds, G.T. Effect of dietary selenium on the metabolism of aflatoxin B1 in turkeys. Food Chem. Toxicol. 1984, 22, 637–642.

252. Diaz, G.J.; Cortés, A.; Botero, L. Evaluation of the ability of a feed additive to ameliorate the adverse effects of aflatoxins in turkey poults. Br. Poult. Sci. 2009, 50, 240–250.

253. Kubena, L.F.; Harvey, R.B.; Huff, W.E.; Corrier, D.E.; Phillips, T.D.; Rottinghaus, G.E. Efficacy of a hydrated sodium calcium aluminosilicate to reduce the toxicity of aflatoxin and T-2 toxin. Poult. Sci. 1990, 69, 1078–1086.

254. Kubena, L.F.; Harvey, R.B.; Phillips, T.D.; Corrier, D.E.; Huff, W.E. Diminution of aflatoxicosis in growing chickens by the dietary

addition of a hydrated, sodium calcium aluminosilicate. Poult. Sci. 1990, 69, 727–735.

255. Kubena, L.F.; Huff, W.E.; Harvey, R.B.; Yersin, A.G.; Elissalde, M.H.; Witzel, D.A.; Giroir, L.E.; Phillips, T.D.; Petersen, H.D. Effects of a hydrated sodium calcium aluminosilicate on growing turkey poults during aflatoxicosis. Poult. Sci. 1991, 70, 1823–1830.

256. Kubena, L.F.; Harvey, R.B.; Phillips, T.D.; Clement, B.A. Effect of hydrated sodium calcium aluminosilicates on aflatoxicosis in broiler chicks. Poult. Sci. 1993, 72, 651–657.

257. Zhao, J.; Shirley, R.B.; Dibner, J.D.; Uraizee, F.; Officer, M.; Kitchell, M.; Vazquez-Anon, M.; Knight, C.D. Comparison of hydrated sodium calcium aluminosilicate and yeast cell wall on counteracting aflatoxicosis in broiler chicks. Poult. Sci. 2010, 89, 2147–2156.

258. Magnoli, A.P.; Monge, M.P.; Miazzo, R.D.; Cavaglieri, L.R.; Magnoli, C.E.; Merkis, C.I.; Cristofolini, A.L.; Dalcero, A.M.; Chiacchiera, S.M. Effect of low levels of aflatoxin B1 on performance, biochemical parameters, and aflatoxin B1 in broiler liver tissues in the presence of monensin and sodium bentonite. Poult. Sci. 2011, 90, 48–58.

259. Coulombe, R.A.; Guarisco, J.A.; Klein, P.J.; Hall, J.O. Chemoprevention of aflatoxicosis in poultry by dietary butylated hydroxytoluene. Anim. Feed Sci. Technol. 2005, 121, 217–225.

260. Guarisco, J.A.; Hall, J.O.; Coulombe, R.A., Jr. Mechanisms of butylated hydroxytoluene chemoprevention of aflatoxicosis-inhibition of aflatoxin B1 metabolism. Toxicol. Appl. Pharmacol. 2008, 227, 339–346.

261. Guarisco, J.A.; Hall, J.O.; Coulombe, R.A., Jr. Butylated hydroxytoluene chemoprevention of aflatoxicosis—Effects on aflatoxin B1 bioavailability, hepatic DNA adduct formation, and biliary excretion. Food Chem. Toxicol. 2008, 46, 3727–3731.

262. Klein, P.J.; van Vleet, T.R.; Hall, J.O.; Coulombe, R.A., Jr. Effects of dietary butylated hydroxytoluene on aflatoxin B1-relevant metabolic enzymes in turkeys. Food Chem. Toxicol.2003, 41, 671–678.

263. Larsen, C.; Ehrich, M.; Driscoll, C.; Gross, W.B. Aflatoxin-antioxidant effects on growth of young chicks. Poult. Sci. 1985, 64, 2287–2291.

264. Gowda, N.K.; Ledoux, D.R.; Rottinghaus, G.E.; Bermudez, A.J.; Chen, Y.C. Antioxidant efficacy of curcuminoids from turmeric (Curcuma longa L.) powder in broiler chickens fed diets containing aflatoxin B1. Br. J. Nutr. 2009, 102, 1629–1634.

265. Bagherzadeh Kasmani, F.; Karimi Torshizi, M.A.; Allameh, A.; Shariatmadari, F. A novel aflatoxin-binding Bacillus probiotic: Performance, serum biochemistry, and immunological parameters in Japanese quail. Poult. Sci. 2012, 91, 1846–1853.

266. Oatley, J.T.; Rarick, M.D.; Ji, G.E.; Linz, J.E. Binding of aflatoxin B1 to Bifidobacteria in vitro. J. Food Prot. 2000, 63, 1133–1136. [PubMed]

267. Peltonen, K.; El-Nezami, H.; Haskard, C.; Ahokas, J.; Salminen, S. Aflatoxin B1 binding by dairy strains of lactic acid bacteria and Bifidobacteria. J. Dairy Sci. 2001, 84, 2152–2156.

268. Shahin, A.A.M. Removal of aflatoxin B1 from contaminated liquid media by dairy lactic acid bacteria. Int. J. Agric. Biol. 2007, 9, 71–75.

269. Topcu, A.; Bulat, T.; Wishah, R.; Boyac, I.H. Detoxification of aflatoxin B1 and patulin byEnterococcus faecium strains. Int. J. Food Microbiol. 2010, 139, 202–205.

270. Deabes, M.M.; Darwish, H.R.; Abdel-Aziz, K.B.; Farag, I.M.; Nada, S.A.; Tawfek, N.S. Protective effects of Lactobacillus rhamnosus GG on aflatoxins-induced toxicities in male albino mice. J. Environ. Anal. Toxicol. 2012, 2, 132.

271. El-Nezami, H.; Salminen, S.J.; Ahokas, J. Biological control of food carcinogens with use ofLactobacillus GG. Nutr. Today 1996, 21, 41S–42S.

272. El-Nezami, H.; Kankaanpää, P.; Salminen, S.; Ahokas, J. Ability of dairy strains of lactic acid bacteria to bind a common food carcinogen, aflatoxin B1. Food Chem. Toxicol. 1998, 36, 321–326.

273. El-Nezami, H.; Kankaanpää, P.; Salminen, S.; Ahokas, J. Physicochemical alterations enhance the ability of dairy strains of

lactic acid bacteria to remove aflatoxin from contaminated media. J. Food Prot. 1998, 61, 466–468. [PubMed]

274. El-Nezami, H.; Mykkänen, H.; Kankaanpää, P.; Salminen, S.; Ahokas, J. Ability ofLactobacillus and Propionibacterium strains to remove aflatoxin B1 from the chicken duodenum. J. Food Prot. 2000, 63, 549–552. [PubMed]

275. Fazell, M.R.; Hajimohammadali, M.; Moshkani, A.; Samadi, N.; Jamalifar, H.; Khoshayand, M.R.; Vaghari, E.; Pouragahi, S. Aflatoxin B1 binding capacity of autochthonous strains of lactic acid bacteria. J. Food Prot. 2009, 72, 189–192.

276. Gratz, S.; Mykkänen, H.; Ouwehand, A.C.; Juvonen, R.; Salminen, S.; el-Nezami, H. Intestinal mucus alters the ability of probiotic bacteria to bind aflatoxin B1 in vitro. Appl. Environ. Microbiol. 2004, 70, 6306–6308.

277. Gratz, S.; Täubel, M.; Juvonen, R.O.; Viluksela, M.; Turner, P.C.; Mykkänen, H.; el-Nezami, H. Lactobacillus rhamnosus strain GG modulates intestinal absorption, fecal excretion and toxicity of aflatoxin B1 in rats. Appl. Environ. Microbiol. 2006, 72, 7398–7400.

278. Gratz, S.; Wu, Q.K.; el-Nezami, H.; Juvonen, R.O.; Mykkänen, H.; Turner, P.C. Lactobacillus rhamnosus strain GG reduces aflatoxin B1 transport, metabolism, and toxicity in Caco-2 cells.Appl. Environ. Microbiol. 2007, 73, 3958–3964.

279. Haskard, C.; Binnion, C.; Ahokas, J. Factors affecting the sequestration of aflatoxin byLactobacillus rhamnosus strain GG. Chem. Biol. Interact. 2000, 128, 39–49.

280. Haskard, C.A.; el-Nezami, H.S.; Kankaanpää, P.E.; Salminen, S.; Ahokas, J.T. Surface binding of aflatoxin B1 by lactic acid bacteria. Appl. Environ. Microbiol. 2001, 67, 3086–3091.

281. Hernandez-Mendoza, A.; Garcia, H.S.; Steele, J.L. Screening of Lactobacillus casei strains for their ability to bind aflatoxin B1. Food Chem. Toxicol. 2009, 47, 1064–1068.

282. Hernandez-Mendoza, A.; Guzman de Peña, D.; Garcia, H.S. Key role of teichoic acids on aflatoxin B1 binding by probiotic bacteria. J. Appl. Microbiol. 2009, 107, 395–403.

283. Hernandez-Mendoza, A.; Rivas-Jimenez, L.; Garcia, H.S. Assessment of aflatoxin B1 binding to Lactobacillus reuteri by microscopy and fluorescence techniques. Food Biotechnol. 2011, 25, 140–150.

284. Kankaanpää, P.; Tuomola, E.; El-Nezami, H.; Ahokas, J.; Salminen, S.J. Binding of aflatoxin B1 alters the adhesion properties of Lactobacillus rhamnosus strain GG in a Caco-2 model. J. Food Prot. 2000, 63, 412–414. [PubMed]

285. Lahtinen, S.J.; Haskard, C.A.; Ouwehand, A.C.; Salminen, S.J.; Ahokas, J.T. Binding of aflatoxin B1 to cell wall components of Lactobacillus rhamnosus strain GG. Food Addit. Contam.2004, 21, 158–164.

286. Lee, Y.K.; el-Nezami, H.; Haskard, C.A.; Gratz, S.; Puong, K.Y.; Salminen, S.; Mykkänen, H. Kinetics of adsorption and desorption of aflatoxin B1 by viable and nonviable bacteria. J. Food Prot. 2003, 66, 426–430. [PubMed]

287. Peltonen, K.D.; el-Nezami, H.S.; Salminen, S.J.; Ahokas, J.T. Binding of aflatoxin B1 by probiotic bacteria. J. Sci. Food Agric. 2000, 80, 1942–1945.

288. Turbic, A.; Ahokas, J.T.; Haskard, C.A. Selective in vitro binding of dietary mutagens, individually or in combination, by lactic acid bacteria. Food Addit. Contam. 2002, 19, 144–152.

289. El-Nezami, H.; Mykkänen, H.; Kankaanpää, P.; Suomalaine, T.; Salminen, S.; Ahokas, J. Ability of a mixture of Lactobacillus and Propionibacterium to influence the faecal aflatoxin content in healthy Egyptian volunteers: A pilot clinical study. Biosci. Microflora 2000, 19, 41–45.

290. El-Nezami, H.S.; Polychronaki, N.N.; Ma, J.; Zhu, H.; Ling, W.; Salminen, E.K.; Juvonen, R.O.; Salminen, S.J.; Poussa, T.; Mykkänen, H.M. Probiotic supplementation reduces a biomarker for increased risk of liver cancer in young men from Southern China. Am. J. Clin. Nutr. 2006, 83, 1199–1203. [PubMed]

291. Lanza, G.M.; Washburn, K.W.; Wyatt, R.D.; Marks, H.L. Genetic variation of physiological response to aflatoxin in Gallus domesticus. Theor. Appl. Genet. 1982, 63, 207–212.

292. Lanza, G.M.; Washburn, K.W.; Wyatt, R.D.; Marks, H.L. Effect of dietary aflatoxin concentration on the assessment of genetic variability of response in a randombred population of chickens. Genetics 1983, 104, 123–131. [PubMed]

293. Manning, R.O.; Wyatt, R.D.; Marks, H.L.; Fletcher, O.J. Effects of dietary aflatoxin in aflatoxin-resistant and control lines of chickens. Poult. Sci. 1990, 69, 922–928.

294. Scott, T.R.; Rowland, S.M.; Rodgers, R.S.; Bodine, A.B. Genetic selection for aflatoxin B1resistance influences chicken T-cell and thymocyte proliferation. Dev. Comp. Immunol. 1991,15, 383–391.

295. Wyatt, R.D.; Marks, H.L.; Manning, R.O. Selection for resistance to aflatoxin in chickens.Poult. Sci. 1987, 66, 1901–1904.

296. Marks, H.L.; Wyatt, R.D. Genetic resistance to aflatoxin in Japanese quail. Science 1979, 206, 1329–1330.

297. Pegram, R.A.; Wyatt, R.D.; Marks, H.L. Comparative responses of genetically resistant and nonselected Japanese quail to dietary aflatoxin. Poult. Sci. 1985, 64, 266–272.

298. Pegram, R.A.; Wyatt, R.D.; Marks, H.L. The relationship of certain blood parameters to aflatoxin resistance in Japanese quail. Poult. Sci. 1986, 65, 1652–1658.

299. Rodeheaver, D.P.; Wyatt, R.D.; Marks, H.L. Relation of serum alpha-amylase to aflatoxin resistance in Japanese quail. Avian Dis. 1986, 30, 568–573.

300. Monson, M.S. Hepatotoxic and Immunomodulatory Transcriptome Responses to Aflatoxin B1 in the Turkey (Meleagris Gallopavo). Ph.D. Thesis, University of Minnesota, Saint Paul, MN, USA, May 2015.

301. Coulombe, R.A. USDA project: Functional Genomics to Enhance Aflatoxin Resistance in Poultry. Utah State University: Logan, UT, USA, 2015.

CHAPTER 7

Protective Effect of Probiotics Strains in Ruminants

Everlon Cid Rigobelo and Fernando Antonio de Ávila

UNESP Animal Science Faculty of Dracena, UNESP Department of Veterinary Pathology, Brazil

1. INTRODUCTION

In last 15 years the use of probiotics strains in animal production has been increased. These probiotics strains can modulate the balance and activities of the gastrointestinal microbiota in which are responsible to gut homeostasis. The intake of probiotics supplemented in ration and provided to the animals, can strongly affect the structure and activities of the gut microbial communities leading to promoting health and improving the performance in livestock, when it is impaired by numerous factors, such as dietary and management constraints. The understanding of the digestive ecosystems in terms of microbial composition and functional diversity is fundamental to modulate the gastrointestinal tract (GIT) of domestic animals providing to them the possibility to maintain the homeostasis of these complex microbial communities, which can be composed of bacteria, protozoa, fungi, archaea, and viruses, thus promoting a reduction of the incidence of diseases. Therefore considerable researchs during 30 years are characterizing the domestic animals ´GIT. The welfare, health and feed efficiency of the animals can be affected by different factors, many of them, environmental factors. Among these factors, feeding practices, composition of animal diets, farms management and productivity constraints can influence the microbial balance in GIT, whose role is fundamental to gut homeostasis and its reduction consequently can affect efficiency digestive When occurs the reduction of microbial in GIT, some reactions as digestion and fermentation of plant polymers are impaired, since the action of the microbiota on gut is strongly related with the realization these reactions, and the animals also are impaired by the fact these polymers to be of particular importance to the herbivorous (Chaucheyras-Durand and Durand, 2010).

2. USE OF ANTIBIOTICS

The problem caused by indiscriminate use of antibiotic as growth promoter in feed to livestock is that this practice has been associated with emergence of resistance to antibiotics in zoonotic bacteria. The use of growth promoter in feed to livestock has been done since 1940 because this practice is correlated with higher health status and improves at performance of animals in terms of feed conversion. The use of antibiotics at animal has had a profound impact on animal health and welfare.

The problems found by this practice require the development of alternative intervention strategies for zoonotic livestock pathogens. Some these strategies could be vaccines in diarrhea in neonates and post weaning animals, limited access to livestock, control of vermin, modifying air flow, high level disinfection regimes, acidification of feed and the supply of probiotic into animals supplemented in ration by example are efficient management to reduce the occurrence of pathogen at the animal production.

3. USE OF PROBIOTICS STRAINS

All additives used in animal feed, including yeasts and bacteria, are strictly regulated within the EU legislative framework. Until May 2003, the risk assessment of animal feed additives for use in European was the responsibility of the Scientific Commitee of Animal Nutrition (SCAN) (Anadon et al., 2006). After this date, the European Food Safety Authority (EFSA) took over the functions of SCAN. While EFSA provide expert scientific advice to the European Commission the approval and risk management of a probiotic product is responsibility of the EC and its constituent member's states. For use of microorganism in

United States as a feed additive is necessary before the product to be outgoing to approval by the Food and Drug Administration (FDA).

The requirements for a novel probiotic product required by EU regulations on animal feed additives are the identification and characterization to species level, and the efficacy data must be provided in support of any claims made for the product. Some characteristics are requested to product such as no adverse effects on the health of performance, the product must be safe for the operator, have no adverse effects upon exposure and also the product must not pose a risk to the safety of the end-consumer (SCAN, 2001).

4. USE OF PROBIOTICS TO CONTROL GASTROINTESTINAL DISEASES IN LIVESTOCK

The intensive production farmed livestock together with the veto of the use of antimicrobial feed supplements in the EU, this situation has increased the risk of contracting gastrointestinal diseases if prophylactic antimicrobial feed supplements are not utilized. The removal of growth promoters has led to a significant increase in the incidence of diseases and also with significant increases in feed costs, the reduced feed weight conversion.

5. USE OF PROBIOTICS IN ANIMALS

Although the mechanisms involved have not been fully elucidated a reduction in pathogen carriage and subsequent clinical disease is one possible mechanisms responsible by reduction of occurrence of disease when the growth promoter is utilized in livestock. After this prohibition many problems arisen and also the need of use of alternatives to resolve this situation. One of these alternatives is the use of probiotics as feed supplement or functional food which may be used for prophylaxis in animals and humans. There are numerous probiotics products commercially available for livestock. Currently commercial livestock probiotic can be separated into two categories, being these, competitive exclusion that are defined and those that are undefined.

6. USE OF PROBIOTIC IN RUMINANTS

In ruminants that have four stomachs being them rumen, omasum, reticle and abomasums when these animals born they have the abomasums extremely big. This situation occurs because the type of food is liquids as milk. Usually the animal becomes ruminants when he from the third or fourth month of age. This development is due the installation of microbiota ruminal in gut and also by distention of organ due the fiber intake. The bacteria from rumen and bowel are acquired through the contact of cattle with the cow or other animal and also by grass intake.

The rumen is as fermentation chamber and it has approximately for 50-85% by use of dry matter from food. The saliva is mixed with food and has a control upon pH of rumen and the papillae existing in inner wall of the rumen increasing the absorption area.

The amount of bacteria from rumen is the approximately 10^{11} CFU/g of counts rumen, the fungi is the 10^3 CFU/g and the protozoa is 10^5 cell /g. There are most of 60 species of bacteria that grow into rumen microbita and this environment has CO_2, CH_4 and N_2 stomach gas maintaining the pH value among 6 – 6.5. The temperature within the rumen is 39°C and the bacteria type living can be characterized according to theirs functions such as cellulolytic, proteolytic, amylolytic.

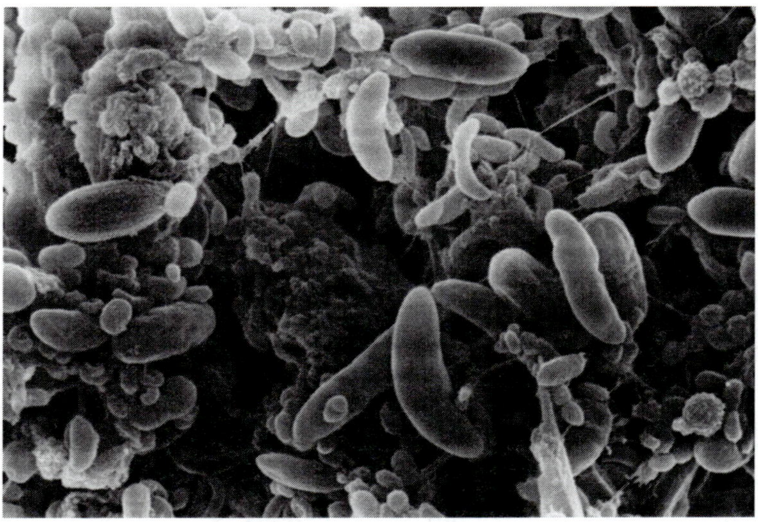

Picture 1. Picture took from Antibiotics and chemotherapeutic and probiotics Avila et al Funep Publisher Brazil 83p.

The proteins and fibrous foods in rumen are converted at ammonia, organic acids and amino acids by microorganism's action. As the majority of amino acids are synthesized of rumen the animals need to be supplied with essential amino acids from ration or injectable. The main factors of stress feed that leaving to a decreasing of ruminal microbiota are dry grasslands, pastures in budding and seasonal changes. The decreasing of ruminal microbiota can be caused by antibiotics use and also environment changes as cccur at auctions, expositions and pre-slaughter. The use of rumen bacteria into ruminants promotes the growth into gut before the establishment of pathogen in these animals causing the prevention of diarrhea occurrence. This situation decreases the weaning time and maintains the balance of rumen microbiota increasing the production of enzymes as cellulase, amylase, urease, protease consequently increasing improving the use fibrous foods. Others benefits to use of probiotics in ruminants are promotes the increasing of weight gain, increasing the milk production and decreasing of diarrhea period.

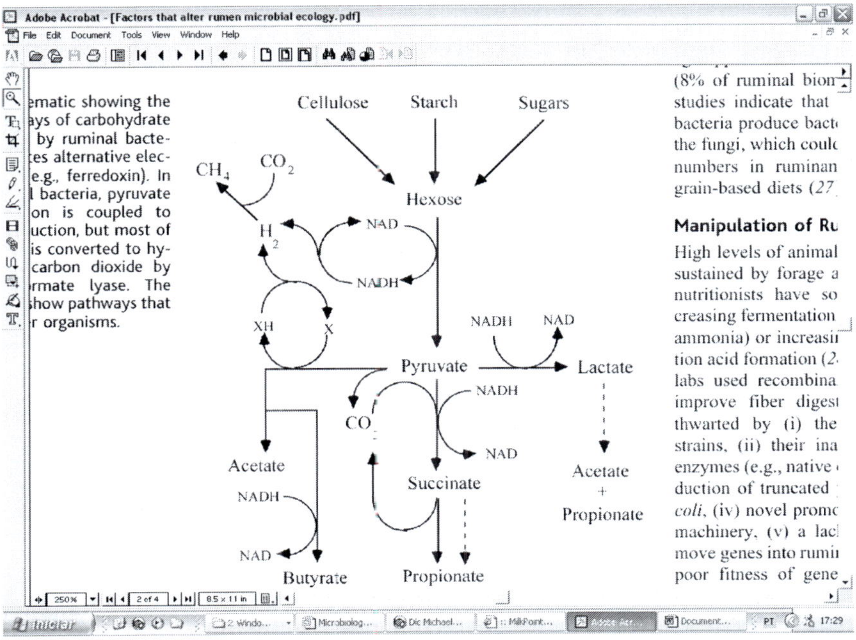

Picture 2. Picture took from Antibiotics and chemotherapeutic and probiotics Avila et al Funep Publisher Brazil 83p.

The advantages of the use of probiotics in livestock are the period of adaptation of animal is not necessary, doesn´t hinder the management on the farm because it can be supplement to ration or mineral salt, and as probiotic is the natural product does not necessary the disposal of milk and also this product can be used during the slaughter of animals as cattle, sheep and buffaloes. According with FERREIRA, (2003) the probiotics microorganisms most used belong to the group of lactic bacteria as*Aerococcus, Atopobium, Bifidobacteirum, Brochothrix, Carnobacterium, Enterococcus, Lactobacillus,Weissella*. The lactic bacteria are positive Gram, anaerobic, negative catalase, presenting of cocos and bacillus way. The probiotics can counts ruminal bacterias as Ruminobacter and Succinovibrio with specifics characteristics that are used in supplementation of ruminants.

Some authors have been showed that some probiotics strains have seen resistant to the antibiotics effects and therefore these strains could be used together the administration of antibiotics in animals. The yeasts are unicellular microorganisms with capacity of survive in several mediums have a great spectrum of pH and many mediums can be saline or without oxygen. The *Saccharomyces boulardii* has been largely tested in human's trials (PENNA et al., 2000). And the *Saccharomyces cerevisiae* in animals showed promising results.

The *Lactobacillus* is constituted by cells that vary long and thin to short and curves with 1.5-6.0μm length and 0.6-0.9 width. The ideal temperature to growth is 45°C and grows in pH 5.5-6.0. The Lactobacillus species known at moment is 56 and the most used as additive are *L. acidophilus, L. rhamnnosis and L. casei.*

The genus *Bifidobacterium* includes 30 species. Many of these 10 are form humans dental caries, vagina and feces, 17 are from animal origin 2 are from wastewater and 1 of fermented milk. These bacteria present optimal growth among 37°C and 41°C and minimal growth among 25° C and 28°C at pH 6-7. The *Bifidobacterium Bifidobacterium animallis, Bifidobacterium lactis, Bifidobacterium longum* species have probiotics characteristics also have capabilities to ferment complex carbon.

Some species of *Bacillus subtilis, Bacillus licheniformis* and *Bacillus cereus* are bacteria positive Gram in rods form. The Bacillus are the only that form spores allowing that these strains to be used in adverse conditions mainly in high temperature.

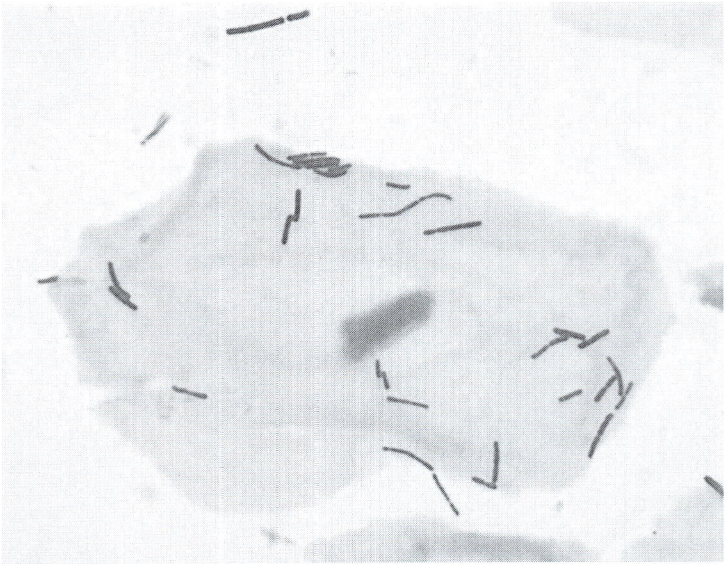

Picture 3. Picture took from Antibiotics and chemotherapeutic and probiotics Avila et al Funep Publisher Brazil 83p.

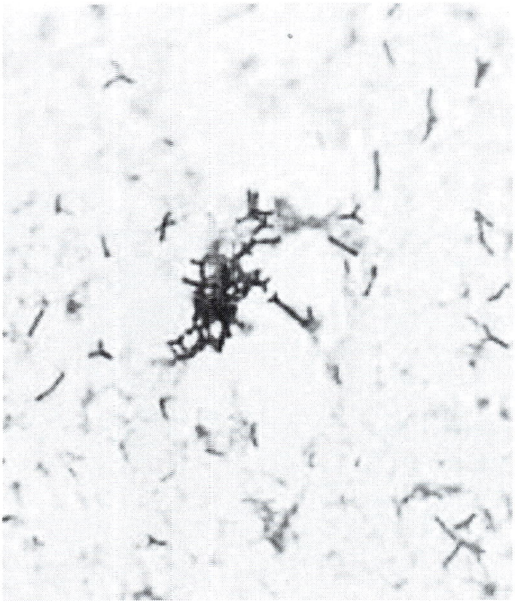

Picture 4. Picture took from Antibiotics and chemotherapeutic and probiotics Avila et al Funep Publisher Brazil 83p.

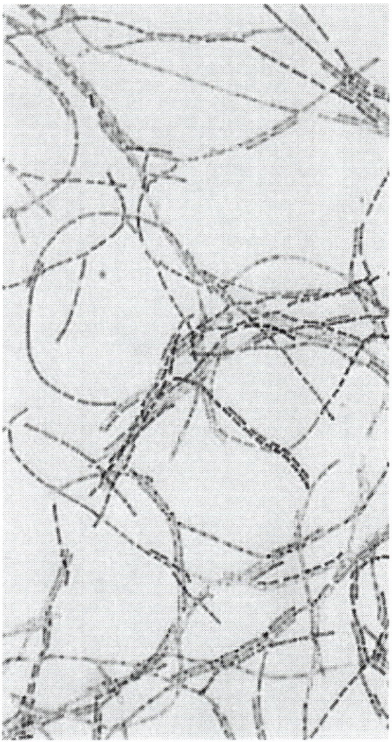

Picture 5. *Enterococcus faecium* is the microorganism belonged to the *Enterococcus* genus belonged to the Lancifield D group. This morphology identification requests the use of coloration by Gram and also catalase test in blade. These bacteria are positive Gram and present the characteristic form of streptococcus (chain cocos), negative catalase and no spore and faculty anaerobic. Through the chemical analysis the strain ferment the lactose, arabinose, mannitol, no ferment the sorbitol. This strain growth into MacConkey medium containing 6.5% of NaCl.

7. OTHERS BENEFITS AND ACTION MODE OF PROBIOTICS STRAINS

7.1. Immune Modulation

The maturation of the humoral immune mechanisms can be conducted by microbial colonization, this events can promote the c circulation of the IgA and IgM secreting cells. The other important factor that can be affected by microbial colonization on the gut of different animals particularly the ruminants are the balance of the different T helper subsets. The memory B and T cells migrate to effectors sites in consequence these events.

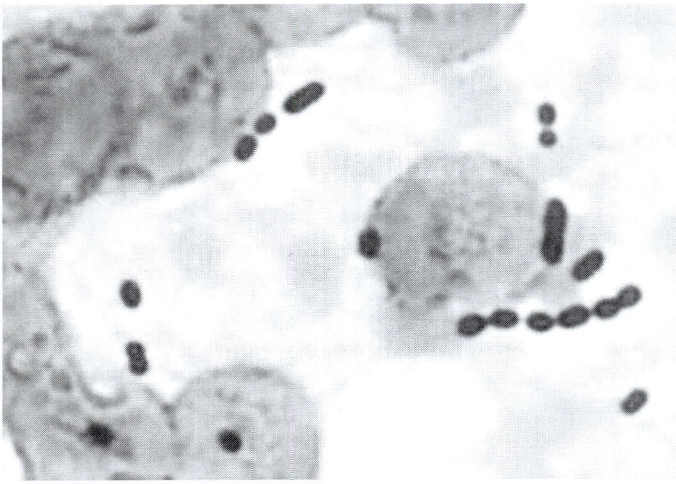

Picture 6. *Ruminobacter amylophilum* is the microorganism belonged to the Ruminobacter genus to its morphology identification is necessary to use Gram coloration this genus present as rods negative Gram. They have motility and no spores. This ferments the cornflour, maltose and liquefy gelatin. They synthesize lactic acid and CO2 from formic acid.

Other mechanisms to immune modulation are followed by active proliferation local induction of certain cytokines and production of secretion antibodies as IgA. When the host is exposure to the antigen, immune cells respond releasing cytokines from host direct the subsequent immune responses. The low-dose tolerance immunity TGF-B associated in via local cytokine is the man mechanisms which the gut associated lymphoid tissue maintains homeostasis. Some lactic acid bacteria can induces the production of proinflammatory cytokines, tumor necrosis factors alpha and interleukin-6 from human peripheral blood mononuclear cells. A strain of *Lactobacillus casei* can inhibit the growth of pathogenic strains as *Pseudomonas aeruginosa* and *Listeria monocytogenes* leading to an increase in the level of macrophages. Others strains as *Lactobacillus acidophilus* and *Bifidobacterium bifidum* could inhance non-specific immunity and concluded that specific lactic acid bacteria could play a role in specific age groups, specific neonates or the elderly. The results can be observed when two groups of animals are compared itself in relation with their age. Usually the positive effect against the colonization by pathogenic bacteria upon the gut occurs most efficiently in neonates than oldest.

Some studies showed a significant increase in IgA immune response. In others, on children with mild to moderate stable Crohn´s Diseases, administration with strain GG improved the gut barrier function and clinical status after six months of therapy.

8. ANTITUMOR ACTIVITY

Some probiotic strains could decrease some enzymes synthesized by many microorganisms may convert procarcinogens into carcinogens and cause colon cancer, some of them azoreductase, β- glucuronidase and nitroreductase. *Lactobacillus acidophilus* could decrease nitroreductase, azoreductase and β glucuronidase activities in carnivorous animals. Another strain as *Lactobacillus rhamnosus* could bacterial β-glucuronidase activity in the large intestine.

Lactobacillus and Lactobacillus bulgaricus suppressed Ehrlich ascitis tumor or Sarcoma 180 in mice. Tumor suppression in associated with intact viable cells, intact dead cells and cell wall fragments or Lactobacilli and Bifidobacteria. When *Lactobaacillus casei* was provided into rats it had effective prevention against the recurrence of superficial bladder cancer.

Nitites used in food processing are converted to carcinogenic nitrosamines in the gastrointestinal tract in several people. Cellular uptake of nitites by *Lactobacillus* and *Bifidobacteria* has been shown in vivo. Also, Lactobacillus has been shown as a great reducer of bile salts. They are implicated in the initiation of colon carcinogens. These strains have been biotransformed of primary to secondary bile salts, this way, there are reduction the possible initiation of cancer. Other authors have been suggested that the decrease of intestinal pH, through metabolic activities of Lactobacillus acid bacteria, could inhibit the growth of putrefactive bacteria, can prevent large bowel cancer.

Many probiotics strains have a positive effect against mould growth and aflatoxin production. These aflatoxins are associated to cause cancer.

Thus the reduction of these moulds decrease the occurrence of cancer caused by this mould.

9. REDUCTION OF CHOLESTEROL

Some studies have showed the effect of fermented milk or milk containing probiotic strains producing lactic acid on serum cholesterol levels. These studies reported that a strain of *Streptococcus thermophilus* and *Lactobacillus acidophilus* reduced cholesterol levels in rats. Milk fermented with lactic acid bacteria and *Streptococcus cerevisae* led to lower serum cholesterol than control group, also phospholipids and bile acids in the fecal samples from mice were lower. When a trial was using rats inoculated with *E. faecium*, they presented a lower cholesterol levels. The same findings were observed in pigs that have been fed a high cholesterol diet.

Another results also, showed that the serum lipoprotein levels of 334 individuals remained unchanged when they were treated with *Lactobacillus acidophilus* and *L. delbrueckii* subsp *bulgaricus* and *E. faecium* administered over six weeks to adults and it resulted in a initial increase in total cholesterol and LDL followed by a sharp decrease two weeks after termination of treatment. The decrease corresponded with an increase in the reduction of iodonitrotetrazolium and superoxide production by peripheral neutrophils and an elevated production of IgG. Several studies don´t explain because there was the reduction in cholesterol levels and suggest that the reduction of cholesterol is not due to assimilation or to a direct interaction between the bacteria and cholesterol. This effect is due to the co-precipitation of cholesterol with deconjugated bile salts at pH value below 6.0. This would not explain the reduction of cholesterol in vivo as the pH of the bower gastrointestinal is neutral to alkaline. Probably there is a physical association between cholesterol and the cell surface.

10. DECREASING OF LACTOSE INTOLERANCE

Some descents from Asia and Africa usually are stricken by lack the intestinal mucosal enzyme β-galactosidase and therefore suffer from reduction in lactase activity. This situation can occur many times after an infection caused rotavirus gastroenteritis. There are much lactic bacteria which are capable to synthesize the enzyme β-galactosidase. Many of them as the bacteria *Streptococcus salivarius* subps *thermophilus* and *Lactobacillus delbrueckii* subps *bulgaricus*. The levels of enzyme produced by these bacteria are high and many products treated with this enzyme presented a low concentration of lactose. These species are sensitive to bile salts. These substances can lead to release of high levels of β-galactosidase in the gastrointestinal tract. Lactose from fermented milk containing the probiotic *Lactobacillus acidophilus* were better absorbed by many people with lower β-galactosidase activity. All symptoms from lactose intolerance were decreased.

11. STOOL TRANSIT

The diarrhea occurrences in neonate are the main cause of death. This disorder affects animals of many species and also the human among them the children. *Lactobacillus* GG had a high decreasing in severity of acute watery diarrhea in young children. Patients treated on erythromycin reacted decreasing the period of diarrhea when they received *Lactobacillus* GG.

The symptoms caused by slow stool transit are diarrhea, stomach pain, abdominal pain and nausea. All symptoms were recovery quickly when the patients received *Lactobacillus* GG. Indeed one of the most severe diarrhea is that caused by *Clostridum difficile*. Usually people stricken by this disease recently passed by treatment with antibiotics. The supply of *Lactobacillus rhamnosus* improved the symptoms of intestinal disorders.

Patients who consumed milk fermented by the strain experienced less diarrhea than those that don´t received. Many of them were patients that were being treated with pelvic radiotherapy. The effect of different LAB n different types of diarrhea has been showed in many studies. Yet are needed others studies to determine which mechanisms the LAB use to relieve diarrhea.

From now on this chapter will present some findings from some trials that were performed with the aim of verifying the protective effect of a probiotic mix that was kindly donated by IMEVE Biotecnology located in Jaboticabal São Paulo State against the colonization caused by STEC in sheep.

Abstract: Shiga toxin-producing *Escherichia coli* (STEC) strains are food-borne pathogens that cause human diseases, and ruminants are usually important reservoirs of STEC. The first step of enteric infection is colonization of the host's gut mucosal surface by pathogenic strains of bacteria. Probiotic bacteria can decrease the severity of infection by competing for receptors and nutrients and by synthesizing an acid that creates an unfavorable environment for the growth of several bacterial species. The aim of this study was to determine whether the inoculation of sheep with a mixture containing 5 x 10^8 (CFU) of *Lactobacillus acidophilus, Lactobacillus helveticus, Lactobacillus bulgaricus, Lactobacillus lactis, Streptococcus thermophilus* and *Enterococcus faecium* per animal decreases the shedding at animals previously inoculated with STEC nonO157. Sheep that received oral inoculums containing 2 × 10^9 viable bacteria of STEC carriers of *stx1*, *stx2* and *eae* genes were compared with others groups that did not receive inoculums. When probiotic was inoculated together with the STEC non-O157, the numbers of these same bacteria in a fecal sample were lower than the group did not receive. It occurred during the 3th, 5th, 6th and 7th weeks post-inoculation. Thus, we conclude that this mixture likely presented a potential protective effect in reducing colonization by STEC non-O157 and can be used as an alternative method to decreases STEC non-157 infection in sheep, thereby reducing transmission to humans.

12. STEC DISEASES

Healthy cattle, sheep and other ruminants can be reservoirs of Shiga-toxin-producing *Escherichia coli*(STEC) strains. STEC have been associated with human diseases such as hemorrhagic colitis and hemolytic uremic syndrome (Hussein 2007; Ramamurthy 2008). These bacteria can be transmitted from person to person (Belongia et al., 1993), but most outbreaks have been associated with the consumption contaminated beef products or a variety of other foods. Before colonization by STEC, it may be possible to determine whether to use the colonization of ruminal mucosa by oral administration of probiotic bacteria as a strategy (Ávila et al., 2000).

Probiotics are live microorganisms that, when administered in the appropriate amount, will benefit the health of the host (Food and Agriculture Organization of the United Nations, 2003; Sanders, 2003). Microbial interference is common to all genera and decreases the severity of infection by mechanisms involving nutrient competition, generation of an unfavorable environment, and competition for attachment or adhesion sites (Chaucheryras-Durand and Durand, 2010). Probiotics bacteria can stimulate the immune system through innate cell surface pattern recognition receptors or via direct lymphoid cell activation. Practical applications for this action of probiotics based on this characteristic include their use in anti-tumor, anti-allergy and immunotherapy treatments, but there is also increasing evidence that some probiotics can sufficiently stimulate a protective immune response to enhance resistance to microbial pathogens (Cross, 2002).

The benefits caused for use of probiotics strains in ruminants are known, however there are few information about the use of probiotics strains to reduction of shedding of STEC non-O157 in sheep.

This study verified the protective effect of probiotic treatment against the colonization of STEC non-O157 in sheep measured the number of STEC recovered from fecal simple.

13. MATERIALS AND METHODS

13.1. Animals and Experimental Locations

The study was performed with 20 sheep of Santa Ines race in the fattening stage, female previously screened by not be carrying of STEC non-O157 strains distributed in four groups with five animals each that were confined at a property located in São Paulo State. The experiment was made January to March 2012. The sheep were selected based on closeness of body weight (41 ± 2) kg and age (9-12) months. Then, all animals were ear-tagged and drenched with Ivomec (MSD- Agvet Merck) for internal parasite control at the rate of 2cc/46kg body weight. During three weeks pre-experimental adaptation period, were offered for all groups of sheep a diet of identical composition *ad libitum* consumption. Group I did not receive the probiotics strains or STEC non-O157 being the control group. Group II received an only oral dose of inoculums containing 2 x 10^9 viable cell of STEC non-O157 per animal. Group III received an only oral dose of inoculums containing 2 x 10^9 viable cell of STEC non-O157 per animal together with daily oral doses at concentration of 5 x 10^8 CFU of*Lactobacillus acidophilus, Lactobacillus helveticus, Lactobacillus bulgaricus, Lactobacillus lactis, Streptococcus thermophilus* and *Enterococcus faecium* per animal lyophilized provided directly in the mouth of animals with help of a cannula of application throughout the experiment. The inoculums were provided with help of a cannula of application and were diluted at 40mL of 0.9% saline solution. Group IV received the probiotics alone at the same number of cells viable and of the same way. During three weeks before of start of experiment always in same hour in the morning were collected feces samples directly of rectum of these animals. The samples were cultured in plate on MacConkey agar then the colonies that grew had their DNA extracted as described by Wani et al. (2003) to verify the absence of STEC non-O157 and *Salmonella*. After the third week the groups of animals were inoculated and monitored by seven weeks with weekly collections of their feces. All animals of present study were not carrying STEC non-O157 before inoculation and were kept in bays separated to avoid cross contamination throughout the experiment in an environmentally controlled building. Each pen had a concrete floor with individual drain, a feeding box and water through and was cleaned once a day and the fecal material deposited was transported to other place where it was composted.

This study was conducted in accordance with the ethical guidelines for investigations involving laboratory animals and was approved by the Ethics in Animal Research Committee (EARC) of UNESP-Univi Estadual Paulista and no adverse effects were observed in the animals receiving the *E. coli* (STEC) and probiotics during the experiment.

14. PROBIOTIC

The probiotics bacteria used were *Bacillus cereus, Lactobacillus acidophilus* and *Enterococcus faecium*all strains in amount of 3 x 10^8 (CFU). These strains were isolated from sheep rumina and intestinal tracts following the recommendations of Hungate (1975) and Wolf et al. (1975). These bacteria have the following features: they are nonpathogenic, enzyme-producing and resistant to lactic acid and low pH. These strains were kindly donated by Imeve Medications Veterinary Industry responsible by all tests realized concerning the quality and conditions of use.

15. STEC NON-O157

To verify the protective effect of probiotics strains reducing the shedding of STEC was used a STEC non-O157 strain isolated from healthy sheep and characterized as described by Possé et al., (2007). It was kindly donated by Laboratory of bacteriological from UNESP Jaboticabal.

16. SAMPLES

For seven weeks, post-inoculation feces samples in same hour in the morning were collected from the sheep and transported to the laboratory, where DNA was extracted. Bacterial strains grown overnight in nutrient broth (Sigma) at 37° C were pelleted by centrifugation at 12,000g for 1 min, resuspended in 200m L of sterile distilled water, and lysed by boiling for 10min. Lysates were centrifuged as described above, and 150m L of the supernatants was used as DNA template for the PCR (Wani et al., 2003). All isolates were subjected to PCR; *stx1*, *stx2,* and *eae* genes were detected using the primers and PCR conditions described by China et al. (1996). Control reference strains were *E. coli* EDL 933 (O157:H7, *stx1*, *stx 2*, *eae*) and *E. coli* K12 (negative control).

17. STECS RECUPERATED

The values of STEC in each sample were determined of two different methods of counting. In both 1 g of each fecal sample was collected, cultured on MacConkey agar, then it was incubated at 37ºC for 24 h. In the first counting, all colonies grown displaying similar genome to STEC non-O157 strain previously inoculated orally were counted. In second counting were selected at least five colonies per sample grown and then separated in STEC non-O157 displaying pattern genome the others isolates from *E. coli* that did not display this specific DNA patterns.

18. *E. COLI* STEC FINGERPRINT BY PULSED-FIELD GEL ELECTROPHORESIS (PFGE) OF CHROMOSOMAL DNA

Genomic DNAs from STEC non-O157 isolates cultured from sheep were prepared as previously described by Barret et al., 1994. The agarose-embedded DNA was digested with 10U of *XbaI*/plug (Gibco BRL) at 37ºC overnight. PFGE was performed in a CHEF-DR II unit (Bio-Rad Laboratories, Hercules, Calif.) using 1% PFGE grade Tris Borate EDTA buffer gels. The DNA was electrophoresed for 20 hours at a constant voltage of 200V (6V/cm) pulse time of 5 to 50 s, an electric field angle of 120° and a temperature of 15°C before being stained with ethidium bromide. Resulting patterns were analyzed on a DNA Pro Scan, ProRFLP program (DNA Proscan, Inc. Nashville, Tenn), and the size of the DNA fragments was used as the criteria for categorizing distinct patterns.

19. RESULTS

The animals received inoculums containing only one isolate of STEC non-O157 carriers of *stx1, stx2*and *eae* genes. After three day post inoculations fecal samples were collected from these animals to make the re-isolating of the strains STEC non-O157 that had been previously inoculated into animals. All strains isolated from fecal samples had their DNA patterns compared with DNA pattern from STEC non-O-157 strain previously inoculated into animals and all those strains had the DNA similar to the strain previously inoculated were counted.

From strains isolated from fecal simples collected during the three weeks prior to inoculation of animals no STEC strain had the similar DNA to the DNA pattern from strains of STEC non-O157 previously inoculated into animals. The results showed that the STEC non-O157 strain previously inoculated into animals was the only strain recovered displaying this specific pattern of DNA. All strains isolated from fecal sample from animals from group I and IV also had no similar DNA patterns to the strain previously inoculated into animals these strains were classified as non-STEC (Table1).

Table 1. Proportion of means of STEC with the means of ordinary *E. coli* grown on plate re-isolated from feces samples from sheep from Groups I to IV during three weeks without inoculation and then during seven weeks post-inoculation.

	Group III	Group IV
Weeks without inoculation		
1	0.0	0.0
2	0.0	0.0
3	0.0	0.0
Weeks post-inoculation		
1	34/134	0.0
2	122/152	0.0
3	133/143	0.0
4	288/119	0.0
5	323/123	0.0
6	129/143	0.0
7	84/138	0.0

Ordinary strain of *E. coli* were all strains that not displayed similar DNA to the strains previously inoculated into animals

The relations among the means values of STEC non-O157 strains displaying the specific pattern of DNA previously inoculated with *E. coli* strains non-STEC from group II and III were respectively as follows: 21/123, 130/142, 146/135, 304/122, 352/132, 190/145 and 90/148; 34/134, 122/152, 133/143, 288/119, 323/123, 129/143 and 84/138 bacteria isolated per gram of feces. (Table1). The means values of STEC non-O157 strains displaying specific pattern of DNA previously inoculated in the animals from groups II and III were compared among itself within the same week to verify the possible reduction of isolates occurred in the animals from group III by administration of probiotics strains (Figure1).

Comparing the means values of isolates of STEC non-157 strains from group II with the means values of isolates of STEC non-O157 strains from Group III within the same weeks verified that the difference was statistically significant among them only the third, fifth, sixth and seventh week post animals´ inoculation (Figure1). There was lowest shedding of STEC non-O157 displaying similar DNA to the pattern of STEC non-O157 previously inoculated into animals belonged to the Group III than Group II, except in the first, second, and fourth week. The Group III had been received probiotic together with the STEC non-O157.

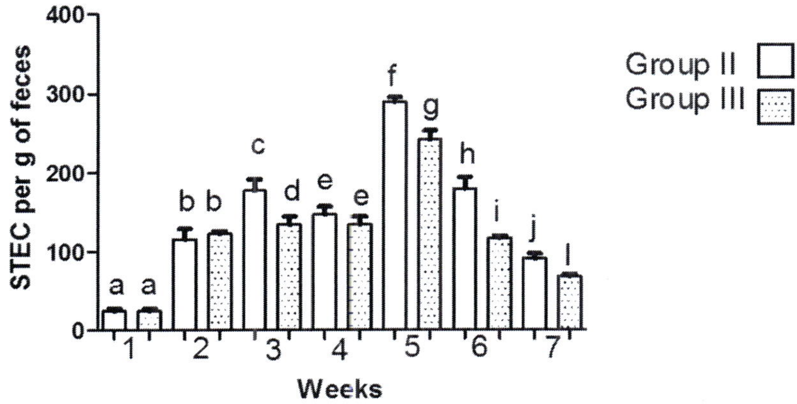

Figure 1. Comparison among the means of STEC from samples feces from Groups II and III.In each week the same letters show that the means not differs among them.

When the quantification was made through the selection at least five colonies from fecal sample during seven weeks of 1 to 5 sheep the results were 24, 26, 29, 30 and 29 in the group II and 20, 15, 19, 18, 16 in the group III (Table.2 and Table.3). The results show that there was no isolating of STEC non-O157 from sheep before the inoculation of bacteria inoculated. The total number of isolates from animals from group III were lowest than from group II. However these values not differ statistically. The aim this second counting was to verify if the reduction of shedding of STEC non-O157 from group III compared with group II would be shown by other way. However, this last counting way did not show statistical difference among the isolates.

Table 2. Total values of STEC re-isolated from feces sample selecting at least five colonies grown per samples from Group II.

	Sheep1	Sheep2	Sheep3	Sheep4	Sheep5
Weeks post-inoculation					
1	2	2	3	3	2
2	2	3	5	5	4
3	5	4	5	4	5
4	5	5	5	5	5
5	4	5	4	5	5
6	3	4	3	4	4
7	3	3	4	4	4
Total	**24**	**26**	**29**	**30**	**29**

Table 3. Total values of STEC re-isolated from feces sample selecting at least five colonies grown per samples from Group III.

	Sheep1	Sheep2	Sheep3	Sheep4	Sheep5
Weeks post-inoculation					
1	3	2	3	3	2
2	3	2	3	2	2
3	4	1	3	2	2
4	3	3	2	3	3
5	1	4	3	3	3
6	4	2	3	2	3
7	2	1	2	3	1
Total	**20**	**15**	**19**	**18**	**16**

20. DISCUSSION

Shiga-toxin-producing *E. coli* (STEC) strains are associated as a foodborne pathogen since 1982 and it has been identified as the cause of several outbreaks (Beutin et al., 2002; Karmali et al., 1989;Willshaw et al., 2001).

Probiotics are live microorganisms taken as food supplements that beneficially affect the host, maintaining a balance in their intestinal microbiota (Fuller, 1989). The ruminants including cattle, sheep and deer are reservoirs of STEC and the fecal shedding of these bacteria forms the vehicle of entry into the human food chain (Lema et al., 2001). The probiotics could be used as strategies to reduction of shedding these pathogens by animals (Chaucheyras-Durand et al. 2010).

In the present study we evaluated the protective effect of a mixture of probiotics strains to decrease the shedding of STEC non-O157 in sheep. The group III that received probiotic had fewer STEC non-O157 recovered from their feces when compared with the group II that did not receive the probiotics being that these differences were significant in 3th, 5th, 6th to 7th weeks. The probiotics strains failed to decrease the shedding of STEC non-O157 by feces during the first, second and fourth week post inoculation. In last three weeks of experiment there was a reduction in the shedding of the STEC non-O157 from feces from group III that received probiotic together with STEC non-O157 compared with the shedding of the STEC non-O157 from feces from group II which received STEC non-O157 only. For unknown reason the shedding of STEC non-O157 from group III was lower than group II during the third week post inoculation. However in the fourth week post inoculation there was no difference among the number of isolates of STEC non-O157 from both group III and II. As the probiotics beneficially affect the host, maintaining a balance in their intestinal microbiota (Fuller, 1989) probably the presence of probiotics strains hindered colonization and consequently the shedding these bacteria by feces.

Several mechanisms have been proposed to explain the beneficial effects of probiotics among them are the production of organic acids by bacterial probiotics can help decrease the gut pH, create more favorable ecological conditions for the resident microbiota and decrease the risk of pathogen colonization (Servin, 2004). The growth of pathogenic bacteria also can be hindered by synthesis of antimicrobial peptides, such as bacteriocins or production of enzymes able to hydrolyze bacterial toxins (Buts, 2004), stimulating the immune system, increasing the absorption of minerals and increasing the syntheses of vitamins (Thuory et al., 2003). Bactericins are produced by many lactic acid bacteria (LAB), including species normally found in the gastrointestinal tract as *L. acidophilus*-group as *L. acidophilus, Lactobacillus amylovorus, L. crispatus, L. crispatus*, *Lactobacillus gallinarum, L. gasseri and L. plantarum*, (De Vuyst et al., 1996 and Dicks & Botes, 2010).

Chaucheyras-Durand et al. (2010) indicated that some strategies may be used in the rumen to decrease the number of viable STEC cells as the use of *Lactobacillus acidophilus* supplemented in the ration, thereby preventing the contamination of food. These strategies are the administration of probiotics in the ruminants. The impact of probiotics and the physicochemical conditions of the rumen digesta on the survival of pathogenic strains could have significant implications for farm management practices and food safety and decrease the risk of food-borne illness.

In our study all sheep belonging to the group that received STEC non-O157 together with daily intake from probiotics strains had lower shedding this STEC non-O157. Some authors as Lema et al., (2001) verified that in lambs, the use of feed supplemented with lactic bacteria such as *Lactobacillusacidophilus* and *Enterococcus faecium* improved meat production. The mixture of probiotic strains used in this study contained strains of lactic bacteria, which probably allowed for the effect cited.Kritas et al., (2006) used *Bacillus licheniformis* and *Bacillus subtilis* supplemented in ration on sheep and verified although the mortality of sheep had not decreased there were beneficial effect on milk yields, fat and protein in milk.

As many bacterial species are present in the intestine, and under normal conditions the majority of these bacteria are strictly anaerobic. This composition makes the gut capable of responding to the possible anatomic and physicochemical variations that occur (Lee et al., 1999). The intestinal microbiota exercises a large influence on many biochemical reactions of the host. The balance maintained by probiotics hinders the growth of pathogenic microorganisms that are present. In contrast, an imbalance in the gut microbiota may cause the proliferation of pathogens and subsequent bacterial infection (Gibson, 1998).

The increased resistance against pathogens is the most important characteristic in developing effective probiotics. The use of probiotics strains excludes potentially pathogenic microorganisms and increases the natural defense mechanisms of the host (Puupponen-Pimiä et al., 2002). The modulation of intestinal microbiota by probiotic microorganisms occurs through a mechanism of competitive exclusion (Guarner and Malagelada, 2003). Also, the probiotics help to reset the intestinal microbiota through adhesion and colonization of the intestinal mucosa. This action hinders the adhesion or invasion of epithelial cells by pathogenic bacteria and decreases the synthesis of toxin. An imbalanced microbiota causes changes, such as the diarrhea associated with infections or treatment with antibiotics, allergic reactions to foods, and intestinal inflammatory diseases. Therefore, correcting an imbalance in the intestinal microbiota constitutes the basis for probiotic therapy (Isolauri et al., 2004). AccordingZhao et al. (1998), probiotics administered prior to exposure to pathogenic *E. coli* may reduce the levels of pathogenic *E. coli* carried in most animals. In this study we observed that concurrent inoculation of probiotics strains with STEC strains probably hindered the colonization of the pathogenic bacteria in the sheep, as compared with the groups that did not receive the probiotics treatment as well as by consequence decreasing thus the shedding by STEC non-O157. According toBatista et al. (2008), the administration *of*

Lactobacillus acidophilus, decreased the number of days the animals displayed symptoms of diarrhea in the group of ruminants that received the probiotic compared with the group that did not receive any probiotic. Roos et al., (2010) verify that the use of*Bacillus cereus* and *Sacharomyces boulardii* enhanced the humoral immune response of lambs to the vaccines.

Some characteristics in probiotics strains are unwanted and much worrisome as well as antimicrobial resistance. Some lactic bacteria could present antibiotic resistance and these bacteria used for food is considered a major danger since this resistance could be transferred to pathogenic bacteria. The probiotics strains used in our study were tested to susceptibility to 27 antibiotics and verified that generally the *Lactobacillus* strains were inhibited to all antibiotics tested (Karapetkov et al., 2011).

In a study with cattle performed in Brazil, the authors used a probiotic contained strains of*Ruminobacter amylophilus, Ruminobacter succinogenes, Succinovibrio dextrinosolvens, Bacillus cereus, Lactobacillus acidophilus* and *Streptococcus faecium*, and these strains were administered at a dose of 3 x 10^8 live cells (CFU) of each strain resuspended in 250 mL of milk and administered orally. This study had many groups of animals. Some animals were vaccinated, others received probiotic and others both were vaccinated and received probiotic. These results showed that the combination of vaccine with the probiotic administered for 15 or 30 days were the most effective treatments for the control of diarrhea and weight gain (Ávila et al., 2000).

Some studies have indicated a higher prevalence of STEC in sheep than in cattle (Beutin et al., 1997; Sidjalat and Bensink, 1997; Urdahl et al., 2003), confirming that sheep are a significant reservoir of STEC. The findings of this study suggest that this probiotic likely presented a potential protective effect in reducing colonization by STEC non-O157 and can be used as an alternative method to decrease STEC non-157 infection in sheep, thereby reducing transmission to humans. Probiotic microorganisms, which benefit from a “natural image”, can expect a promising future in animal nutrition (Chaucheyras-Durand and Durand, 2010).

ACKNOWLEDGEMENT

The authors would like to thank FAPESP by financial support that permitted the realization of study. Process: 2009/14923-8

REFERENCES

1. F. Ávila, A. , A. Paulillo, C. Schocken-Iturrino, R. , P. et, al, 2000Evaluation of efficiency of a probiotic in the controlo f diarrhea and weight gain in calves. Arquivos Brasileiros de Medicina Veterinaria e Zootecnia 4146

2. C. Batista, G. , S. Coelho, G. , E. Rabelo, et al.2008Performance and health of calves fed milk without antimicrobials residue or milk from mastitis treated cows with or without probiotic. Arquivos. Brasileiro de Medicina Veterinária e Zootecnia. 185191

3. E. A. Belongia, N. T. Osterholm, J. T. Soler, et al.1993Transmission of Escherichia coli O157: H7 infection in Minnesota child day-care facilities. Journal American. Medicine. Association. 269, 883888

4. L. Beutin, D. Geier, S. Zimmermann, et al.1997Epidemiological relatedness and clonal types of natural populations of Escherichia coli strains producing Shiga toxins in separate populations of cattle and sheep. Applied Environmental Microbiology 6321752180

5. L. Beutin, S. Kaulfuss, T. Cheasty, B. Brandenburg, S. Zimmermann, K. Gleier, G. A. Willshaw, H. R. Smith, 2002Characteristics and association with disease of two major subclones of Shiga toxin (Verocytotoxin)-producing strains of Escherichia coli (STEC) O157 that are present among isolates from patients in Germany. Diagnostic Microbiology Infect Diseases, 44337346

6. J. P. Buts, 2004Exemple dún medicament probiotique: Sacchamoryces boulardii lyophilize. In Rambaud, J.C., Buts, J.P., Corthier, G. and Flourié, B. (eds) Flore microbienne intestinale. John Libbey Eurotext, Montrouge, France, 221244

7. F. Chaucheyras-Durand, F. Fahima, A. Ameilbonne, et al.2010Fates of acid-resistant and non-acid-resistant shiga toxin-producing

Escherichia coli strains in rumiant digestive contents in the absence and presence of probiotics. Applied. Environmental. Microbiology, 640647

8. F. Chaucheyras-Durant, H. Durant, 2010Probiotics in animal nutrition and health. Beneficial Microbes, 139

9. B. China, V. Pirson, J. Mainil, 1996Typing of bovine attaching and effacing Escherichia coli bymultiplex amplification of virulence-associated genes. Applied. Environmental. Microbiology. 8234623463

10. Cross, M., 2002Microbes versus microbes: immune signals generated by probiotic lactobacilli and their role in protection against microbial pathogens. FEMS. 245253

11. L. De Vuyst, R. Callewart, B. Pot, 1996Characterization of the antagonistic activity of Lactobacillus amylovorus DCE 471 and large scale isolation of its bacteriocin amylovorin L471. Systematic and Applied Microbiology 19920

12. l. M. T. Dicks, M. Botes, Probiotic lactic acid bacteria in the gastro-intestinal tract: health benefits, safety and mode of action. 2010Beneficial Microbes, 11129

13. Food And Agriculture Organization Of The United Nations, World Health Organization.Evaluation of health and nutritional properties of probiotics in food including powder milk with live lactic acid bacteria. Córdoba, 2003. 34p. Available at:http://ftp.fao.org/ es/esn/ food/probioreport_ en.pdf>. Accessed 03 Fev 2005

14. R. Fuller, 1989Probiotics in man and animals. Journal Applied Bacteriology Oxford, 66365378

15. G. Gibson, R. , M. Roberfroid, B. , 1998Dietary modulation of the human colonic microbiota: introducing the concept of prebiotics. Journal Nutrition, 12514011412

16. F. Guarner, J. Malagelada, R. , 2003Gut flora in health and disease. Lancet, London, 360512518

17. H. S. Hussein, 2007Prevalence and pathogenicity of Shiga toxin producing Escherichia coli in beef cattle and their products. Journal of Animal Science, 85, E63E73.

18. E. Isolauri, S. Salminen, A. Ouwehand, C. , 2004Probiotics. Best Pract. Research. Clinical of Gastroenterology, 2299313

19. N. Karapetkov, Rumyan. N. Geogieva, E. Karaivanova, 2011Antibiotic susceptibility of different lactic acid bacteria strains, Beneficial Microbes, 2335339

20. M. A. Karmali, 1989Infection by verocytotoxin-producing Escherichia coli. Clinical Microbiology. Review, 21538

21. S. K. Kritas, A. Govaris, G. Christodoulopopoulos, A. R. Burriel, 2006Effect of Bacillus licheniformis and Bacillus subtilis supplementation of Ewe´s feed on sheep milk production and young lamb mortality. Journal of Veterinary Medicine,53170173

22. Y. Lee, K., K. Nomoto, S. Salminen, S. Gorbach, L. , 1999Handbook of probiotics. New York, Wiley, 211pp.

23. M. Lema, L. Williams, D. R. Rao, 2001Reduction of fecal shedding of enterohemorrhagic Escherichia coli O157:H7 in lambs by feeding by microbial feed supplement. Small Ruminant Research 393139

24. B. Posse, L. D. Zutter, M. Heyndrickx, L. Herman, 2007Metabolic and genetic profiling of clinical O157 and non-O157 Shiga-toxin-producing Escherichia coli, Institut Pasteur, 158591599

25. R. Puupponen-Pimiä, A. Aura, M. , K. Oksmancaldentey, M. , et al.2002Development of functional ingredients for gut health. Trends Food Science Technological, Amsterdam, 13311

26. T. Ramamurthy, 2008Shiga toxin-producing Escherichia coli (STEC): the bug in our backyard. Indian Journal of Medical Research, 128233236

27. T. B. Roos, V. C. Tabeleao, L. A. Dummer, et al.2010Effect of Bacillus cereus var Toyoi and Saccharomyces boulardii on the immune response of sheep to vaccines. Food and Agricultural Immunology, 21113118

28. M. Sanders, E. , T. Klaenhammer, R. , 2003Invited review: the scientific basis of Lactobacillus acidophilus NCFM functionality as a probiotic. Journal. Dairy Science, 84319331

29. A. I. Servin, 2004Antagonistic activities of lactobacilli and bifidobacteria against microbial pathogens. FEMS Microbiology Reviews 28405440

30. H. Sidjabat, J. Bensink, C., 1997Verotoxin-producing Escherichia coli from the faeces of sheep, calves and pigs. Australian Veterinary Journal 75292293

31. K. Thuory, M. , H. Probert, M. , C. Smejkal, W. , et al.2003Using probiotics and prebiotics to improve gut health. Drug Discovery Today, Haywards Heath,15692700

32. A. Urdahl, M. , L. Beutin, E. Skjerve, S. Zimmermann, et al.2003Animal host associated differences in Shiga toxin-producing Escherichia coli isolated from sheep and cattle on the same farm. Journal Applied Microbiology. 92101

33. S. Wani, A. Bhat, M., A. Samanta, I. Nishikawa, Y. , A. Buchh, S. , 2003Isolation and characterization of Shiga toxinproducing Escherichia coli (STEC) and enteropathogenic Escherichia coli (EPEC) from calves and lambs with diarrhea in India. Letters. Applied. Microbiology. 37121126

34. G. A. Willshaw, T. Cheasty, H. R. Smith, Brien. S. J. O′, G. K. Adak, 2001Verocytotoxin-producing Escherichia coli (VTEC) O157 and other VTEC from human infections in England and Wales: 1995-1998. Journal Medicine Microbiology, 50135142

35. T. Zhao, M. Doyle, B. Harmon, et al.1998Reduction of carriage of enterohemorrhagic Escherichia coli O157:H7 in cattle by inoculation with probiotic bacteria. Journal Clinical Microbiology, 641647

CHAPTER 8

Effect of Fatty Acids on Reproductive Performance of Ruminants

[1], José Herrera-Camacho[1], Alejandra Soberano-Martínez[3], Karlos Edmundo Orozco Durán[2], Carlos Aguilar-Pérez[2] and Juan Carlos Ku-Vera [2]

[1] Instituto de Investigaciones Agropecuarias y Forestales Universidad Michoacana de San Nicolás de Hidalgo, México

[2] Campus de Ciencias Biológicas y Agropecuarias Universidad Autónoma de Yucatán, México

1. INTRODUCTION

Different types of fats have been utilized in an attempt to improve reproductive function in ruminant animals. Fatty acids derived from plants and oil seeds have exerted a major impact on reproductive performance, some of the most common sources include sunflower, linseed, cottonseed, rapeseed and soyabean. Animal fat (tallow) and calcium salts of saturated fatty acids may escape in a significant percentage rumen hydrogenation to be incorporated into adipose tissue and milk. Fish by-products contain a high proportion of polyunsaturated fatty acids (PUFAs) and pass without being altered in the rumen exerting no effects on rumen fermentation. Each dietary source of fat varies regarding composition of specific fatty acids (Table 1).

Early studies of the effect of fat in the ration on reproductive performance were carried out by Burr & Burr (1930), who observed that fat deficiency in the ration of growing rats induced alterations in ovulation rate and on the onset of oestrus, while lipid supplementation reestablished reproductive performance of the females, coining the concept of essential fatty acids. In later studies, research was aimed at evaluating the effect of fat supplementation in different animal species both ruminant and non-ruminant, on reproductive aspects such as the establishment of puberty (Smith et al., 1989), semen production (Castellano et al., 2010), maternal recognition of pregnancy (Abayasekara & Wathes, 1999,Filley et al., 2000, Lopes et al., 2009) by means of the suppression of luteolytic signals (Mattos et al., 2000), restart of ovarian activity after parturition (de Fries et al., 1998), follicle development, quality of oocytes (Staples & Thatcher, 2005; Bilby et al., 2006c), and of the embryo (Cerri et al., 2009), modification in the mechanism of synthesis and secretion of hormones involved in reproductive processes (Staples et al., 1998) and on production aspects such as quality of milk (Rego et al., 2004;Bernal et al., 2010) or meat (Wood et al., 2003). Due to the fact that some fatty acids (FA) are essential for mammals and to the role of fatty acids on reproductive processes, it is possible that cattle reproduction will be influenced more by the type of lipids consumed than for the total lipid intake. This is particularly important since ruminants hydrogenate PUFAs in the rumen, limiting the amount of PUFAs that are absorbed from the small intestine (Thatcher & Staples, 2007, Santos et al., 2008,Doreau et al., 2011). However, it is possible that some specific PUFAs may pass intact the reticulo-rumen and be absorbed from the small intestine, allowing in this way the improvement of reproductive efficiency directly on the target tissue of the reproductive system of the female (autocrine or paracrine) or by an indirect effect mediated by the endocrine system (Staples & Thatcher, 2005).

Several studies have shown that PUFAs of the ω-3 family such as eicosapentaenoic acid (C20:5, ω-3) and docosahexaenoic acid (C22:6, ω-3) suffer insignificant biohydrogenation in the rumen (Thatcher & Staples, 2007). These fatty acids are usually found in feedstuffs derived from fish (and other marine products), such as the oil and the meal which are considered non essential since they can be synthesized from linoleic acid (ω-3), and apparently play an important role in animal performance (Thatcher & Staples, 2007). Furthermore, lipid supplements partially resistant to biohydrogenation in the rumen have been developed such as calcium salts of long chain fatty acids (Ca-LCFA) with the aim of increasing the amount of unsaturated FA which can be absorbed by limiting biohydrogenation (Mattos et al., 2000). PUFAs act as mediators in a series of processes in several reproductive tissues, including fluidity of cell membrane, intracellular signaling and susceptibility to oxidative damage (Wathes et al., 2007). Changes in chain length, degree of unsaturation and position of the double bonds in the acil chain of fatty acids may have a major impact on reproductive function and play a role in livestock reproduction (Mattos et al., 2000). Potential mechanisms may include increment of energy density of the ration (Ferguson et al., 1990), even when for some workers (Williams & Amstalden, 2010), the effect of fat supplementation on reproduction is independent of the energy density of ration or of changes in live weight of animals. Considering all the above described, the aim of this review is to examine some of the reproductive processes in the bovine and ovine females which could be regulated or modified by the inclusion of lipids in the ration.

Table 1. Fatty acid composition of selected dietary fat sources. [1]Commercial preparations considered partially inert in the rumen; [2]Also contain 14% C20:5 and 9% C22:6 (**Thatcher & Staples, 2007**).

Fat source	Fatty acid						
	C14:0	C16:0	C16:1	C18:0	C18:1	C18:2	C18:3
	Myristic	Palmitic	Palmit-oleic	Stearic	Oleic	Linoleic	Linolenic
Tallow	3	25	3	18	43	3.8	<1
Yellow grease	2	21	4	11	44	14	<1
Energy Booster 100[1]	3	40	1	41	10	2	<1
Megalac: EnerG-II[1]	1	50	<1	4	36	8	<1
Megalac- R[1]	1	36	<1	4	26	29	3
Canola oil	<1	4	<1	2	63	19	9
Cottonseed oil	1	23	<1	3	18	54	1
Flaxseed oil	<1	5	1	3	20	16	55
Extruded Linseed	<1	7.6		5.2	20	14.5	51.3
Rapeseed oil	<1	5	<1	2	54	22	11
Safflower oil	<1	7	<1	2	12	78	<1
Soyabean oil	<1	11	<1	4	23	54	8
Sunflower oil	<1	7	<1	5	19	68	1
Menhaden fish oil[2]	7	16	8	3	12	1	2

2. EFFECT OF LIPID SUPPLEMENTATION ON PREGNANCY RATE

Incorporation of lipids in rations for dairy cattle usually increases energy density of ration and improves lactation and reproductive performance

(Funston, 2004). However, when they are supplied in early lactation, frequently there is a reduction in feed intake due to a reduction in dry matter digestibility and to an increase in energy of greater availability, so when lipids are supplied in the early postpartum period, there is little alteration in the energy status of the animal even when a higher energy density ration is consumed (Santos et al., 2008).

Then, if dietary fat does not alter the energy status of dairy cows, reproductive response results more from the supply of some fatty acids, than from the effect of the energy supply *per se* (Staples et al., 1998, Williams & Amstaldem, 2010).

There are several studies that report a better reproductive performance in cows fed supplementary lipids. In this respect, Staples et al. (1998), showed that lipid consumption exerted a positive effect on reproductive aspects in dairy cows (Table 2). In beef cattle, the same trend has been observed. It is in this context that, de Fries et al. (1998) reported that Brahman cows consuming 5.2% lipids in the ration showed a trend towards an increase in pregnancy rate than those cows which consumed only 3.7% lipids in the ration. Ferguson et al. (1990) observed a 2.2 times increase in the possibility of pregnancy in lactating cows consuming 0.5 kg lipids per day. In another study, it was demonstrated that grazing cows supplemented with fat, pregnancy rate at first service was 16% higher than in cows which did not receive fat in the ration (Bader et al., 2000).

Bellows et al., (2001) observed that supplementation with safflower seed, soyabeans, or sunflower seed (4.7, 3.8 and 5.1% fat in the ration, respectively) for the last 65 days before calving increased subsequent pregnancy rates (94%, 90% and 91%, respectively) of first-calf beef heifers compared with the control (79 %) that received only 2.4 % fat in the ration. In another study Bellows et al., (2001), using good quality forage and a higher amount of fat in the ration (6.5%) during 68 days before calving, was unable to improve pregnancy rates relative to a control ration (2.2% fat), this result indicates that when adequate nutrients are available, the effect of supplemental fat may be masked.

Grazing Holstein cows which were supplemented for 103 days, as from day 10 post-partum, with two sources of bypass fat Megalac plus 3% (MP; 0.4 kg/day, containing Ca salts of palm fatty acids and Ca salts of methionine hydroxy analogue) and Megapro Gold (MPG; 1.5 kg/day, containing Ca salt of palm fatty acids, extracted rapeseed meal and whey permeate), MPG increased ($P<0.05$) conception rate to first service compared to control group (CG). Conception rate to first service was similar (P=0.14) on MP compared to CG. For pregnancy to second service, CG had a higher (P<0.05) conception rate than MP. There were no significant differences between treatments in overall pregnancy rate, services per conception, number of cows served in the first three weeks of the breeding season or the 6-weeks in-calf rate. Comparing the combined fat treatments to CG resulted in a higher (P<0.05) conception rate to first service for the fat treatments but no significant difference in overall pregnancy rate (McNamara et al., 2003).

Table 2. Reproductive performance of beef cows supplemented with fat in the ration. [1]First insemination (Adapted from **Thatcher & Staples, 2007**).

Reference	Fat source	Percent inclusion	Pregnancy rate
Fergunson et al., 1990	Ca-Palm oil	2.0 %	59[1]
Sklan et al.,1991	Ca-Palm oil	2.6 %	82
Scott et al., 1995	Ca-Palm oil	1 lb d^{-1}	98
Garcia-Bojalil et al., 1998	Ca-Palm oil	2.2 %	86
Son et al., 1996	Tallow	3 %	62
Espinoza et al., 2010	Tallow	9.5%	Herd A 70; Herd B 55%
Frajblat and Butler, 2003	Energy Booster	1.7 %	86
Petit et al., 2001	Flaxseed	17%	87
Ambrose et al., 2006a	Flaxseed	9%	48[1]
Ambrose et al., 2006b	Flaxseed	9%	26[1]
Fuentes et al., 2007	Extruded Linseed	1.7 kgd^{-1}	39
McNamara et al., 2003	MegaPro Gold	3.3 lb d^{-1}	54
Juchem et al., 2004	Soy + Trans C18:1	1.5%	34[1]
Cullens, 2005	Megalac-R	2%	58[1]
Aguilar-Pérez et al., 2009	ByFat ®	1.8%	33
Espinoza et al., 2010	Megalac®	9.6%	Herd A 80; Herd B 58%
Castañeda-Gutierrez et al., 2005	Ca-CLA	0.3 lb d^{-1}	81
Bernal-Santos et al., 2003	Ca-CLA	0.3 lb d^{-1}	42
Bruckental et al., 1989	Fish meal	7.3%	72
Armstrong et al., 1990	Fish meal	1.8 lb d^{-1}	64
Carroll et al., 1994	Fish meal	3.5 %	89
Burke et al., 1997	Fish meal	2.8	41

In another study, Aguilar-Pérez et al., (2009) observed that pregnancy rate of F_1 (Holstein x Zebu) cows grazing under tropical conditions in Mexico, was not affected by supplementation with bypass fat (33.3%), relative to a control group (41.7%) at 90 days postpartum. In conclusion, fat supplementation increased conception rate to first service but did not significantly affect the proportion of cows pregnant at the end of the breeding season, these results suggest that the higher quality of the forage supplied in the different seasons that the trial lasted, may have been a factor that masked the effect of fat supplementation. Juchem et al. (2010) evaluated the effect of supplementation before and after parturition with Ca-LCFA of palm oil or with a mixture of linoleic and *trans*-octadecenoic acids and observed that cows fed PUFAs showed a higher pregnancy rate than those supplemented with palm oil at 27 and 41 days (37.9 *vs* 28.6% and 35.5 *vs* 25.8%, respectively) after artificial insemination. Recently, Lopes et al. (2009) fed Nellore cows with 0.1 kg $cow^{-1}day^{-1}$ of PUFAs and observed that pregnancy rate was improved. In this respect, Mattos et al. (2000) suggested that the positive effect of fat on fertility may not be due to improvement in energy balance of the cows but rather to the specific effect of some dietary fatty acids on the physiology of the hypothalamus-hypophysis-ovary axis and even the uterus.

In a review of previous studies in which conjugated linoleic acids (CLA) were supplemented to dairy cows during early lactation, de Veth et al. (2009) demonstrated that the probability of pregnancy increases in 26% when CLA are increased in the ration and that the optimum CLA amount is 10.0 g d^{-1}, after which the beneficial effects are reduced. It is possible that the positive effect of lipid supplementation may be due to specific fatty acids (Staples & Thatcher, 2005), and the absorption of unsaturated FA in ruminants is limited due microbial biohydrogenation in the rumen (Lopes et al., 2009). Some studies have evaluated the possibility that unsaturated FA intake, particularly those of the*n*-6 (linoleic acid) and *n*-3 (α-linolenic, eicosapentaenoic, docosahexaenoic acids) families, may have some influence on reproduction in cows, even when reports in the literature are not always consistent (Santos et al., 2008). In this respect, when cows were fed 0.75 kg of linseed rich in α-linolenic acid (*n*-3), or sunflower rich in linoleic acid (*n*-6), pregnancy rate tended to increase in cows of the first treatment (Ambrose et al., 2006a).

In other studies, no response was observed with linseed (Fuentes et al., 2008). Similarly, feeding *n*-3 fatty acids from fish oil in the form of Ca-LCFA did not improve pregnancy rate postpartum at first service in beef cows when compared to supplementation with beef tallow (Juchem, 2007) or with Ca-LCFA from palm oil (Silvestre, quoted by Santos et al., 2008), even when pregnancy rate at second service postpartum was higher in cows fed *n*-3 fatty acids (Silvestre, quoted by Santos et al., 2008). In grazing F_1 (*Bos taurus* x *Bos indicus*) cows, Aranda-Ávila et al. (2010) observed a 15.4% increase in pregnancy rate when cows were supplemented with corn oil, relative to a control group (54.5 *vs* 41.7 %

respectively) after 35 day supplementation; however, differences were not statistically significant. It is possible that the poor response observed in those studies may be due to an increase in milk production along with a loss of body weight, which occurs in greater or lesser degree in cows during the early postpartum period (Sklan et al., 1994).

3. EFFECT OF LIPID SUPPLEMENTATION ON THE HYPOTHALAMUS-HYPOPHYSIS-OVARY AXIS

The major objective of cow-calf enterprises is to produce one calf per cow annually. Thus, management strategies that enhance reproductive performance of milk and beef cows are beneficial to the productivity of cow-calf operations. Previous studies reported that utilization of dietary fat as a nutraceutical, particularly PUFAs, positively influenced reproductive function in both milk and beef cows (Williams & Stanko, 2000). Furthermore, these positive effects were independent of the additional energy contribution from the PUFAs sources (Funston, 2004). Different mechanisms have been proposed by means of which fat supplementation may affect functioning of the hypothalamus-hypophysis-ovary axis. Early work in this respect suggested that fat supplementation may affect secretion of reproductive and metabolic hormones and further research demonstrated that fat addition to the ration modified ovarian activity in heifers and adult cows postpartum.

The mechanism (or mechanisms) by which dietary fat improves reproductive performance has not been elucidated. Several hypotheses have been proposed: 1) an amelioration of a negative energetic balance, thus leading to an earlier return to oestrus postpartum and, therefore, improved fertility; 2) an increase in steroidogenesis favorable to improved fertility; 3) manipulation of insulin so as to stimulate ovarian follicle development; and 4) a stimulation or inhibition of the production and release of $PGF_{2\alpha}$, which influences the persistence of the corpus luteum (Staples et al., 1998)

3.1. Hormonal Secretion And Lipid Metabolites

Some studies showed that dietary fat supplementation in dairy heifers increased circulating concentrations of progesterone (Talavera et al., 1985), and enhanced lifespan of induced corpus luteum during early postpartum in beef cows (Williams, 1989; Ryan et al., 1995). Other studies suggest that when lipids are included in the ration of cows to increase energy density, caloric balance is improved which directly influences hypophysis-gonadal activity postpartum (Harrison et al., 1995), increasing, in principle, the amplitude and frequency of secretion of luteinizing hormone (LH) in animals (Sklan et al., 1994). In this respect, de Luna et al. (1982), reported an increase in the secretion of luteinizing hormone in ovariectomized cows treated with GnRH and supplemented with beef tallow. In sheep, secretion of luteinizing hormone in response to the injection of GnRH at day 10 of the oestrus cycle was greater in Pelibuey sheep supplemented with Ca-LCFA from palm oil during 30 days than in the control group (Espinoza et al., 1997).

Other studies, using isocaloric and isonitrogenous diets in cows of poor body condition indicated that the increase in dietary fat consumption augmented the number of follicles of medium-size by 1.5- to 5-fold within 3 to 7 weeks and these changes occurred coincident with changes in serum insulin, growth hormone and intraovarian insulin-like growth factor (IGF-1) (Wehrman et al., 1991; Ryan et al., 1992;Thomas et al., 1997). Table 3 summarizes the effects of dietary fat supplementation on follicular physiology and growth as observed in different studies.

On the other hand, it has been shown that hiperlipidic rations supplied both to dairy as well as to beef cows, induced and increase in the levels of blood cholesterol, as it was observed by Hightshoe et al. (1991) in cows supplemented postpartum with Ca-LCFA from palm oil. Similar results, were reported in Angus and Hereford cows which consumed a supplement which contained 125 g of Ca-LCFA from palm oil (Espinoza et al., 1995), in Chinampas (*Bos taurus*) cows consuming Ca-LCFA from palm oil or beef tallow (Espinoza-Villavicencio et al., 2010); this is particularly important since cholesterol is the main precursor for progesterone synthesis in the corpus luteum as well as of other steroid hormones at the follicular level (Childs et al., 2008c). Thus, when supplementation with protected fats from rumen biohydrogenation is augmented, it is possible to increase concentration of plasma progesterone in cows (Lopes et al., 2009), which is associated positively with pregnancy rate (Santos et al., 2008). When endometrial secretions are modified (Gray et al., 2001) changes are induced in endometrial architecture which are fundamental for the appropriate development of the embryo (Wang et al., 2007). Nonetheless, in other studies, no response has been observed of fat supplementation on progesterone synthesis. In this respect, Robinson et al. (2002), using cows fed a ration rich in linoleic acid showed a reduction of up to 50% in the serum levels of progesterone between days 4 and 8 of the oestrus cycle, while those cows consuming a ration rich in linolenic acid showed a lower concentration of this hormone, but in days 4 and 15 of the oestrus cycle (Robinson et al., 2002). Recent work (Huante, 2010), demonstrated that secretion of progesterone in Brahman cows with low body condition at calving and during the postpartum (2.0 points, in the scale 1-5) was unaffected by supplementation with oilseeds with 60% of unsaturated fatty acids, relative to the control group (0.065±0.013 *vs* 0.054±0.013 ng/ml, respectively). These results suggest that due to the low body condition, the cow uses feed energy for its maintenance and not for reproductive processes. Those results suggest that the above mentioned fatty acid, exert different effects in the synthesis of ovarian steroids (Hinckley et al., 1996).Childs et al. (2008b) fed beef heifers a ration enriched with fish oil (*n*-3) and even when no increase in serum progesterone was achieved at day 7 of the oestrus cycle, they postulated that there was evidence of a greater synthesis of progesterone during the whole oestrus cycle due to the increase in the serum concentration of cholesterol and to the greater size of the corpus luteum.

Table 3. Summary of the effect of dietary fat supplementation in cattle on ovarian follicular growth and steroid production. (Modified from Williams & Amstalden, 2010)

Reference	Effect
Wehrman et al., 1991; Ryan et al., 1992; Hightshoe et al., 1991; Lucy et al., 1991; Thomas & Williams, 1996; Thomas et al., 1997; Lammoglia et al., 1996; Stanko et al., 1997; de Fries et al., 1998	Increased number of medium-sized follicles (polyunsaturated fat > saturated and highly polyunsaturated fat effects)
Lucy et al., 1989, 1991	Milk cows supplemented with Ca-LCFA palm oil, the basal level of LH was increase
Wehrman et al., 1991; Ryan et al., 1992	Increased granullosa cell progesterone production *in vitro*, increased follicular fluid progesterone
Lopes et al., 2009, Salas-Razo et al., 2011	Cows supplemented with rumen inert polyunsaturated fat had greater mean serum progesterone concentrations compared with control
Ryan et al., 1992; Thomas & Williams, 1996	No effect on superovulation rate
de Fries et al., 1998; Bilby et al., 2006a; Garnsworthy et al., 2008	Increased number of large follicles; increased size of largest follicle

While in sheep, concentration of progesterone in the follicular fluid was greater than in sheep which consumed the ration enriched with *n-3* fatty acids, than those which were fed with a greater amount of*n*-6 fatty acids (Wonnacott et al., 2010). Some *in vitro* studies have demonstrated that progesterone metabolism can be inhibited by high concentrations (300 μM) of α-linolenic acid in the culture media (Piccinato et al., 2010). This is perhaps related to the fact that certain PUFAs may also increase serum concentration of insulin and this in turn, reduce hepatic expression of some enzymes of the cythocrome P_{450} complex which catabolize progesterone (Lemley et al., 2008). Oestradiol, another hormone derived from cholesterol has stimulatory effects on the uterine secretion of $PGF_{2\alpha}$ (Knickerbocker et al. 1986) and can increase sensibility of the corpus luteum to prostaglandins by intensifying its regression (Howard et al., 1990). Previous studies demonstrate that intra-abomasal infusion of yellow fat induced a reduction in blood levels of oestradiol between days 15 and 20 of the oestrus cycle, compared to cows which were infused with glucose intra-abomasally. The relationship between dietary fatty acids and synthesis of steroid hormones may be direct through a direct effect on steroidogenesis (acute regulatory protein of steroids; STAR, cythocrome P450 protein, family 11; subfamily A polypeptide enzyme 1; CYP11A1) etc. or indirectly through prostaglandins. STAR plays a fundamental role in the synthesis of steroid hormones (Stocco et al., 2005). In previous work, it has been observed that in beef heifers there was a reduction in the concentration of arachidonic acid in the endometrial tissue when they were fed *n*-3 PUFAs (Childs et al., 2008a; Childs et al., 2008b). On the other hand, there is evidence that supplementation with linoleic acid (*n*-6) increased uterine secretion of $PGF_{2\alpha}$ (Cullens, 2005), which could derive in a greater availability of arachidonic acid in the endometrium. Under such conditions it could be speculated that a possible way to act of the *n*-3 and *n*-6 fatty acids in steroidogenesis is by reducing or increasing the availability of arachidonic acid, respectively and as a consequence, the expression of RNAm for STAR synthesis.

3.2. Lipids On Ovarian Activity

These results suggest that another of the mechanisms by means dietary lipids may improve reproductive performance of cattle is influencing follicular development and ovulation. In this, respect, Lucy et al. (1991), replaced corn with Ca-LCFA from palm oil in the ration of dairy cows at calving, and increased the number of medium size follicles (6-9 mm) and of follicles greater than 15 mm within 25 days postpartum. Furthermore, the diameter of the greatest follicle was superior in cows fed Ca-LCFA from palm oil (18.2 *vs* 12.4 mm). When this study was repeated with isocaloric diet, observed effects were similar (Lucy et al. 1993).

The greatest increase in medium follicle populations occurred in response to plant oil consumption, which is likely a direct result of the effects of high levels of linoleic acid in the rumen. Maximum follicular growth responses to plant oil supplementation have occurred when plant oils were fed at 4 to 6% of dietary dry matter, with lesser increases observed at lower levels of added fat. Animal tallow, calcium salts of saturated fatty acids or fish oil have been shown to have less clear effects on follicular growth than plant-derived oils. Moreover, postpartum beef cows which calved in a very poor body condition (BCS of 3; 1-9 scale) were unable to develop medium or large follicles at a rate equal to those with a body condition score of 4 or greater after 3 weeks of fat consumption (Ryan et al., 1994).

The number of medium size follicles (5 to 10 mm) was higher in beef cows which consumed feed with a greater content of PUFAs (Thomas et al., 1997) and in dairy cows which consumed a diet enriched with 5% *n*-3 fatty acids derived from fish oil (Heravi-Moussavi et al., 2007). Similar results were observed in cows fed with diets enriched with *n*-3 or *n*-6 fatty acids (Robinson et al, 2002). Staples & Thatcher (2005) summarized the effect of increasing lipids in the ration on the size of the dominant follicle: on average, the size of the dominant follicle was 3.2 mm greater than in females fed with some source of fat, which represents a 23% increase. Other studies (Bilby et al., 2006d; Garnsworthy et al., 2008) showed that the size of the dominant follicle is increased in cows fed diets rich in PUFAs. In sheep, the use of Ca-LCFA from palm oil (El-Shahat & Abo-El maaty, 2010) or from corn oil (Herrera et al., 2008) in the feed improved the number and size of the preovulatory follicles, as well as rate of ovulation and the superovulatory response in Pelibuey sheep (Herrera et al., 2008). These evidences demonstrate that consumption of lipids accelerate follicular growth (de Fries et al., 1998), which may influence the restart of ovulation postpartum, as it was shown by Marín-Aguilar et al. (2007) who observed that Holstein cows supplemented with plant oil (60% PUFAs) reduced by 7 days the restart of ovarian activity relative to a control group. de Veth et al. (2009) observed that time at first ovulation in dairy cows was reduced by 8 days when they were supplemented with 8 g/d of *trans*-10,*cis*-12 CLA.

4. LIPIDS AND ITS EFFECT ON ENDOMETRIAL SECRETION OF PROSTAGLANDINS

Studies in a variety of species have shown that dietary PUFAs can modulate prostaglandin synthesis and metabolism. Eicosanoids, comprising prostaglandins, thromboxanes, leukotrienes and lipoxins, are all synthesized from C20 fatty acids (Mattos et al., 2000). The most biologically active two series prostaglandins are derived from arachidonic acid, but the less active three series prostaglandins can be produced from eicosapentaenoic acid by the action of the same enzymes (Robinson et al., 2002).

Prostaglandins play an important role in reestablishing oestrus cycles both immediately after parturition and thereafter until conception. Prostaglandin $F_{2\alpha}$ ($PGF_{2\alpha}$) is responsible for uterine involution after parturition. The uterus releases $PGF_{2\alpha}$ during each oestrus to regress each new corpus luteum if the cow is not pregnant and initiate a new oestrus cycle. During the period of corpus luteum regression, concentrations of $PGF_{2\alpha}$ and progesterone are inversely related. If the cow does conceive, release of $PGF_{2\alpha}$ from the uterus is prevented in order to preserve the corpus luteum and maintain pregnancy (Funston & Filley, 2002).

Linoleic acid is a substrate for the synthesis of $PGF_{2\alpha}$. Linoleic acid can be desaturated and elongated to arachidonic acid (C20:4, *n*-6), which is a precursor for $PGF_{2\alpha}$. Regulatory enzymes for this conversion include delta 6 desaturase and cyclooxygenase. Linoleic acid can inhibit prostaglandin synthesis by competitive inhibition with these key enzymes. Arachidonic, and two fatty acids found in fishmeal, eicosapentaenoic (C20:5) and docosahexanoic (C22:6), have been shown to inhibit cyclooxygenase activity as well. It is important to note that linolenic acid (C18:3) was also present in the endometrial prostaglandin synthesis inhibitor isolated by Thatcher et al. (1994), and that linolenic acid has been shown to be a strong inhibitor of prostaglandin synthesis (Mattos et al., 2000). The amount and probably type of particular fatty acids reaching the target tissues likely influence if prostaglandin synthesis is going to be stimulated or inhibited (Thatcher & Staples, 2000).

Figure 1 shows the schematic metabolic pathway of dietary n-6 and n-3 PUFAs and potential mechanisms for regulation of $PGF_{2\alpha}$ secretion. Absorbed PUFAs are desaturated and elongated in organs such as the mammary gland, adipose tissue, testis, brain, placenta and the liver (of non-ruminants). Dietary PUFAs and their desaturation and elongation products are incorporated into phospholipids of the plasma membrane. The amount of each fatty acid incorporated depends on the amount of precursor present in the diet. External stimuli such as the binding of oxytocin (OT) to the oxytocin receptor (OTr) stimulates the activity of phospholipase A2 (PLA2) and phospholipase C (PLC), which cleave phospholipids from the plasma membrane and ultimately increase availability of diacylglycerol (DAG) and fatty acids for processing by prostaglandin H synthetase (PGHS). Eicosapentaenoic acid (EPA; C20:5, n-3) is processed by PGHS to generate prostaglandins of the 3 series. Arachidonic acid (AA; C20:4 *n*-6) can be processed by PGHS, epoxygenase and lipoxygenase to generate prostaglandins of the 2 series, epoxyeicosatrienoic acids (EETs), leukotrienes and hydroxyeicosatetraenoic acids (di-HETEs), respectively (Mattos et al., 2000).

In vitro studies have demonstrated that some *n*-3 (eicosapentaenoic and docosahexaenoic) fatty acids reduce biosynthesis of prostaglandins of the series 2 in cells and tissues (Mattos et al., 2003). Similarly, some isomers from conjugated linoleic acid inhibited synthesis of $PGF_{2\alpha}$ and the effect was independent of the concentration of linolenic acid and the ratio *n*-6:*n*-3 (Harris et al., 2001). The above results are interesting due to the fact that most embryonic losses in cattle occur during days 8-16 after artificial insemination (Sreenan et al., 2001), which leads to believe that some embryos may not reach the appropriate size at that moment to inhibit synthesis of $PGF_{2\alpha}$ for luteolysis to occur (Thatcher et al., 1994), showing the inability to inhibit luteolytic action by $PGF_{2\alpha}$ during the critical period of maternal recognition of pregnancy (Childs et al., 2008a). In this context, inhibition of the synthesis of $PGF_{2\alpha}$ could increase the rates of embryo survival and pregnancy (Binelli et al., 2001). PUFAs (*n*-3) such as eicosapentaenoic acid may inhibit uterine synthesis of $PGF_{2\alpha}$ by competing with arachidonic acid by means of COX, or in the case of docosahexaenoic acid, by competence with arachidonic acid by phospholipase A2 enzymes (Mattos et al., 2000).

Fish meal has relatively high concentrations of eicosapentaenoic and docosahexaenoic acids, in such a way that their incorporation in the ration of cattle may reduce the synthesis of $PGF_{2\alpha}$ and delay regression of the corpus luteum, improving embryo survival and herd fertility (Staples et al., 1998)

Previous studies showed that the infusion of a fat source rich in linoleic acid (17%) into the abomasum of lactating dairy cows resulted in a significant reduction in the release of PGFM, as measured in peripheral plasma, in response to an injection of oxytocin on day 15 of a synchronized oestrous cycle (Oldick et al., 1997). These results indicate that high concentrations of PUFAs in the diet can decrease endometrial secretion of prostaglandins.

R. Mattos et al.

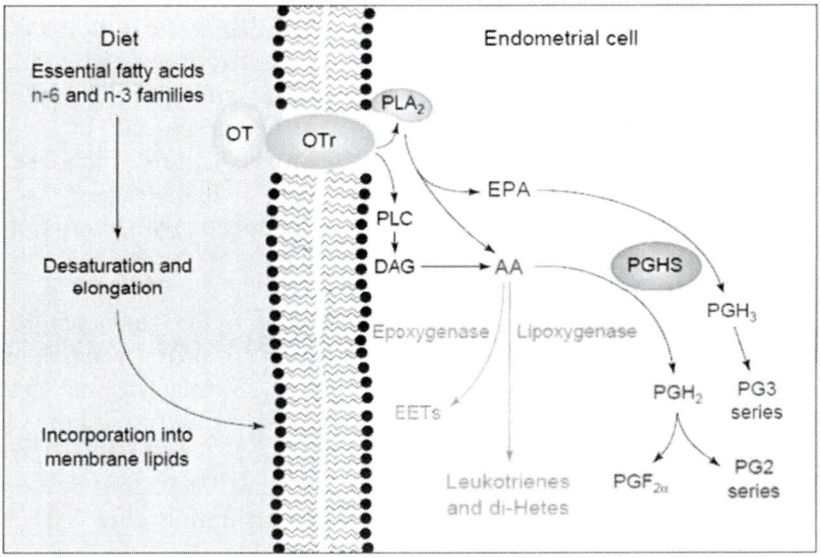

Figure 1. Schematic representation of the metabolism of dietary *n*-6 and *n*-3 polyunsaturated fatty acids and potential mechanisms for regulation of $PGF_{2\alpha}$ secretion (**Mattos et al., 2000**).

In this respect, in cows, fed with *n*-3 fatty acids derived from fish meal, a reduction was observed in endometrial secretion of arachidonic acid, increasing in this same tissue the levels of eicosapentaenoic acid and total *n*-3 fatty acids (Burns et al., 2003; Bilby et al., 2006b). Similar effects were observed in dairy cows which consumed increasing amounts of fish meal or Ca-LCFA from fish oil (Bilby et al. 2006a). Due to the addition of *n*-3 and *n*-6 in the phospholipid component of endometrium, it is possible that changes in the content of fatty acids in the endometrial tissue may modulate secretion of $PGF_{2\alpha}$ in cows (Santos et al., 2008). In further studies, Burns et al. (2003) reported an increase in eicosapentaenoic acid at day 18 postpartum, as well as a reduction in arachidonic acid in the caruncules of beef cows fed fish meal. In another study, supplementing the diet with fish oil during the periparturient period reduced uterine secretion of $PGF_{2\alpha}$ in lactating dairy cows. In this sense, beef heifers fed *n*-3 PUFAs reduced production of arachidonic acid in endometrial tissue, apparently due to the fact that linoleic acid must elongate and desaturate to form arachidonic acid using the same enzymes than for the synthesis of eicosapentaenoic acid, starting from α-linolenic acid (Coyne et al., 2008).

There is evidence that during the prepartum period, lipid supplementation with 30% fatty acids as linoleic acid (*n*-6) increased uterine secretion of $PGF_{2\alpha}$ (Cullens, 2005). Increase in the synthesis of $PGF_{2\alpha}$ when cows are fed *n*-6 fatty acids during the transition period before calving or during the early postpartum (puerperium), can increase the potential of the uterus and of the immune system to secrete eicosanoids, which have influence in postpartum uterine health and in the inmunocompetence of the cow (Santos et al., 2008).

Childs et al. (2008b) fed heifers with a diet rich in *n*-6 fatty acids (entire soyabeans) or *n*-3 fatty acids (fish oil) and observed that on day 15 of the oestrous cycle, serum concentration of PGFM metabolite was higher than in the first group. These results are interesting if it is considered that by stimulating secretion of $PGF_{2\alpha}$ during early postpartum uterine involution can be favored (Mattos et al., 2004) and reproductive efficiency is improved, since when cows were supplemented before calving with calcium salts rich in *n*-6 fatty acids the occurrence of postpartum diseases such as membrane retention, metritis and mastitis, was reduced (Cullens, 2005). Thus, both arachidonic acid, but as well precursors of this fatty acid, such as linoleic acid increase the production of series 2 prostaglandin, while *n*-3 fatty acids compete with arachidonic acid and therefore reduce synthesis of such prostaglandin. In studies carried out *in vitro*, addition of eicosapentaenoic acid to the culture media of endometrial cells of cows, reduced production of endometrial $PGF_{2\alpha}$ from 88 to 40%, but this was reverted as the ratio of *n*-6:*n*-3 in the culture media was increased from 0 to 19% (Caldari-Torres et al., 2006). This observation is consistent with previous reports and suggest that the net inhibition of uterine $PGF_{2\alpha}$ synthesis by *n*-3 fatty acids may depend on the amount of *n*-6 fatty acids that reach the secretory tissue (Achard et al., 1997).

However, recent studies (Meier et al., 2009) showed that the bovine endometrial and trophoblastic tissues during short-term culture, incubated in a media supplemented with fatty acids: eicosapentaenoic (20:5-3; EPA), docosahexaenoic acids (22:6-3; DHA) or linoleic acids (C18:2-6; LIN), the release of PGE_2 from 'pregnant' endometrium was higher ($P=0.094$) than from 'non-pregnant' endometrium, while $PGF_{2\alpha}$ concentrations were similar. Treatment with fatty acids had no effect on

$PGF_{2\alpha}$ or PGE_2release from either pregnant or non-pregnant endometrium. The individual fatty acid treatments had no effect on the ratio of $PGF_{2\alpha}$ to PGE_2 from trophoblast tissues, but when the data from the three fatty acid treatments were combined (EPA, DHA and LIN treatment groups) the ratio of $PGF_{2\alpha}$ to $PGE_2$was reduced (P=0.026) when compared to the medium only. This result indicated that the ability of exogenous fatty acids to modify embryonic prostaglandin release needs to be examined in the context of supplementing dairy cows with different sources of fats.

On other hand, the dynamics of bovine corpus luteum regression in response to exogenous $PGF_{2\alpha}$ can also be altered by dietary fish meal. In this respect, Burke et al., (1997) fed cows (n=56) with fish meal at either 0 or 2.8% of ration dry matter from 24 to109 days post-partum. On day 58 postpartum, all cows were injected with a luteolytic dose of $PGF_{2\alpha}$. Two days after injection of $PGF_{2\alpha}$, the proportion of cows with plasma concentrations of progesterone that were >1 ng ml^{-1} was greater when fish meal was included in the ration than when a control ration was fed (29.0 *vs* 4%). Thus, it is possible that fatty acids present in fish oil reduce the sensitivity of the corpus luteum to $PGF_{2\alpha}$.

5. LIPIDS THEIR EFFECT ON EMBRYO DEVELOPMENT

Establishment of pregnancy in the ruminant requires the ovulation of a competent oocyte, of insemination at the appropriate time and of a correct pattern of secretion of oestradiol and progesterone during the follicular and luteal phase of oestrus. The embryo must develop in an appropriate way and avoid luteolysis producing enough interferon τ which stimulates the expression of genes in the endometrium to inhibit the synthesis of oxytocin receptors and consequently final production of $PGF_{2\alpha}$, allowing the establishment of a corpus luteum (Bott et al., 2010). In dairy cows there is a significant loss of embryos during this period, it is considered that only 40% of cows remain pregnant at day 28 after artificial insemination (Santos et al., 2008). There is evidence that such events can be influenced by PUFAs consumed in the ration (Wathes et al., 2007). Fatty acids play an important role in the modification of the biophysical properties and in the activity of biological membranes, including fluidity and cell proliferation (Bilby et al., 2006d). The competence and quality of the ovocyte and of the embryo are related to the type of fatty acid, specifically, with the content of particular fatty acids en the phospholipids of cell membrane which play a role in development and during and after fertilization (Santos et al., 2008).

The amount of lipids in the ovocyte of ruminants is about 76 ng approximately and has around 58% triglycerides, 20% phospholipids, 20% cholesterol and 10% free fatty acids (McEvoy et al., 2000). Fatty acids found in greater amounts in the phospholipid fraction of the membrane of cattle ovocyte are palmitic (16:0) and oleic (18:1) acids. PUFAs represent less than 20% of the total, being linoleic acid the most abundant of them (Santos et al., 2008). Marei et al. (2010) pointed out that linoleic acid (*n-6*) is the most abundant fatty acid in follicular fluid of cattle and has an important role in the regulation of the process of maturation of the ovocyte, since when the cell complex of the cumulus and ovocyte were treated with linoleic acid there was a delay in the growth of the latter. Some studies have shown the fatty acid profiles in follicular fluid are affected by the estrogenic activity of the follicle (Renaville et al., 2010). Table 4 shows saturated and unsaturated fatty acids in plasma and in some reproductive tissues.

Ratio of saturated fatty acids to PUFA in granulose cells (Adamiak et al., 2005) and in the ovocyte (Wonnacott et al., 2010) is greater than in plasma. This suggests the presence of a mechanism of selective uptake in the ovarian follicles or *de novo* synthesis of saturated fatty acids from acetate (Wonnacott et al., 2010). Fatty acids can be oxidized as an energy source during maturation of the ovocyte and during early embryo development before implantation, in such a way that cattle ovocytes exposed to metal palmoxirate to block oxidation of fatty acids, showed a lower ability to form blastocysts after *in vitro* fertilization (Ferguson & Leese, 2006). Cetica et al. (2002) reported a significant increment in the activity of the enzyme lipase during maturation of cattle ovocyte, which releases fatty acids from triglycerides for later oxidation. Apart from the role of fatty acids, particularly those saturated, as sources of energy, the content of PUFAs in the ovocyte may affect maturation, cryopreservation and its capacity for further development (Wathes et al., 2007). From the *n-3* fatty acids, linolenic acid has been implicated in the growth and differentiation of the ovocyte, regulation of meiotic arrest during the germinal vesicle stage and in avoiding the breaking of this structure (Kim et al., 2001).

In sheep, Zeron et al. (2002) showed that supplementation with Ca-LCFA from fish oil during 13 weeks, resulted in better quality ovocytes and better integrity of their membrane, compared to that of sheep which were not fed lipid supplements (74.3% and 57.0%, respectively), which increased the ratio of long chain fatty acids in plasma of cells from the cumulus, although these changes were not observed in the ovocytes, suggesting selective uptake by the ovocyte or a highly regulated uptake, which could limit potential impact of cow nutrition on the proportion of fatty acids in their gametes. While in beef cattle, Fouladi-Nashta et al. (2007) fed cows with 200 or 800 g per day of Ca-LCFA from palm oil, which resulted in a greater percentage of ovum which developed up to the blastocyst stage and had a greater amount of cells due to an increment in the number of cells of the trophectoderm. By influencing the molecular mechanisms which control nucleus maturation of the ovocyte, *in vitro* treatment of the cumulus-oocyte complex with linolenic acid (*n-3*), increased the percentage of gametes which reached the meiotic second division, increased the division of the embryo as well as the rate of blastocyst (Marei et al., 2009).

Table 4. Saturated and unsaturated fatty acids in plasma and some reproductive tissues.Adapted from**Wonnacott et al. (2010)** and **Palmquist (2010)**. [a]6.4 % oil in the dietsupplemented from linseed oil, [b]5.7% oil in the diet supplemented from sunflower oil.[c]Important effects due dietary fatty acids are in bold.

Fatty acid group	Plasma (µg/ml)		Granulosa cell (µg/pellet)		Oocytes (ng/oocyte)	
	Dietary treatment [ab]					
	n-3	n-6	n-3	n-6	n-3	n-6
Saturated	30.9	42.1	39.1	39.7	75.7	71.1
Unsaturated	59.3	49.4	49.7	52.1	20.9	25.8
Monounsaturated fatty acids	21.5	18.0	20.9	23.9	7.8	12.9
Polyunsaturated fatty acids	37.8	31.4	28.9	28.2	13.0	12.9
n-6 PUFA	25.4	28.9	8.5[c]	24.1	5.2	10.9
n-3 PUFA	**12.4**	**2.5**	**20.4**	**4.1**	**7.8**	**2.1**
Ratio n6:n3	**2.1**	**11.5**	**0.4**	**6.2**	**0.7**	**8.3**

When a group of lactating superovulated cows were fed with rich sources of saturated fatty acids (*n*-6 or *n*-3), rate of fertilization and the number of transferable embryos was not different; however, embryo development was increased in cows which consumed unsaturated fatty acids, compared to the cows which consumed saturated fatty acids (Thangavelu et al., 2007). Cerri et al. (2009) supplemented dairy cows with Ca-LCFA from palm oil or linoleic + octadecaenoic acids between days 25 prepartum and 80 days postpartum observing in the second group a greater proportion of excellent and good quality embryos, apart from a higher number of blastomere.

In hair ewes, Herrera et al. (2008) showed that PUFAs in the ration increased superovulatory response, registering increased ($P<0.05$) numbers of corpus luteum (14.73±1.87 *vs* 10.73±1.42); total cells collected, including ova plus embryos (9.18±2.16 *vs* 4.18±1.36) and embryos (6.72±1.78 *vs* 3.09±1.36) in the PUFAs treatment than in the control treatment, respectively.

On the contrary, Childs et al. (2008b) fed cows with a ration enriched with *n*-3 PUFAs and did not observe any effect on the number of normal embryos nor in the amount of good quality embryos (grade 1 and 2); however, these cows showed a lower number of degenerated embryos. Similarly, Bilby et al. (2006d) did not find an effect of fatty acids in the ration of dairy cows on the quality of embryos after maturation and *in vitro* fertilization. In a similar way, Thangavelu et al. (2007) did not establish a difference in the total number of transferable embryos of dairy cows supplemented with either PUFAs or saturated fat. Marques et al. (2007) did not observe any effect of the addition of arachidonic or eicosapentaenoic acids to the culture media for the *in vitro* maturation of ovocytes on the subsequent embryo development. This agrees with observations by Lawson et al. (2007) who added increasing amounts of eicosapentaenoic acid to the culture media for the maturation of ovocytes *in vitro*. In a recent study, Wakefield et al. (2007) suggested that contrary to the possible beneficial effects, supplementing a ration with *n*-3 PUFAs during the period just before or immediately after conception, may reduce normal development of the embryo, since this seems to disturb mitochondrial metabolism.Leroy et al. (2010) cultivated bovine zygotes in media supplemented with serum from heifers fed rations with a high content of lipids protected from rumen biohydrogenation and observed a lower production of blastocysts compared to a control treatment, noting that in the first there was a greater expression of genes related to apoptosis.

Even when *in vivo* and *in vitro* studies have shown a better embryo development with rations supplemented with lipids, results are not consistent and it is important to ascertain in particular, which fatty acids are the most beneficial for embryo survival (Santos et al., 2008). Oocytes of all mammals contain an endogenous lipid reserve. This feature reflects their common ancestral origin, the yolk-rich amniote egg. However, lipids are species-specific in terms of their apparent abundance and utilization. Despite the significant role of the lipid reserves in cell structure and function, very few studies have provided detailed descriptions of its nature and composition in mammalian oocytes. Table 5 gives the fatty acid composition of total lipid extracted from zone-intact oocytes of cattle and sheep.

6. CONCLUSION

Data reviewed shows that supplementation with different sources of lipids and fatty acids improve reproductive performance of the female ruminant. However, it is important to consider that the optimum response will be achieved when undernutrition status of the female is not extremely sever. A nutrient balance (protein:energy) in the ration consumed by the animal is fundamental to obtain maximum benefit from supplementation with fat, since fatty acids do not supply nitrogen for amino acid synthesis and consequently for the correct functioning of the hypothalamus-hypophysis axis. Improvements in reproductive performance may be a result of increased energy density of the ration or of the direct effects of specific fatty acids on reproductive processes. As is the case for any technology or management strategy that improves specific aspects of ovarian physiology and cyclic activity, actual improvements in pregnancy rate or total weight of calf weaned are dependent on a variety of management practices and environmental conditions. Until these interrelationships are better understood, livestock producers are recommended to attempt to formulate low cost/balanced rations. If a source of supplemental fat is available locally and can be incorporated with little or no change in the cost of the ration, it would be wise for farmers to do so. Research studying the role of fat supplementation on reproductive responses has not been that consistent, therefore, adding fat to the ration would be advised when the risk of low reproductive performance (young, growing animals and limiting nutrients [protein, energy] in the basal ration) is the greatest.

Table 5. Fatty acid composition of total lipid extracted from zone-intact oocytes of cattle and sheep (adapted from McEvoy et al., 2000). Each sample (*n* = 2–4) represents 1000 oocytes. [a]Percentage (w/w) of the total fatty acids in oocyte lipid. [b]One of the four original cattle oocyte samples was excluded because of uncertainty about the validity of the assay result involving cholesterol ester component of total lipid. nd: not detected.

		Mean (± SEM) distribution of fatty acids (%, w/w)[a]	
Name	Formula	Cattle (*n* = 3)[b]	Sheep (*n* = 2)
Lauric	12:0	0.23 ± 0.15	nd
Myristic	14:0	2.48 ± 1.02	0.39±0.032
Palmitic	16:0	32.0 ± 1.64	24.7±0.74
Palmitoleic	16:1 n-7	2.24 ± 0.45	4.38±0.20
Heptadecanoic	17:0	0.76 ± 0.14	0.41±0.407
Stearic	18:0	14.2±2.47	16.2±0.30
Oleic	18:1 n-9	25. ±1.75	26.2±0.23
Vaccenic	18:1 n-7	3.71± 0.12	3.64±0.32
Linoleic	18:2 n-6	5.17± 0.12	6.98±0.10
γ-Linolenic	18:3 n-6	0.75 ± 0.16	1.01±0.06
α-Linolenic	18:3 n-3	0.49±0.09	2.01±0.41
Stearidonic	18:4 n-3	nd	1.68±0.30
Arachidic	20:0	1.35±0.70	3.11±0.45
Eicosenoic	20:1 n-9	0.27±0.13	0.19±0.187
Eicosadienoic	20:2 n-6	0.54±0.20	0.91±0.476
Eicosatrienoic	20:3 n-6	0.27±0.15	0.52±0.070
Arachidonic	20:4 n-6	1.13±0.57	1.50±0.60
Eicosapentaenoic	20:5 n-3	1.15±1.15	3.03±1.09
Behenic	22:0	1.23±0.63	3.03±1.09
Erucic	22:1n-9	0.20±0.10	nd
Docosatetraenoic	22:4 n-6	0.27±0.15	nd
Docosapentaenoic	22:5 n-3	0.88±0.28	1.41±0.33
Docosahexaenoic	22:6 n-3	0.50±0.25	1.74±0.06
Lignoceric	24:0	2.30±1.21	nd

REFERENCES

1. D. R. E. Abayasekara, C. D. Wathes, 1999 Effects of altering dietary fatty acid composition on prostaglandin synthesis and fertility. Prostaglandins, Leukotrienes and Essential Fatty Acids, 61 5 (November 1999), 275 287 . 0952-3278

2. D. Achard, M. Gilbert, C. Benistant, S. B. Slama, D. L. De Witt, W. L. Smith, M. Lagarde, 1997 Eicosapentaenoic and docosahexaenoic acids reduce PGH synthase 1 expression in bovine aortic endothelial cells. Biochemical Biophysical Research Communications, 241 2 (December 1997), 513 518 . 0000-6291X

3. S. J. Adamiak, K. Mackie, R. G. Watt, R. Webb, K. D. Sinclair, 2005 Impact of nutrition on oocyte quality: cumulative effects of body composition and diet leading to hyperinsulinemia in cattle. Biology of Reproduction, 73 5 (November 2005), 918 926 . 0006-3363

4. C. Aguilar-Pérez, J. Ku-Vera, C. P. Garnsworthy, 2009 Effects of bypass fat on energy balance, milk production and reproduction in grazing crossbred cows in the tropics. Livestock Science, 121 1 (March 2009), 64 71 . 1871-1413

5. D. J. Ambrose, C. T. Estill, M. G. Colazo, J. P. Kastelic, R. Corbett, 2006b Conception rates and pregnancy losses in dairy cows fed a diet supplemented with rolled flaxseed. Proceedings of the 7th International Ruminant Reproduction Symposium, Wellington, New Zealand. Abstract 50.

6. D. K. Ambrose, J. P. Kastelic, R. Corbett, P. A. Pitney, H. V. Petit, J. A. Small, P. Zalkovic, 2006a Lower pregnancy losses in lactating dairy cows fed a diet enriched in α-linolenic acid. Journal of Dairy Science, 89 8 (August 2006), 3066 3074 . 1811-9751

7. I. Aranda-Ávila, J. Herrera-Camacho, J. R. Aké-López, R. A. Delgado-León, J. C. Ku-Vera, 2010 Effect of supplementation with corn oil on postpartum ovarian activity, pregnancy rate, and serum concentration of progesterone and lipid metabolites in F1 (Bos taurus × Bos indicus) cows. Tropical Animal Health and Production, 42 7 October 2010), 1435 1440 . 1573-7438

8. J. D. Armstrong, E. A. Goodall, F. J. Gordon, D. A. Rice, W. J. Mc Caughey, 1990 The effects of levels of concentrate offered and inclusion of maize gluten or fish meal in the concentrate on reproductive performance and blood parameter of dairy cows. Animal Production, 50 1 (September, 1990), 1 10 . 1357-7298

9. J. F. Bader, E. E. D. Felton, M. S. Kerley, D. D. Simms, D. J. Patterson, 2000 Effects of postpartum fat supplementation on reproduction in primiparous 2-year-old and mature cows. Journal of Animal Science, 78 Suppl. 1 (July 2000), 224 1525-3163 Journal of Dairy Science, Vol.83, Suppl. No.1, (July 2000), pp. 224. ISSN 0022-0302

10. S. G. Bernal, A. M. O'Donnell, J. L. Vicini, G. F. Hartnell, D. E. Bauman, 2010 Enhancing omega-3 fatty acids in milk fat of dairy cows by using stearidonic acid-enriched soybean oil from genetically

modified soybeans. Journal of Dairy Science, 93 1 (January 2010), 32 37 . 0022-0302

11. G. Bernal-Santos, I. I. J. W. Perfield, D. M. Barbano, D. E. Bauman, T. R. Overton, 2003 Production responses of dairy cows to dietary supplementation with conjugated linoleic acid (CLA) during the transition period and early lactation. Journal of Dairy Science, 86 10 (October 2003), 3218 3228 . 0022-0302

12. T. R. Bilby, J. Block, B. C. Amaral, Filho. O. Sa, F. T. Silvestre, P. J. Hansen, C. R. Staples, W. W. Thatcher, 2006d Effects of dietary unsaturated fatty acids on oocyte quality and follicular development in lactating dairy cows in summer. Journal of Dairy Science, 89 10 October 2006), 3891 3903 . 0022-0302

13. T. R. Bilby, A. Guzeloglu, Laren. L. A. Mac, C. R. Staples, W. W. Thatcher, 2006a Pregnancy, bST and omega-3 fatty acids in lactating dairy cows: II. Gene expression related to maintenance of pregnancy. Journal of Dairy Science, 89 9 September 2006), 3375 3385 . 0022-0302

14. T. R. Bilby, T. Jenkins, T. R. Staples, W. W. Thatcher, 2006b Pregnancy, bovine somatotropin, and dietary n-3 fatty acids in lactating dairy cows: III. Fatty acid distribution. Journal of Dairy Science, 89 9 (September 2006), 3386 3399 . 0022-0302

15. T. R. Bilby, A. Sozzi, M. M. Lopez, F. T. Silvestre, A. D. Ealy, C. R. Staples, W. W. Thatcher, 2006c Pregnancy, bovine somatotropin, and dietary n-3 fatty acids in lactating dairy cows: I. ovarian, conceptus, and growth hormone-insulin-like growth factor system responses. Journal of Dairy Science, 89 9 September 2006), 3360 3374 . 0022-0302

16. M. Binelli, W. W. Thatcher, R. Mattos, P. S. Baruselli, 2001 Antiluteolytic strategies to improve fertility in cattle. Theriogenology, 56 9 December 2001), 1451 1463 . 0009-3691X

17. R. C. Bott, R. L. Ashley, L. E. Henkes, A. Q. Antoniazzi, J. E. Bruemmer, G. D. Niswender, F. W. Bazer, T. E. Spencer, N. P. Smirnova, R. V. Anthony, T. R. Hansen, 2010 Uterine vein infusion of interferon Tau (IFNT) extends luteal life span in ewes. Biology of Reproduction, 82 4 April 2010), 725 735 . 0006-3363

18. I. Bruckental, D. Dori, M. Kaim, H. Lehrer, Y. Folman, 1989 Effects of source and level of protein on milk yield and reproductive performance of high-producing primiparous and multiparous dairy cows. Animal Production, 48 2 April 1989), 319 329 . 1357-7298

19. J. M. Burke, C. R. Staples, C. A. Risco, R. L. De La Sota, W. W. Thatcher, 1997 Effect of ruminant grade menhaden fish meal on reproductive and productive performance of lactating dairy cows. Journal of Dairy Science, 80 12 December 1997), 3386 3398 . 0022-0302

20. P. D. Burns, T. E. Engle, M. A. Harris, R. M. Enns, J. C. Whittier, 2003 Effect of fish meal supplementation on plasma and endometrial fatty acid composition in nonlactating beef cows. Journal of Animal Science, 81 11 November 2003), 2840 2846 . 1525-3163

21. G. O. Burr, M. M. Burr, 1930 On the nature and role of the fatty acids essential in nutrition. Journal of Biological Chemistry, 86 2 April 1930), 587 621 . 0021-9258

22. C. Caldari-Torres, C. Rodriguez-Sallaberry, E. S. Greene, L. Badinga, 2006 Differential effects of n-3 and n-6 fatty acids on prostaglandin F2α production by bovine endometrial cells. Journal of Dairy Science, 89 3 March 2006), 971 977 . 0022-0302

23. D. J. Carroll, F. R. Hossain, M. R. Keller, 1994 Effect of supplemental fish meal on the lactation and reproductive performance of dairy cows. Journal of Dairy Science, 77 10 Octiber 1994), 3058 3072 . 0022-0302

24. E. Castañeda-Gutierrez, T. R. Overton, W. R. Butler, D. E. Bauman, 2005 Dietary supplements of two doses of calcium salts of conjugated linoleic acid during the transition period and early lactation. Journal of Dairy Science, 88 3 March 2005), 1078 1089 . 0022-0302

25. Audet. I. Castellano, J. L. Bailey, P. Y. Chouinard, J. P. Laforest, J. J. Matte, 2010 Effect of dietary n-3 fatty acids (fish oils) on boar reproduction and semen quality. Journal of Animal Science, 88 7 July 2010), 2346 2355 . 1525-3163

26. R. L. A. Cerri, S. O. Juchem, R. C. Chebel, H. Rutgliano, R. G. S. Bruno, K. N. Galvaõ, W. W. Thatcher, J. E. Santos, 2009 Effect of fat source differing in fatty acid profile on metabolic parameters,

fertilization, and embryo quality in high-producing dairy cows. Journal of Dairy Science, 92 4 April 2009), 1520 1531 . 0022-0302

27. P. Cetica, L. Pintos, G. Dalvit, M. Beconi, 2002 Activity of key enzymes involved in glucose and triglyceride catabolism during bovine oocyte maturation in vitro. Reproduction, 124 5 November 2002), 675 681 . 1741-7899

28. S. Childs, F. Carter, C. O. Lynch, J. M. Sreenan, P. Lonergan, A. A. Hennessy, D. A. Kenny, 2008b Embryo yield and quality following dietary supplementation of beef heifers with n-3 polyunsaturated fatty acids (PUFA). Theriogenology, 70 6 October 2008), 992 1003 . 0009-3691X

29. S. Childs, A. A. Hennessy, J. M. Sreenan, C. Wathes, Z. Cheng, C. Stanton, M. G. Diskin, D. A. Kenny, 2008a Effect of level of dietary n-3 polyunsaturated fatty acid supplementation on systemic and tissue fatty acid concentrations and on selected reproductive variables in cattle. Theriogenology, 70 4 September 2008), 595 611 . 0009-3691X

30. S. Childs, C. O. Lynch, A. A. Hennessy, C. Stanton, D. C. Wathes, J. M. Sreenan, M. G. Diskin, D. A. Kenny, 2008c Effect of dietary enrichment with either n-3 or n-6 fatty acids on systemic metabolite and hormone concentration and ovarian function in heifers. Animal, 2 6 May 2008), 883 893 . 1751-7311

31. G. S. Coyne, D. A. Kenny, S. S. Childs, J. M. Sreenan, S. M. Waters, 2008 Dietary n-3 polyunsaturated fatty acids alter the expression of genes involved in prostaglandin biosynthesis in the bovine uterus. Theriogenology, 70 5 September 2008), 772 782 . 0009-3691X

32. F. M. Cullens, 2005 Effects of the timing of initiation of fat supplementation on productive and reproductive responses of periparturient dairy cows during summer. MSc. Thesis. University of Florida, USA. 146

33. C. A. de Fries, D. A. Neuendorff, R. D. Randel, 1998 Fat supplementation influences postpartum reproductive performance in Brahman cows. Journal of Animal Science, 76 3 March 1998), 864 870 . 1525-3163

34. C. J. de Luna, W. H. Brown, D. E. Ray, T. N. Wegner, 1982 Effects of protected fat supplement on GnRH induced LH release in

ovariectomized and early postpartum beef cow. Journal of Animal Science, 55 No.Suppl. 1, 348 Abstr) 1525-3163

35. M. J. de Veth, D. E. Bauman, W. Koch, G. E. Mann, A. M. Pfeiffer, W. R. Butler, 2009 Efficacy of conjugated linoleic acid for improving reproduction: A multi-study analysis in early-lactation dairy cows. Journal of Dairy Science, 92 6 June 2009), 2662 2669 . 0022-0302

36. M. Doreau, D. Bauchart, Y. Chilliard, 2011 Enhancing fatty acid composition of milk and meat through animal feeding. Animal Production Science, 51 1 December 2010), 19 29 . 1836-0939

37. K. H. El -Shahat, maaty. A. M. Abo El, 2010 The effect of dietary supplementation with calcium salts of long chain fatty acids and/or l-carnitine on ovarian activity of Rahmani ewes. Animal Reproduction Science, 117 1-2 (January 2010), 78 82 . 0378-4320

38. J. L. Espinoza, J. A. Ramírez-Godínez, J. A. Jiménez, A. Flores, 1995 Effects of calcium soaps of fatty acids on postpartum reproductive activity in beef cows and growth of calves. Journal of Animal Science, 73 10 October 1995), 2888 2892 . 1525-3163

39. J. L. Espinoza, J. A. Ramirez-Godinez, S. S. Simental, J. Jiménez, R. Ramirez, A. Palacios, R. de Luna, 1997 Effects of calcium soaps of fatty acids on serum hormones and lipid metabolites in Pelibuey ewes. Small Ruminant Research, 26 1 December 1997), 61 68 . 0921-4488

40. J. L. Espinoza-Villavicencio, R. Ortega-Pérez, A. Palacios-Espinosa, A. Guillén-Trujillo, 2010 Efecto de la suplementación de grasas sobre características productivas, tasas de preñez y algunos metabolitos de los lípidos en vacas para carne en pastoreo. Archivos de Medicina Veterinaria, 42 1 Enero 2010); 25 32 . 0030-1732X

41. E. M. Ferguson, H. J. Leese, 2006 A potential role for triglyceride as an energy source during bovine oocyte maturation and early embryo development. Molecular Reproduction and Development, 73 9 September 2006), 1195 1201 . 0104-0452X

42. J. D. Ferguson, D. Sklan, W. V. Chalupa, D. S. Kronfeld, 1990 Effects of hard fat on in vitro and in vivo rumen fermentation, milk production, and reproduction in dairy cows. Journal of Dairy Science; 73 10 October 1990), 2864 2879 . 0022-0302

43. S. J. Filley, H. A. Turner, F. Stormshak, 2000 Plasma fatty acids, prostaglandin F2α metabolite, and reproductive response in postpartum heifers fed rumen bypass fat. Journal of Animal Science, 78 1 January 2000), 139 144 . 1525-3163

44. A. A. Fouladi-Nashta, C. G. Gutierrez, J. G. Gong, P. C. Garnsworthy, R. Webb, 2007 Impact of dietary fatty acids on oocyte quality and development in lactating dairy cows. Biology of Reproduction, 77 1 July 2007), 9 17 . 0006-3363

45. M. C. Fuentes, S. S. Calsamiglia, C. Sánchez, A. González, J. Newbold, J. E. P. Santos, L. M. Rodríguez-Alcalá, J. Fontecha, 2008 Effect of extruded linseed on productive and reproductive performance of lactating dairy cows. Livestock Science, 113 2 February 2008), 144 154 . 1871-1413

46. R. Funston, S. Filley, 2002 Effects of fat supplementation on reproduction in beef cattle. Proceedings of The Applied Reproductive Strategies in Beef Cattle Workshop, September 5-6, 2002, Manhattan, Kansas. USA. 1 8 p.

47. R. N. Funston, 2004 Fat supplementation and reproduction in beef females. Journal of Animal Science, 82 13 January 2004), 154 161 . E suppl. 1525-3163

48. C. M. Garcia-Bojalil, C. R. Staples, C. A. Risco, J. D. Savio, W. W. Thatcher, 1998 Protein degradability and calcium salts of long chain fatty acids in the diets of lactating dairy cows: productive responses. Journal of Dairy Science, 81 5 May 1998), 1374 1384 . 0022-0302

49. P. C. Garnsworthy, A. Lock, G. E. Mann, K. D. Sinclair, R. Webb, 2008 Nutrition, metabolism, and fertility in dairy cows: 2. Dietary fatty acids and ovarian function. Journal of Dairy Science, 91 10 October 2008), 3824 3833 . 0022-0302

50. C. A. Gray, K. M. Taylor, W. S. Ramsey, J. R. Hill, F. W. Bazer, F. F. Bartol, T. E. Spencer, 2001 Endometrial glands are required for preimplantation conceptus elongation and survival. Biology of Reproduction 64 6 June 2001), 1608 1613 . 0006-3363

51. M. A. Harris, R. A. Hansen, P. Vidsudhiphan, J. L. Koslo, J. B. Thomas, B. A. Watkins, K. G. Allen, 2001 Effects of conjugated linoleic acids and docosahexaenoic acid on rat liver and reproductive tissue fatty acids, prostaglandins and matrix metalloproteinase

production. Prostaglandins Leukotrienes and Essential Fatty Acids, 65 1 July 2001), 23 29 . 0952-3278

52. J. H. Harrison, R. L. Kincaid, J. P. Mc Namara, S. Waltner, K. A. Loney, R. E. Riley, J. D. Cronrath, 1995 Effect of whole cottonseeds and calcium salts of long-chain fatty acids on performance of lactating dairy cows. Journal of Dairy Science, 78 1 January 1995), 181 193 . 0022-0302

53. A. R. Heravi-Moussavi, R. O. Gilbert, T. R. Overton, D. E. Bauman, W. R. Butler, 2007 Effect of feeding fish meal and n-3 fatty acids on ovarian and uterine responses in early lactating dairy cows. Journal of Dairy Science, 90 1 January 2007), 145 154 . 0022-0302

54. J. Herrera-Camacho, J. R. Aké-López, J. C. Ku-Vera, G. L. Williams, J. A. Quintal-Franco, 2008 Respuesta ovulatoria, estado de desarrollo y calidad de embriones de ovejas Pelibuey superovuladas suplementadas con ácidos grasos poliinsaturados. Tecnica Pecuaria en Mexico, 46 2 Abril-Junio 2008), 107 117 . 0040-1889

55. R. B. Hightshoe, R. C. Cochran, L. R. Corah, G. H. Kiracofe, D. L. Harmon, R. C. Perry, 1991 Effects of calcium soaps of fatty acids on postpartum reproductive function in beef cows. Journal of Animal Science, 69 10 October 1991), 4097 4103 . 1525-3163

56. T. Hinckley, R. M. Clark, S. L. Bushmich, R. A. Milvae, 1996 Long chain polyunsaturated fatty acids and bovine luteal cell function. Biology of Reproduction, 55 2 August 1996), 445 449 . 0006-3363

57. J. J. Howard, R. G. Scott, J. H. Britt, 1990 Associations among progesterone, estradiol-17b, and prostaglandin in cattle treated with hCG during diestrus to extend corpus luteum function. Prostaglandins, 40 1 July 1990), 51 70 . 1098-8823

58. Huante CR. 2010 Efecto de la suplementación de dos tipos de ácidos grasos sobre el reinicio de la actividad ovárica y metabolitos de lípidos en vacas F1 (Bos taurus x Bos indicus) durante el posparto temprano. Maestria Institucional en Ciencias Biológicas. Instituto de Investigaciones Agropecuarias y Forestales. Universidad Michoacana de San Nicolás de Hidalgo. Morelia, Michoacán, México.

59. S. O. Juchem, 2007 Lipid Digestion and Metabolism in Dairy Cows: Effects on Production, Reproduction and Health. PhD thesis, University of California Davis, USA.

60. S. O. Juchem, R. L. A. Cerri, R. Bruno, K. N. Galvao, E. W. Lemos, M. Villasenor, A. C. Coscioni, H. M. Rutgliano, W. W. Thatcher, D. Luchini, J. E. P. Santos, 2004 Effect of feeding Ca salts of palm oil (PO) or a blend of linoleic and monoenoic trans fatty acids (LTFA) on uterine involution and reproductive performance in Holstein cows. Journal of Animal Science, 82 Journal of Dairy Science, Vol.87 / Poultry Science, Vol.83, Suppl.1 310 Abstract 633). 1525-3163 / ISSN 0022-0302 / ISSN: 0032-5791

61. S. O. Juchem, R. L. A. Cerri, M. Villaseñor, K. N. Galvão, R. G. S. Bruno, H. M. Rutigliano, E. J. De Peters, F. T. Silvestre, W. W. Thatcher, J. E. P. Santos, 2010 Supplementation with calcium salts of linoleic and trans-octadecenoic acids improves fertility of lactating dairy cows. Reproduction in Domestic Animals, 45 1 February 2010), 55 62 . 1439-0531

62. J. Y. Kim, M. Kinoshita, M. Ohnishi, Y. Fukui, 2001 Lipid and fatty acid analysis of fresh and frozen-thawed immature and in vitro matured bovine oocytes. Reproduction, 122 1 July 2001), 131 138 . 1470-1626

63. J. J. Knickerbocker, W. W. Thatcher, D. B. Foster, D. Wolfenson, F. F. Bartol, D. Caton, 1986 Uterine prostaglandin and blood flow responses to estradiol-17β in cyclic cattle. Prostaglandins, 31 4 April 1986), 757 776 . 1098-8823

64. M. A. Lammoglia, S. T. Willard, J. R. Oldham, R. D. Randel, 1996 Effects of dietary fat and season on steroid hormonal profiles before parturition and on hormonal, cholesterol, triglycerides, follicular patterns and postpartum reproduction in Brahman cows. Journal of Animal Science, 74 9 September 1996), 2253 2262 . 1525-3163

65. C. Lawson, M. Wade, D. A. Kenny, P. Lonergan, 2007 Effect of addition of eicosapentaenoic acid (EPA) to culture medium on development of bovine embryos in vitro. In: Annual Meeting of the European Embryo Transfer Association. 190 p.

66. C. O. Lemley, S. T. Butler, W. R. Butler, M. E. Wilson, 2008 Short communication: Insulin alters hepatic progesterone catabolic

enzymes cytochrome P450 2C and 3A in dairy cows. Journal of Dairy Science, 91 2 February 2008), 641 645 . 0022-0302

67. J. L. Leroy, V. Van Hoeck, M. Clemente, D. Rizoz, A. Gutierrez-Adan, A. Van Soom, M. Uytterhoeven, P. E. Bols, 2010 The effect of nutritionally induced hyperlipidemia on in vitro bovine embryo quality. Human Reproduction, 25 3 March 2010), 768 778 . 0268-1161

68. N. Lopes, A. B. Scarpa, B. I. Cappellozza, R. F. Cooke, J. L. M. Vasconcelos, 2009 Effects of rumen-protected polyunsaturated fatty acid supplementation on reproductive performance of Bos indicus beef cows. Journal of Animal Science, 87 12 December 2009), 3935 3943 . 1525-3163

69. M. C. Lucy, R. L. De La Sota, C. R. Staples, W. W. Thatcher, 1993 Ovarian follicular populations in lactating dairy cows treated with recombinant bovine somatotropin (sometribove) or saline and fed diets differing in fat content and energy. Journal of Dairy Science; 76 4 April 1993), 1014 1027 . 0022-0302

70. M. C. Lucy, C. R. Staples, F. M. Michel, W. W. Thatcher, 1991 Effect of feeding calcium soaps to early postpartum dairy cows on plasma prostaglandin F2α, Luteinizing hormone, and follicular growth. Journal of Dairy Science, 74 2 February 1991), 483 489 . 0022-0302

71. M. C. Lucy, W. W. Thatcher, F. J. Michel, C. R. Staples, 1989 Effect of dietary calcium soaps of long chain fatty acids (Megalac) on plasma prostaglandin F metabolite (PGFM), LH, energy balance and follicular population in early postpartum dairy cattle. Journal of Animal Science, 67 No. Suppl. 1 389 (Abstr). 1525-3163

72. F. W. Marei, D. C. Wathes, A. A. Fouladi-Nashta, 2010 Impact of linoleic acid on bovine oocyte maturation and embryo development. Reproduction, 139 6 June 2010), 979 988 . 1470-1626

73. W. F. Marei, D. C. Wathes, A. A. Fouladi-Nashta, 2009 The effect of linolenic acid on bovine oocyte maturation and development. Biology of Reproduction, 81 6 December 2009), 1064 1072 . 0006-3363

74. M. A. Marín-Aguilar, J. C. Tinoco-Magaña, J. Herrera-Camacho, L. G. Sánchez-Gil, V. M. Sánchez-Parra, J. L. Solorio-Rivera, A.

García-Valladares, 2007 Reinicio de la actividad ovárica y nivel de metabolitos de lípidos en vacas lecheras suplementadas con aceite vegetal durante el posparto temprano. Interciencia, 32 3 (March 21007), 180 184 . 0378-1844

75. C. C. Marques, M. C. Baptista, M. I. Vasques, A. E. M. Horta, R. M. Pereira, 2007 Effect of polyunsaturated fatty acids (PUFA) on bovine oocyte in vitro maturation and subsequent embryo development and freezability. Reproduction in Domestic Animals, 109 No.Suppl. 2 (September 2007), 116 0936-6768

76. R. Mattos, A. Guzeloglu, L. Badinga, C. R. Staples, W. W. Thatcher, 2003 Polyunsaturated fatty acids and bovine interferon-s modify phorbol ester-induced secretion of prostaglandin F2α and expression of prostaglandin endoperoxide synthase-2 and phospholipase-A2 in bovine endometrial cells. Biology of Reproduction, 69 3 September 2003), 780 787 . 0006-3363

77. R. Mattos, C. R. Staples, A. Arteche, M. C. Wiltbank, F. J. Diaz, T. C. Jenkins, W. W. Thatcher, 2004 The effects of feeding fish oil on uterine secretion of PGF2α, milk composition, and metabolic status of periparturient Holstein cows. Journal of Dairy Science, 87 4 April 2004), 921 932 . 0022-0302

78. R. Mattos, C. R. Staples, W. W. Thatcher, 2000 Effects of dietary fatty acids on reproduction in ruminants. Reviews of Reproduction, 5 1 January 2000), 38 45 . 1470-1626

79. T. G. Mc Evoy, G. D. Coull, P. J. Broadbent, J. S. M. Hutchinson, B. K. Speake, 2000 Fatty acid composition of lipids in immature cattle, pig and sheep oocytes with intact zona pellucida. Journal of Reproduction and Fertility, 118 1 January 2000), 163 170 . 0022-4251

80. S. Mc Namara, T. Butler, D. P. Ryan, J. F. Mee, P. Dillon, F. P. O'Mara, S. T. Butler, D. Anglese, M. Rath, J. J. Murphy, 2003 Effect of offering rumen-protected fat supplements on fertility and performance in spring-calving Holstein-Friesian cows. Animal Reproduction Science, 79 1-2 ; (November 2003); 45 56 . 0378-4320

81. S. Meier, A. M. Ledgard, T. A. Sato, A. J. Peterson, M. D. Mitchell, 2009 Polyunsaturated fatty acids differentially alter PGF2α and PGE2 release from bovine trophoblast and endometrial tissues

during short-term culture. Animal Reproduction Science, 111 2-4 (April 2009), 353 360 . 0378-4320

82. B. S. Oldick, C. R. Staples, W. W. Thatcher, P. Gyawu, 1997 Abomasal infusion of glucose and fat-effect on digestion, production, and ovarian and uterine function of cows. Journal of Dairy Science, 80 7 July 1997), 1315 1328 . 0022-0302

83. D. L. Palmquist, 2010 Essential fatty acids in ruminant diets. Proceedings of the 21nd Annual Ruminant Nutrition Symposium, February 2-3, 2010. Gainesville, Florida, USA. 127 141

84. C. A. Piccinato, R. Sartori, S. Sangsritavong, A. H. Souza, R. R. Grummer, D. Luchini, M. C. Wiltbank, 2010 In vitro and in vivo analysis of fatty acid effects on metabolism of 17β-estradiol and progesterone in dairy cows. Journal of Dairy Science, 93 5 May 2010), 1934 1943 . 0022-0302

85. A. O. Rego, V. P. Portugal, B. M. Sousa, H. J. D. Rosa, M. C. Vozuela, E. S. A. Borba, R. J. B. Bessa, 2004 Effect of diet on the fatty acid pattern of milk from dairy cows. Animal Research, 53 3 May-June 2004), 213 220 . 1627-3591

86. B. Renaville, N. Bacciu, A. Comin, M. Motta, I. Poli, G. Vanini, A. Prandi, 2010 Plasma and follicular fluid fatty acid profiles in dairy cows. Reproduction in Domestic Animals, 45 1 February 2010), 118 121 . 0936-6768

87. R. S. Robinson, P. G. A. Pushpakumara, Z. Cheng, A. R. Peters, D. Abayasekara-E-E, D. C. Wathes, 2002 Effects of dietary polyunsaturated fatty acids on ovarian and uterine function in lactating dairy cows. Reproduction, 124 1 July 2002); 119 131 . 1470-1626

88. D. P. Ryan, R. A. Spoon, M. K. Griffith, G. L. Williams, 1994 Ovarian follicular recruitment, granulosa cell steroidogenic potential, and growth hormone/insulin-like growth factor-1 relationships in beef cows consuming high lipid diets: Effects of graded differences in body condition maintained during the puerperium. Domestic Animal Endocrinology, 11 2 April 1994), 161 174 . 0739-7240

89. D. P. Ryan, B. Bao, M. K. Griffith, G. L. Williams, 1995 Metabolic and luteal sequelae to heightened dietary fat intake in postpartum

beef cows. Journal of Animal Science, 73 7 July 1995), 2086 2093 . 1525-3163

90. D. P. Ryan, R. A. Spoon, G. L. Williams, 1992 Ovarian follicular characteristics, embryo recovery, and embryo viability in heifers fed high fat diets and treated with follicle-stimulating hormone. Journal of Animal Science, 70 11 November 1992); 3505 3513 . 1525-3163

91. G. Salas-Razo, J. Herrera-Camacho, E. Gutiérrez-Vázquez, J. C. Ku-Vera, J. R. Aké-López, 2011 Postpartum ovarian activity resumption and plasma concentration of lipid metabolites and progesterone in Indobrasil cows in the dry tropics of Michoacan supplemented with bypass fat. Tropical and Subtropical Agroecosystems (in press). 1870-0462 1870 0462

92. J. E. P. Santos, T. R. Bilby, W. W. Thatcher, C. R. Staples, F. T. Silvestre, 2008 Long chain fatty acids of diet as factors influencing reproduction in cattle. Reproduction in Domestic Animals, 43 No. Suppl. 2 (July 2008), 23 30 . 1627-3591

93. T. A. Scott, R. D. Shaver, L. Zepeda, B. Yandell, T. R. Smith, 1995 Effects of rumen-inert fat on lactation, reproduction, and health of high producing Holstein herds. Journal of Dairy Science, 78 11 Novembre 1995), 2435 2451 . 0022-0302

94. D. Sklan, M. Kaim, U. Moallem, Y. Folman, 1994 Effect of dietary calcium soaps on milk yield, body weight, reproductive hormones, and fertility in first parity and older cows. Journal of Dairy Science, 77 6 June 1994), 1652 1660 . 0022-0302

95. D. Sklan, U. Moallem, Y. Folman, 1991 Effect of feeding calcium soaps of fatty acids on production and reproductive responses in high producing lactating cows. Journal of Dairy Science, 74 2 February 1991), 510 517 . 0022-0302

96. S. S. Smith, M. Neuringer, S. R. Ojeda, 1989 Essential fatty acid deficiency delays the onset of puberty in the female rat. Endocrinology, 125 1650 1659 . 0013-7227

97. J. Son, R. J. Grant, L. L. Larson, 1996 Effects of tallow and escape protein on lactational and reproductive performance of dairy cows. Journal of Dairy Science, 79 5 May 1996), 822 830 . 0022-0302

98. J. M. Sreenan, M. G. Diskin, D. G. Morris, 2001 Embryo survival rate in cattle: a major limitation to the achievement of high fertility. In: Fertility in the high producing dairy cow. BSAS Occasional Publication, 1 93 105 .

99. R. L. Stanko, P. Fajersson, L. A. Carver, G. L. Williams, 1997 Follicular growth and metabolic changes in beef heifers fed incremental amounts of polyunsaturated fat. Journal of Animal Science, 75 No. Suppl. 1 (July-August 1997), 223 Abstr 376). 1525-3163

100. C. R. Staples, J. M. Burke, W. W. Thatcher, 1998 Influence of supplemental fats on reproductive tissues and performance of lactating cows. Journal of Dairy Science, 81 3 March 1998), 856 871 . 0022-0302

101. C. R. Staples, W. W. Thatcher, 2005 Effects of fatty acids on reproduction of dairy cows. In: Garnsworthy PC, Wiseman, J. (Eds.), Recent Advances in Animal Nutrition. Nottingham University Press, Nottingham, UK, 229 256 p.

102. D. M. Stocco, X. Wang, Y. Jo, P. R. Manna, 2005 Multiple signaling pathways regulating steroidogenesis and steroidogenic acute regulatory protein expression: more complicated than we thought. Molecular Endocrinology, 19 11 November 2005), 2647 2659 . 0888-8809

103. F. Talavera, C. S. Park, G. L. Williams, 1985 Relationships among dietary lipid intake, serum cholesterol and ovarian function in Holstein heifers. Journal of Animal Science, 60 4 April 1985); 1045 1051 . 1525-3163

104. G. Thangavelu, M. G. Colazo, D. J. Ambrose, M. Oba, E. K. Okine, M. K. Dyck, 2007 Diets enriched in unsaturated fatty acids enhance early embryonic development in lactating Holstein cows. Theriogenology 68 949 957 . 0009-3691X

105. W. W. Thatcher, K. L. Macmillan, P. J. Hansen, F. W. Bazer, 1994 Embryonic losses: causes and prevention. In: Fields, M.J.S., Sand, R.S. Editors. Factors affecting calf crop. Boca Raton, FL: CRC Press; 135 153 p.

106. W. W. Thatcher, C. R. Staples, 2000 Effects of dietary fat supplementation on reproduction in lactating dairy cows. Advanced in Dairy Technology, 12 12 December 2000), 213 232 . 1184-0684

107. W. W. Thatcher, R. C. Staples, 2007 Using fats and fatty acids to enhance reproductive performance. Proceedings of the 5th Mid-Atlantic Nutrition Conference.. 115 Zimmermann, N.G., ed., University of Maryland, College Park, MD 20742. USA. 116 129 p.

108. M. G. Thomas, G. L. Williams, 1996 Metabolic hormone secretion and FSH-induced superovulatory responses of beef heifers fed dietary fat supplements containing predominantly saturated or polyunsaturated fatty acids. Theriogenology, 45 January), 451 458 . 0009-3691X

109. M. G. Thomas, B. Bao, G. L. Williams, 1997 Dietary fats varying in their fatty acid composition differentially influence follicular growth in cows fed isoenergetic diets. Journal of Animal Science, 75 9 September 1997), 2512 2519 . 1525-3163

110. S. L. Wakefield, M. Lane, S. J. Schulz, M. L. Hebart, J. G. Thompson, M. Mitchell, 2007 Maternal supply of omega 3 polyunsaturated fatty acids alter mechanisms envolved in oocyte and early embryo development in the mouse. American Journal of Physiology, Endocrinology and Metabolism, 294 2 February 2008); E425 E434 . 0193-1849

111. C. K. Wang, R. S. Robinson, A. P. Flint, G. E. Mann, 2007 Quantitative analysis of changes in endometrial gland morphology during the bovine oestrous cycle and their association with progesterone levels. Reproduction, 134 2 August 2007); 365 371 . 1470-1626

112. D. C. Wathes, D. R. Abayasekara, R. J. Aitken, 2007 Polyunsaturated fatty acids in male and female reproduction. Biology of Reproduction, 77 2 August 2007), 190 201 . 0006-3363

113. M. E. Wehrman, T. H. Welsh, Jr , G. L. Williams, 1991 Diet-induced hyperlipidemia in cattle modifies the intrafollicular cholesterol environment, modulates ovarian follicular dynamics, and hastens the onset of postpartum luteal activity. Biology of Reproduction, 45 3 September 1991), 514 522 . 0006-3363

114. G. L. Williams, R. L. Stanko, 2000 Dietary fats as reproductive nutraceuticals in cattle. Proceedings of the American Society for Animal Science 1998/1999. Journal of Animal Science, 77 No.ESuppl (January 2000), 1 12 . 1525-3163

115. G. L. Williams, 1989 Modulation of luteal activity in postpartum beef cows through changes in dietary lipid. Journal of Animal Science, 67 3 March 1989); 785 793 . 1525-3163

116. G. L. Williams, M. Amstalden, 2010 Understanding postpartum anestrus and puberty in the beef female. Proceedings, Applied Reproductive Strategies in Beef Cattle. January 28 29 ; San Antonio, TX. USA. 55-71 p.

117. K. E. Wonnacott, W. Y. Kwong, J. Hughes, A. M. Salter, R. G. Lea, P. C. Garnsworthy, K. D. Sinclair, 2010 Dietary omega-3 and-6 polyunsaturated fatty acids affect the composition and development of sheep granulosa cells, oocytes and embryos. Reproduction, 139 1 January 2010), 57 69 . 1470-1626

118. J. D. Wood, R. I. Richardson, G. R. Nute, A. V. Fisher, M. M. Campo, E. Kasapidou, P. R. Sheard, M. Enser, 2003 Effects of fatty acids on meat quality: a review. Meat Science, 66 1 January 2004), 21 32 . 0309-1740

119. Y. Zeron, D. Sklan, A. Arav, 2002 Effect of polyunsaturated fatty acid supplementation on biophysical parameters and chilling sensitivity of ewe oocytes. Molecular Reproduction and Development, 61 2 February 2002), 271 278 . 1098-2795

CHAPTER 9

Vitamin-Like Supplementation in Dairy Ruminants: The Case of Choline

Luciano Pinotti[1]

Department of Health, Animal Science and Food Safety, Università di Milano, Italy

1. INTRODUCTION

Recent developments in nutrition have established that choline is an essential nutrient for mammals when a sufficient supply in methionineand folates are not available in the diet. Vitamin B12 is also involved in this process. The dynamic interactions between these components introduced the concept of choline as vitamin-like compound. Two types of choline functions are known: as choline *per se*, for which the choline moiety is required, and functions as a methyl donor. Choline per se plays a major role in lipid metabolism, particularly in lipid transport, as lipotropic agent. Choline is also an important source of labile methyl groups for the biosynthesis of other methylated compounds. Based on this second function choline and methionine are interchangeable, as sources of methyl groups. Accordingly choline occupies a key position between energy and protein metabolism in mammals.

Choline and methyl group metabolism in ruminants however are different. In adult ruminants, choline is extensively degraded in the rumen; for this reason dietary choline contributes insignificantly to the choline body pool and methyl group metabolism is generally conservative with a relatively low rate of methyl catabolism and an elevated rate of de novo synthesis of methyl groups via the tetrahydrofolate (THF) system. This can be exacerbated in lactating dairy ruminant, in which the dietary availability of choline is nearly non-existent, but the output of methylated compounds in milk is high, while methionine as well as other sources of methyl groups are likely to be in short supply, especially at the onset of lactation. In light of this, the hypothesisthat choline can be a limiting nutrientfor milk production has been formulated and tested in several studies. Accordingly this chapter will focus on the effects of rumen-protected choline (RPC) supplementation to transition and early lactating dairy ruminants (cows and goats) on milk production and on metabolic health. Experimental data will be discussed in a systematic analysis, in order to define possible recommendations for high yielding dairy ruminants.

2. WHAT IS CHOLINE?

In 1849 Adolph Strecker, a German chemist, isolated a compound from pig bile, to which he subsequently (in 1862) applied the name choline (from the Greeke, *chole*, bile). In 1867, Bayer, determined the chemical structure of choline (McDowell, 1989).

Choline is a beta-hydroxyethyltrimethylammonium hydroxide (figure 1), which is widely distributed in nature as free choline, acetylcholine, and more complex phospholipids and their metabolic intermediates. Its role in the body is complex. It is needed for neurotransmitter synthesis (acetylcholine), cell-membrane signaling (phospholipids), lipid transport (lipoproteins), and methyl-group metabolism (homocysteine reduction) (Zeisel & Da Costa, 2009).

2.1. Vitamin Or Not?

In mammals choline has been classified as one of the B-complex vitamins but it does not satisfy the standard definition of a vitamin (Pinotti et al., 2002): it is synthesised endogenously and there is no evidence that it is an enzyme co-factor; furthermore, unlike other water-soluble vitamins, it is difficult to identify a deficiency syndrome for choline in healthy mammals because of its interrelation with methionine, folic acid, and vitamin B12 (Scott, 1999; Zeisel, 1988). Finally, choline is a vital component of tissues and is required in the diet of non-ruminant species at much higher levels than the water-soluble vitamins (g vs. mg) (Whitehead & Portsmouth, 1989). However, the presence of an endogenous synthetic pathway does not render choline dispensable and deficiency results in several dysfunctions when other nutrients are limiting (Zeisel et al., 1991). For this reason it has been suggested that choline may be an essential nutrient (or vitamin like compound) for mammals when excess methionine and folic acid are not available in the diet (Zeisel & Da Costa, 2009). Consequently choline was officially recognized as an essential nutrient for humans by the Institute of Medicine in 1998 (Food and Nutrition Board, 1998).

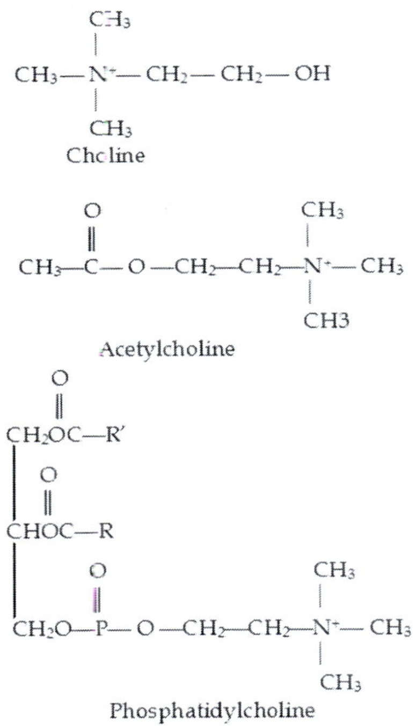

Figure 1. Structural formula for free choline, acetylcholine, and phosphatidylcholine (lecithin). R and R’ refers to any fatty acids (adapted from McDowell, 1989).

2.2. How Requirement For Choline Can Be Met

The metabolic need for choline can be met in two ways: either by dietary choline and via de novo biosynthesis through the methylation of phosphatidylethanolamine to phosphatidylcholine.From the point of view of animal nutrition, relatively rich sources of choline are soybean, soybean meal, rapeseed meal, fish meal and dried yeast, even though its bioavailability is considered "moderate" (Pinotti et al., 2002; Whitehead & Portsmouth, 1989).Dietary choline from a variety of choline-containing feed and food is absorbed by the intestine and uptake is mediated by choline transporters nutrient (McDowell, 1989). *De novo* synthesis of choline occurs by the sequential methylation of phosphatidylethanolamine, the methyl groups being supplied by S-adenosyl-L-methionine (SAM) (Mato et al., 1994).However, *de novo* synthesis of choline alone is not sufficient or rapid enough to satisfy all the animal's need. Methyl groups needed in this pathway may be derived from exogenous sources such as methionine, and betaine but can arise de novo in the body from the tetrahydrofolate (THF) system (Zeisel, 1988; Zeisel, 1992); vitamin B12 is involved in this process (Kennedy et al., 1995). Thus, dietary factors such us methionine, betaine, myo-inositol, folic acid, and vitamin B12 or combination of different levels of fat, carbohydrates, and protein in the diet, as well as the physiological state, all have influence on the "requirements" of choline (McDowell, 1989; Zeisel & Da Costa, 2009).

2.3. Functions Of Choline

Choline is considered a vitamin-like compound with two main functions: as choline *per se*, for which the choline moiety is required, and as methyl donor, although the two roles overlap.

Choline *per se* is an essential constituent of all cell membranes, where it is required to make the phospholipids phosphatidylcholine, lysophosphatidylcholine, choline plasmalogen, and sphingomyelin. Phosphatidylcholine, one of the most abundant phospholipids in higher plants and animals, is the predominant phospholipid (>50%) in most mammalian cell membranes (Kuksis & Mookerjea, 1978;Ruiz et al., 1983; Zeisel, 1988; Zeisel, 1992). Choline *per se* plays a major role in lipid metabolism, particularly in lipid transport, as lipotropic agentbecause of its ability to prevent or correct excess fat deposition in the liver generally arising as a result of its dietary deficiency (Kuksis &

Mookerjea, 1978;Zeisel, 1988). Impaired triglyceride secretion to very low density lipoproteins (VLDL) is considered a major cause of fatty liver in dietary choline deficiency (Zeisel, 1988). In this context, it is noteworthy that bovine VLDL phospholipids are mainly phosphatidylcholine, with smaller proportions of sphingomyelin and phosphatidylethanolamine (Moore & Christie, 1981). Thus, when massive mobilisation of fatty acids – as at the onset of lactation in dairy ruminants- is associated with lipotropic factor deficiency (e.g. choline) triglycerides accumulate in the liver and may lead to the development of fatty liver (Gruffat et al., 1996).Choline *per se* is required to prevent hemorrhagic kidney lesions in rats (Kuksis & Mookerjea, 1978); and together with other nutrients particularly manganese salts, is required to prevent perosis, a bone disease of poultry (Ruiz et al., 1983). In the prevention of perosis, choline is needed as constituent of phospholipids required for normal maturation of the cartilage matrix of the bone, whereas in the prevention ofhemorrhagic kidney lesions choline seems involved in the renal phospholipids turnover (Kuksis & Mookerjea, 1978).

Choline *per se* is also essential for the synthesis of the neurotransmitter acetylcholine (Kuksis & Mookerjea, 1978; Zeisel, 1988). It plays important roles in brain and memory development in the fetus and appears to decrease the risk of the development of neural tube defects (Zeisel & Da Costa, 2009).

As methyl donor, choline, like methionine, is an important source of labile methyl groups for biosynthesis. Actually, the two principal methyl donors in animal metabolism are betaine, a choline metabolite, and S-adenosyl-L-methionine (SAM) a metabolite of methionine (Pinotti et al., 2002 for references). At least 50 SAM-dependent reactions have been identified in mammals, and it is likely that the number is much higher. Such methylation reactions play major roles in biosynthesis of lipids, the regulation of several metabolic pathways, and detoxification in the body (Zeisel & Da Costa, 2009). Accordingly it has been suggested that choline, and methionine are closely interrelated metabolically (Mato et al., 1994); it is also clear that choline has only little capacity to reduce the requirement for methionine, even though choline seems important to spare methionine as a methyl donor.

Although the choline-methionine interrelationship has been studied extensively (see Pinotti et al, 2002for reference), less emphasis has been placed on the link between choline and its oxidation product, betaine. Betaine serves as an osmoregulator and is a substrate in the betaine–homocysteine (HCy) methyltransferase reaction, which links choline and betaine to the folate-dependent one-carbon metabolism (Kempson & Montrose, 2004; Ueland, 2011). The availability of choline and its metabolite betaine appear to influence the conversion or recycling of the HCy moiety of methionine, through betaine-HCy methyltransferase activity (Stipanuk, 1986).However, betaine fails to prevent fatty livers and haemorrhagic kidney (McDowell, 1989), indicating that the requirement for choline per se must be met as choline, and that betaine can substitute only the methyl donor function of choline, probably because betaine cannot be reduced to choline (Zeisel, 1988). Thus, choline *per se* must provide at least 50% of the total choline requirement, while the remaining portion of the choline requirement can be replaced by betaine (Dilger et al., 2007).

3. CHOLINE IN RUMINANTS

Ruminants differ from most other mammals in regard to their choline and methyl group metabolism. In adult ruminants, choline is extensively degraded in the rumen. Quantitative studies in sheep showed that 76% of [14C]-choline injected into the rumen was expired as methane over 6h, whereas approximately 15% accumulated as trimethylamine. Under such conditions, less than 10% of choline escapes degradation by incorporation, as phosphatidylcholine, into the structural membranes of ciliate protozoa (Neill et al., 1979). Nevertheless, the concentration of phosphatidylcholine in ruminal digesta was higher than in abomasal digesta, suggesting that protozoa are selectively retained in the rumen (Neill et al., 1979). In sheep with a defaunated rumen, the concentration of phosphatidylcholine in the abomasal digesta was higher than in the rumen (Dawson et al., 1981), leading to the suggestion that some of this abomasal Phosphatidylcholine was derived from non-dietary sources (regurgitation of bile from the lower digestive tract) (Robinson et al., 1984). In fact, intravenous injection of labelled choline in sheep indicated that the small amount of phosphatidylcholine present in abomasal digesta is largely (69%) of non-dietary or ruminal origin (Dawson et al., 1981).

The physiological state however, is important (Table 1). In the pre-ruminant lamb the activities of liver Cho-oxidase and betaine-HCy methyltransferase increase markedly after birth, but subsequently decrease as the animals reach the ruminant state (Xue & Snoswell, 1986a). The increasing activities of liver choline oxidase and betaine-HCy methyltransferase in pre-ruminant lambs are probably related to the abundance of choline-containing compounds in the milk (Xue & Snoswell, 1986a- Figure 2). By contrast, in adult ruminants only small quantities of methyl group nutrients are available from the diet and methionine synthase assumes a much more important role (Dawson et al., 1981; Neill at al., 1978; Neill et al., 1979; Kennedy et al., 1995). Methionine synthase is responsible for the de novo synthesis of methionine methyl groups from one-carbon units furnished by tetrahydrofolate (THF) (figure 1). The enzyme utilises methyl-THF as methyl donor and methylcobalamin as tightly-bound coenzyme. De novo synthesis of methyl groups via this system appears to be minimal when labile methyl group intake is sufficient or excessive (Stipanuk, 1986).

Dietary choline therefore contributes insignificantly to the choline body pool in adult ruminants (Table1; Figure 3). Methyl group metabolism is generally conservative with a relatively low rate of methyl catabolism and an elevated rate of de novo synthesis of methyl groups via the tetrahydrofolate (THF) system (Girard et al., 2010; Henderson et al., 1983; Robinson et al., 1984; Snoswell & Xue, 1987; Xue & Snoswell, 1986a; Xue & Snoswell, 1986b). In dairy ruminants the situation can be even worse (Table 1): the dietary availability of choline is still low while the output of methylated compounds to milk is particularly high, and methionine, as well as other methyl group sources - one-carbon units for methylneogenesis via the THF system-, is likely to be in short supply, especially at the onset of lactation (Pinotti et al., 2002). In dairy ruminants producing large quantities of milk in fact, methionine is also the first limiting amino acid. This means that the elevated requirement for methioninefor transmethylation reactions and milk protein synthesis may lead to altered methyl group metabolism (Girard et al., 2010; LaCount et al., 1995; Lobley et al., 1996). The effects of lactation on methyl group metabolism have been examined in sheep (Xue & Snoswell, 1985); it was found that the activities of hepatic phospholipid methyltransferase (for phosphatidylcholine synthesis) and methionine synthase were significantly higher (+ 33% and +34%, respectively) in lactating than non-lactating ewes (Xue & Snoswell, 1985).

4. CHOLINE IN DAIRY COWS NUTRITION

The earliest investigations were interested in determining the effects of added dietary choline on milk production. For this purpose supplementation experiments using unprotected choline (Erman et al., 1984; Atkins et al., 1988; Sharma & Erdman, 1988b), were performed. In these studies using unprotected choline, neither milk production nor milk composition were affected, due, it was suggested, to rapid choline degradation in the rumen (Atkins et al., 1988; Sharma & Erdman, 1988b). In fact the work of Atkins et al. (1988) indicated that choline chloride was more degradable than naturally occurring choline in feed, while increasing choline chloride intake from 23 to 326 g/d only raised duodenal choline flow from 1.2 to 2.5 g/d, so that recovery was low indeed (Sharma & Erdman, 1988b). The mean estimates of rumen degradable choline (measured *in vitro*) actually were 80-98% for common feedstuffs and supplements (Table 2), indicating that use of choline-rich feedstuffs or additives in diets can only marginally increase the post ruminal flow of choline in ruminants (Sharma and Erdman, 1989a).

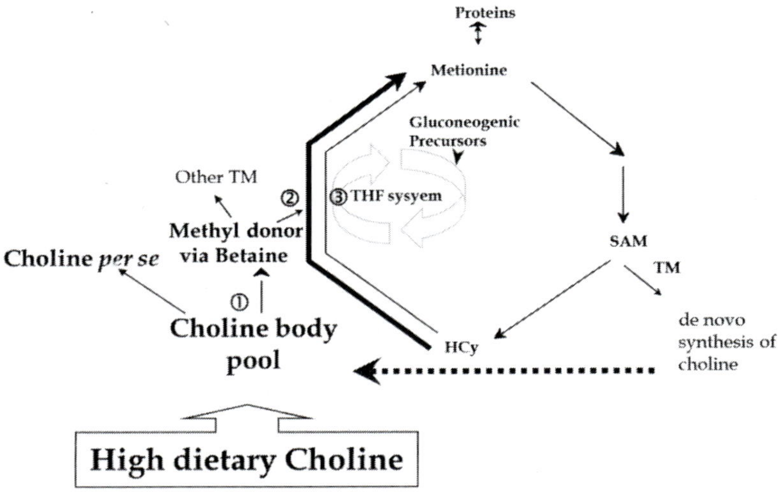

Figure 2. Choline and methyl groups metabolism in pre-ruminants. SAM, S-adenosyl-L-methionine. HCys, Homocysteine. TM, Transamethylation pathway. THF Tetrahydrofolate system. ①Choline oxidase (EC 1.1.3.17). ② Betaine-homocysteine methyltransferase (EC 2.1.1.5). ③ Methionine synthase (EC 2.1.1.13).(adapted from **Pinotti et al., 2002**).

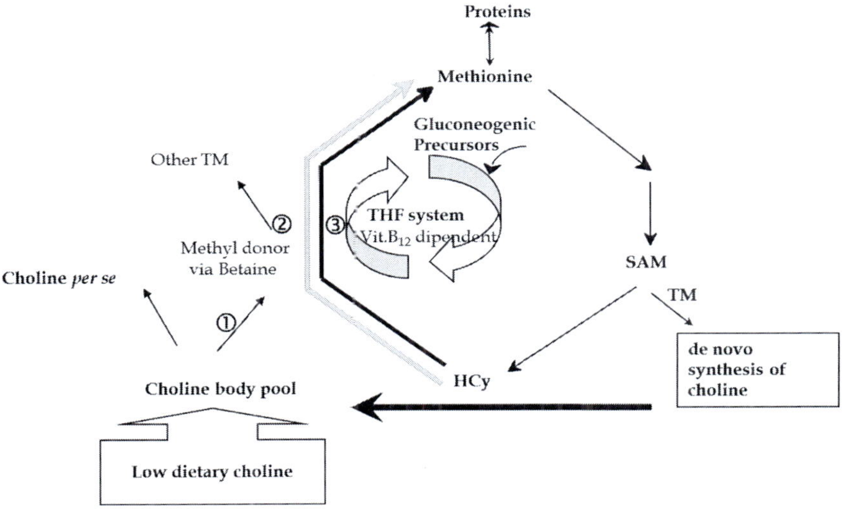

Figure 3. Choline and methyl groups metabolism in adult ruminants in positive energy balance (adapted from**Pinotti et al., 2002**). SAM, S-adenosyl-L-methionine. HCys, Homocysteine. TM, Transamethylation pathway. THF Tetrahydrofolate system. ①Choline oxidase (EC 1.1.3.17). ② Betaine-homocysteine methyltransferase (EC 2.1.1.5). ③ Methionine synthase (EC 2.1.1.13).(adapted from **Pinotti et al., 2002**).

Table 1. Table 1.Evolution of metabolism in methyl groups in ruminants in three physiological stages[*at the onset of lactation, the substrates of this path often become limiting](adapted from **Pinotti et al., 2002**).

Physiological Stage	Availability	Enzymes Activities
Pre-Ruminants	⇧ choline ⇧ betaine	⇧ Cho-oxidase ⇧Betaine-HCy-methyltransferase
Adult Ruminants	⇩ choline ⇩ betaine ⇧ methyl syntheses via THF	⇧ Methionine synthetase
Lactating Ruminants	⇩ choline ⇩ betaine ⇩methionine ⇩gluconeogenetic precursors ⇧ methyl syntheses via THF*	⇧ Methionine synthetase

However, others studies (Sharma & Erdman, 1988a; Sharma & Erdman, 1989b) on the effects of abomasally infused choline, suggested a possible requirement for supplementing choline in lactating dairy cows.For example, Sharma & Erdman (1988a) investigated abomasal infusion of choline or methionine in dairy cows, either alone or with the choline synthesis inhibitor 2-amino-2-methyl-1-propanol (2AMP). They found that abomasal infusion of 30 g/d of choline was more effective than abomasal infusion of 45.6 g/d methionine (the molar equivalent of 30 g choline) in increasing milk yield and milk fat content. Additionally, milk production, milk fat percentage, milk fat yield, milk protein percentage, and milk protein yield of cows infused with 2AMP plus methioninewere all lower than in cows infused with 2AMP plus choline (Sharma & Erdman, 1988a). These data suggest not only that supplemental choline is required to achieve maximal performance, but also that choline formed from methionine (Emmanuel & Kennelly, 1984) may be partially responsible for these stimulatory effects on milk production. Form a technical/feeding point of view these studies also evidenced that choline must be rumen-protected to be effective in dairy cow nutrition.

Table 2. Table 2.Rumen stability: In vitro choline degradation (**Sharma & Erdman, 1989a**). Data obtained during four individual fermentation runs, samples were incubated in vitro with an inoculum mixture containing rumen fluid obtained from a rumen-fistulated dairy cow fed 17.5% corn silage and 28.7% grass silage and 53.8% concentrate diet.

Feedstuffs	Effective degradation, %
Barley	79.4
Cottonseed	84.7
Soybean meal	82.9
Fish-meal	83.8
Supplements	
Choline stearate	98.0
Choline chloride	98.6

4.1. Recent Advance In Choline Supplementation In Dairy Cow

Although the choline requirement of dairy cows is still unknown (National Research Council, 2001), higher choline availability (by feeding rumen-protected choline, RPC) can have a favorable effect on milk production (Erdman and Sharma 1991; Erdman, 1994;Hartwell et al. 2000; Piepenbrink and Overton 2003; Pinotti et al. 2003; Scheer et al. 2002). In dairy ruminants, in fact choline has been proposed as limiting nutrient, especially at the onset of lactation (Erdman and Sharma 1991; Pinotti et al., 2002). Based on those considerations, the effects of rumen-protected choline (RPC) supplementation to transition cows have been investigated in several studies (figure 4).

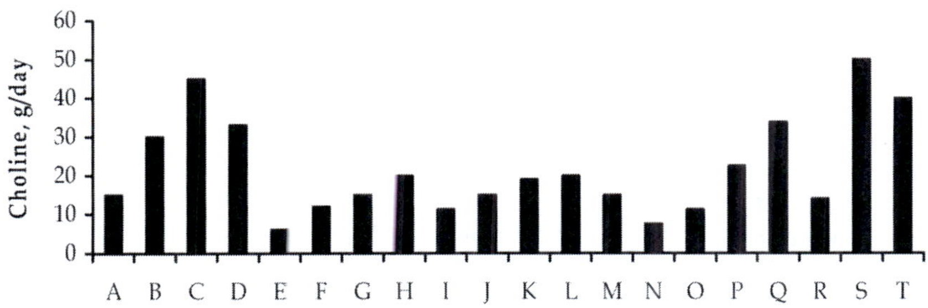

Figure 4. Levels of choline chloride supplementation [provided in a rumen-protected (RP) form] to transition dairy cows in published studies. Citations keys are reported on the x axis: (A)–(C) Erdman & Sharma, 1991; (D) Erdman, 1994; (E)–(F)Hartwell et al., 2000; (G) Scheer et al., 2002; (H) Pinotti et al., 2003; (I)–(K) Piepenbrink & Overton, 2003; (L) Pinotti et al., 2004; (M) Janovick Guretzky at al., 2006; (N)-(Q)Xu et al., 2006; (R) Zahra et al., 2006; (S)Elek et al., 2008; (T) Davidson et al., 2008. Choline chloride supplementation is shown on the y axis.

Findings in transition and early lactating dairy cows suggest that greater choline availability can improve not only milk production (Erdman and Sharma, 1991; Hartwell et al., 2000; Pinotti et al., 2003), but also lipid (Piepenbrink Overton, 2003; Pinotti et al., 2003), methyl group metabolism (Baldi & Pinotti, 2006) and choline secretion in milk (Deuchler et al., 1998; Elek et al., 2008; Pinotti et al., 2003; Pinotti et al, 2004). Furthermore, in the Pinotti et al. (2003) study, although plasma concentrations of vitamin E declined after parturition in choline-supplemented animals, the reduction was less than in controls, suggesting improved vitamin E status. However, the mechanisms of this effect have not been elucidated with certainty, even though an improved fat absorption and transport induced by choline supplementation has been proposed (Pinotti et al., 2003). Janovick Guretzky et al. (2006)

found that choline supplementation did not affect milk production in either Holstein or Jersey transition cows, but in discussing their results they noted that calculated methionine balance was negative postpartum, so that supplemental choline might not have spared enough methionine to produce a physiological benefit. This conclusion is in line with another study (Brüsemeister & Südekum, 2006) that, in reviewing several published works on choline supplementation in dairy cows, concluded that negative methionine balance – and therefore a basal diet of adequate quality and composition (Baldi & Pinotti, 2006) – is essential for obtaining a response to rumen-protected choline. A furtherstudy of Zahra et al. (2006), reported that choline supplementation improves milk production (1.2 kg/day) in the first 60 d of lactation, although supplementation was only from 21 d prepartum to 28 days in milk. However, choline's effect on milk yield was attributable mainly to increased milk production (4.4 kg/day) in animals with body condition score ≥4 three weeks before calving, that were also consuming more feed. Supplementation of rumen-protected choline would therefore seem to be essential for optimising high-quality milk production in high-yielding dairy cows, as also reported in subsequent studies (Davidson et al., 2008; Elek et al., 2008; Xu et al., 2006), particularly in animals fed basal diets that limit post-ruminal methionine supply. The magnitude of the production response is likely to be affected by basal diet composition, the dose and mode of administration of the rumen-protected choline, and the stage of lactation, as discussed elsewhere (Baldi & Pinotti, 2006; National Research Council, 2001).

With regard to metabolic health, fewstudies in dairy cow also investigated the effects of choline supplementation on lipid metabolism (Piepenbrink & Overton, 2003; Pinotti et al. 2003; Pinotti et al. 2004), and hepatic lipidosis (Grummer, 2006; van Vuuren et al., 2010). In dairy ruminants it is known that the necessity to increase the lipolysis of adipose fat stores is most critical during the first stage of lactation (Grummer, 1993; McCarthy et al., 1968). During this stage, fat mobilisation leads to increased blood levels of non-esterified fatty acids (NEFA) which are taken up by the liver and oxidised to ketone bodies or carbon dioxide or esterified to triglycerides. The liver normally packages the triglycerides in VLDLs and secretes them, but sudden increases in plasma NEFA may not be adequately processed by the ruminant liver (Gruffat et al., 1996; Pullen et al., 1990). Choline serves as methyl donor in the synthesis of carnitine (Griffith, 1987) which is essential for fatty acid oxidation. The esterification of NEFA to triglycerides, and their secretion to VLDL also involves choline (Gruffat et al., 1996). In this context the NEFA/cholesterol ratio can be considered a measure of the fat retained in or metabolized by the liver and its lowered value observed in RPC cows suggested a lower risk of fatty liver in these animals (Holtenius, 1989). Thus the lowered plasma NEFA levels and the NEFA/cholesterol ratio in RPC-supplemented cows at parturition suggest more efficient liver function and improved lipid metabolism in general (Pinotti et al. 2003).

A subsequent study (Pinotti et al. 2004) provided similar results, in that plasma NEFA was reduced by 18% on day 20 of lactation in supplemented animals. Likewise, Chung et al. (2009) found that RPC have a dose dependent effect on reducing plasma NEFA concentration by 11% and 23% in cows receiving 25g and 50 g of rumen-protected choline. These findings are also consistent with the evidence that RPC supplementation to dairy cows throughout the periparturient period decreased liver accumulation of lipids (stored as intracellular triglycerides- van Vuuren et al., 2010) and increased liver glycogen content (Piepenbrink & Overton, 2003). The effects of choline at liver level have been recently confirmed in transition dairy cows (van Vuuren et al., 2010), in which RPC supplementation resulted in lower TAG concentration, by 36% and 32%compare to untreated animals, in week 1 and 3 post-partum, respectively. Results obtained in these studies indicated that not only hepatic fatty acid metabolism and cow performance are responsive to increasing the supply of choline (Piepenbrink & Overton, 2003), but also that choline supplementation can reduce cellular lipid accumulation in the liver especially during the transition period.

A further aspect investigated in some of these studies was the effect of choline on methyl group status of lactating cows. Studies carried out by Pinotti et al. (2004) showed that the administration of 50 g of (RPC, providing 20 g of choline chloride) to transition dairy cows increased plasma levels of choline-containing phospholipids and folates by the end of the transition period (20 days in milk). These findings indicate that that RPC supplementation is effective both in increasing choline availability and in optimizing methyl group metabolism: the increase in plasma folates may be attributable to a sparing effect of choline on methyl group metabolism. As noted, choline is an important source of labile methyl groups, but when it is in short supply, methyl groups (e.g. for methionine resynthesis from homocysteine) must be synthesized *de novo* by the THF system or derived from other sources. The THF system consumes gluconeogenic precursors, so in periods of glucose imbalance (e.g. early lactation) such precursors can also be limiting (Armentano 1994; Rukkwamsuk et al. 1999). An unexpected finding of studies in this area is that plasma vitamin B12 concentrations in RPC-supplemented cows were lower than in controls (Baldi & Pinotti 2006). Vitamin B12 is considered a marker of methyl group status, and the lower concentrations in treated animals appear inconsistent with the observed increases in plasma folates and choline-containing phospholipids. However, vitamin B12 is required as a coenzyme in the biosynthesis of methionine from homocysteine and in gluconeogenesis from propionate. Increased gluconeogenesis may therefore deplete vitamin B12, as indicated by the finding of a tendency toward higher concentrations of plasma glucose in RPC-supplemented cows. Nevertheless, these results were obtained from a limited number of animals, and therefore they should considered with caution since our knowledge of the interaction of choline with vitamins and other essential nutrients are incomplete.

4.2. Systematic Analysis Of Choline Supplementation In Dairy Cows

In the recent past different models (linear and logistic) were developed to permit the calculation of changes in milk yield and milk composition from amounts of rumen-protected choline in dairy cow diets. For example, Pinotti et al. (2010) investigated the effects of RPC administration on milk production by a regression model based on the results of 11 different studies (42 experimental groups) published between 1991 and 2008 (Davidson et al., 2008; Deuchler et al., 1998; Elek et al., 2008;Erdman et al., 1991; Hartwell et al., 2000; Janovick Guretzky et al., 2006; Pipenbrink and Overton, 2003; Pinotti et al., 2003; Pinotti et al., 2004; Xu et al., 2006; Zahra et al., 2006). The analysis included RPC supplementation (control/RPC) as fixed effect, the variability among experiments as random effect and their interaction. Treatments schedule, dry matter intake and dietary composition, accounted for most of the variability among experiments and were highly correlated in a preliminary data analysis; therefore, these variables were considered as experimental effects and treated as random components in the mixed model, assuming that the results of different experiments are affected by different experimental conditions. For each study, data were also weighted by the reciprocal of the squared standard error of the mean value, in order to consider the unequal variance among studies.

Table 3. Results from mixed model when the different amounts of RPC administrated were grouped into three supplementation levels (adapted from Pinotti et al., 2010).

Choline amount administered	Milk yield estimate (kg/d)	RPC effect
Control (0 g/d)	32.35	-
RPC low level (<10 g/d)	33.40	0.15
RPC high level (> 10 g/d)	35.18	<0.001

The data reviewed in this analysis were consistent with the fact that choline supplementation significantly increased milk yield in dairy cows, even though, as expected, the dose of choline administered is also relevant (table 3). Considering three supplementation classes, i.e. control (no choline: 0 g/d), low level (<10 g/d of choline chloridein a rumen protected form) and high level (>10 g/d of choline chloridein a rumen protected form), the meta-analysis indicated that a low level (less than 10 g per day) of RPC supplementation tended to increase milk yield and a high level of supplementation of RPC significantly increased milk yield. This results were in line with another meta-analysis, performed using a nonlinear model (Sales et al., 2010), on the effects of rumen-protected choline in dairy cows,in which choline supplementation has a positive effect on milk yield and milk protein content, but not on milk fat content.

5. OTHER DAIRY RUMINANT

In the case of small ruminants the effects of choline infusion on transmethylation reactions have been studied in sheep by Lobley et al. (1996), while Emmanuel and Kennely (1984) investigated methionine and choline incorporation into plasma and milk of lactating goats. More recently Banskalieva et al. (2005) assessed the efficacy of rumen-protected choline supplementation in meat goats. However, studies on choline supplementation and its effects on milk production, and lipid and methyl group status in periparturient dairy goats have been published only recently (Pinotti et al., 2008; Savoini et al., 2010). In a production trial (Pinotti et al., 2008) a supplementation of 4g/d of choline in rumen protected form was given to periparturient dairy goats, starting 4 weeks prior to expected kidding and continuing for 5 weeks after parturition. The quantity of choline given was based on experiments in dairy cows (Pinotti et al., 2005) and metabolicBody weight ($BW^{0.75}$) of the goats at the beginning of the experiment. During the first 6 weeks of lactation, milk yield and four percent fat-corrected milk (FCM) yield were increased by 7 and 12%, respectively, in RPC supplemented than non-supplemented goats. Milk fat concentration, fat and protein yield were also increased by RPC treatment.The link between choline supplementation and milk response has been mainly attributed to the metabolic interchangeability of choline and methionine, in the sense that both can furnish labile methyl groups. This interchangeability has been investigated by Emmanuel and Kennelly (1984) in goats, who estimated that 6% of the choline pool was derived from methionine and that approximately 28% of methionine is used for choline synthesis via the pathway for the de novo biosynthesis of the choline moiety involving sequential methylation of phosphatidylethanolamine. Reduction of methyl group demands from the methionine pool may therefore offer production benefits to dairy ruminants and might be achieved by nutritional manipulation other than alteration of the methionine supply (Lobley et al., 1996). This was probably the situation in the study by Pinotti and coworkers (2008), in which during the early stages of lactation of dairy goats dietary methionine was probably limiting (goats received about 80% of the optimal level of methionine). These considerations imply that feed composition, mainly protein supply and the availability of methionine (National Research Council, 2001) influence the effects of choline supplementation not only in dairy cows (Baldi & Pinotti, 2006; Brüsemeister & Südekum, 2006), but also in goats.

The milk production response to choline supplementation in dairy goats, was obtained without any detrimental effect on methyl groups status. In fact, folate availability at kidding was higher in choline-supplemented than in non-supplemented goats, while during lactation lack of a drop in plasma folate and vitamin B12 in the choline-supplemented animals, suggests that good methyl group status was maintained in choline-treated goats (Pinotti et al., 2008; Savoini et al., 2010). This results was obtained, even though RPC receiving goats were producing more milk than the choline un-suplemented ones.

In a parallel study (Baldi et al. 2011) the effect of rumen-protected choline administration on metabolic profile, selected liver constituents, and on mRNA expressions of selected enzymes, transcription factors, and nuclear receptors involved in mammary lipid metabolism in dairy goats, have been investigated. It was evident from this study that at comparable milk yield during the first month of lactation, choline supplemented goats showed a lower level of plasma β-hydroxybutyrate, whereas no other differences were observed for other metabolites. The reasons for this difference are uncertain: authors conjecture that it might be due to the role of choline in lipid metabolism as already reported in dairy cows (Baldi & Pinotti, 2006; Chung et al., 2009). With regard to liver variables, it has been reported that total liver lipid/ DNA content was significantly lower (-35%) in RPC than CTR goats. As already reported above, dairy cows receiving choline have also been reported to have reduced hepatocellular lipid accumulation (Cooke et al., 2007; Grummer, 2006).

Within the mammary tissue many selected enzymes [e.g. lipoprotein lipase (LPL)], transcription factors and nuclear receptors that can be affected by choline availability are involved in lipid uptake and milk fat secretion. Furthermore, choline is actively secreted into mammalian milk. The major choline-containing compounds in bovine milk are unesterified choline, phosphatidylcholine, and sphingomyelin (Pinotti et al., 2003). Kinsella (1973) reported that a bovine mammary gland yielding 25 litres milk secretes 10±3g phospholipids per day, corresponding on average to 5% of the phospholipids of the mammary tissue. The phospholipids of the membrane of the milk fat globule constitute the major choline-containing component of bovine milk (McPherson & Kitchen, 1983). The above data suggest that choline is an important metabolite in lactating mammary tissue, that it is used avidly when available (Kinsella, 1973), mainly in the lipid metabolism. Thus, it has been proposed that choline supplementation can have positive effects on lipid trafficking particularly lipid transport toand within extra-hepatic tissues (Baldi & Pinotti, 2006; Cooke et al., 2007; Piepenbrink & Overton, 2003) including the mammary gland. However, at udder level in lactating goats, choline did not affect the expression of several mammary gland transcripts involved in lipid metabolism (Baldi et al. 2011). The effects of nutrition on the expression of mammary lipogenic genes in dairy ruminants (cows, ewes, goats) have been investigated mainly by feeding milk-fat depressing diets, characterized by high levels of concentrates, or conjugated linoleic acids (CLA), or other lipids (fish oil, vegetable oils, full fat seeds, etc.) (Bauman, 2008; Bernard et al., 2006; Harvatine and Bauman 2006; Lock et al., 2008). Absence of significant effects on gene expression observed in goats could be due to the dose or to the type of supplementation. Thus, while in Baldi et al. (2011) were supplied 4 g/d choline (about 0.2% of DMI), other studies used higher levels of supplementation (for fats 3.6-11.2% of DMI) or more potent nutritional factors (e.g. CLAs are potent inhibitors of mammary gland fat synthesis) (see Bernard et al., 2006 for review). Nevertheless, these results were obtained from a limited number of animals, and choline effects and function in lactating mammary tissue deserve further investigation.

6. CONCLUSIONS

It can be concluded from this outline that dietary supply of choline may not always be sufficient to maximize milk production in both dairy cows and goat; and although the requirement for choline can in theory be satisfied by other nutrients, it is unlikely that this happens in practice especially at the onset of lactation. The magnitude of the production response is likely to be affected by basal diet composition, the dose of the rumen-protected choline, and the stage of lactation (Baldi & Pinotti, 2006;National Research Council 2001). In particular with respect to lactation cycleseveral studies provided convincing evidence that protected choline positively affects lactation performance especially when feeding is initiated prior to calving and is continued during early lactation (Grummer, 2011). In fact,assuming that choline is limiting in high yielding dairy ruminants, the greatest demand for choline should occur at lactation onset. It is also at this stage that the availability of other choline-related nutrients (i.e. methionine, folates, and vitamin B12) is often low (Baldi & Pinotti, 2006; Girard et al.,2010). From a metabolic and hephatic point of view, since choline is a lipotropic factor, it may be particularly beneficial at this time in view of the adipose and liver metabolism changes that occur during the transition from late pregnancy to early lactation (Piepenbrink and Overton 2003): choline may optimize the balance between fat retained and fat metabolized by the liver, thereby improving lipid metabolism in general. These roles of choline however, do not exclude that RPC supplementation is also effective in optimizing methyl group metabolism, which might be an added benefit of adequate choline supplementation. With regard to dose, even though an estimate of the minimum choline needed in dairy cows for maintenance functions (based on metabolic body size), is approximately 4 to 6 g/day (Donkin, 2011), production response were abstained when higher doses were fed (Pinotti et al., 2010). Consequently, although insufficient data are available to establish the choline requirement, it can be suggested that supplementation doses in this range are likely to be adequate for transition dairy ruminants.

However, it is also evident that our knowledge of the interaction of choline with vitamins and other essential nutrients is incomplete in the dairy ruminants. For example there are indications that both methyl group precursors (including choline) and appropriate co factors (folate and vitamin B12) are important for the optimal metabolic support of milk production (Pinotti et al., 2008), even though methionine may not always be involved in this scenario (Preynat et al, 2009).These aspects not only reflect the inadequacy of a nutritional approach based on the supply and utilisation of individual nutrients, but also suggest that the requirements of B-complex vitamins in dairy ruminants should be reconsidered.

ACKNOWLEDGMENT

This chapter has been developed in the frame of the COST Action FA0802 Feed for Health. The author wish also to thanks Dr. Gerwin Meijer from the Animal Science Group, Wageningen University, for his advisor in the early literature study on this topic, Prof. Antonella Baldi and Prof. Vittorio Dell'Orto, as well as all the member of the Animal Nutrition Research Group from the Veterinary Medicine Faculty of the Università di Milano (Italy), for their collaboration and support in several experiments carried out in this area.

REFERENCES

1. L. Armentano, 1994Impact of metabolism by extragastrointestinal tissues on secretory rate of milk proteins.Journal of Dairy Science779September 1994), 280928200022-0302

2. K. B. Atkins, R. A. Erdman, J. H. Vandersall, 1988Dietary choline effects on milk yield and duodenal choline flow in dairy cattle.Journal of Dairy Science711January 1988, 1091160022-0302

3. A. Baldi, R. Bruckmaier, F. D'ambrosio, A. Campagnoli, C. Pecorini, R. Rebucci, L. Pinotti, Rumen-protected choline supplementation in periparturient dairy goats: effects on liver and mammary gland. Journal of Agricultural Science, 150No. FirstView,Jenuary 201117

4. A. Baldi, L. Pinotti, (2006, 2006Choline metabolism in high-producing dairy cows; metabolic and nutritional basis.Canadian Journal of Animal Scicience,862June 2006, 2072121918-1825

5. V. Banskalieva, R. Puchala, A. L. Goetsch, J. Luo, T. Sahlu, 2005Effects of ruminally protected betaine and choline on net flux of nutrients across the portal-drained viscera and liver of meat goat

wethers consuming diets differing in protein concentrationSmall Ruminant Research, 572-3March 2005), 1932020921-4488

6. D. E. Bauman, J. W. Perfield, K. J. Harvatine, L. H. Baumgard, (2008, 2008Regulation of Fat Synthesis by Conjugated Linoleic Acid, Lactation and the Ruminant Model.Journal of Nutrition, 1382February, 2008), 4034091541-6100

7. L. Bernard, C. Leroux, Y. Chilliard, (2006, 2006Characterisation and nutritional regulation of the main lipogenic genes in the ruminant lactating mammary gland. In, Ruminant physiology, digestion, metabolism and impact of nutrition on gene expression, immunology and stress, Sejrsen, K., Hvelplund, T. & Nielsen, M.O. (Ed.), 295326Wageningen Academic Publishers, 978-9-53761-981-7Wageningen, The Netherlands.

8. F. Brüsemeister, K. H. Sudekum, 2006Rumen-protected choline for dairy cows: the in situ evaluation of a commercial source and literature evaluation of effects on performance and interactions between methionine and choline metabolismAnimal Research552 March-April 2006), 931041627-3591

9. Y. Chung, H. , N. E. Brown, C. M. Martinez, T. W. Cassidy, G. A. Varga, (2009, 2009Effects of rumen-protected choline and dry propylene glycol on feed intake and blood parameters for Holstein dairy cows in early lactation.Journal of Dairy Science, 926June 2009), 272927360022-0302

10. R. F. Cooke, Río. N. Silva Del, D. Z. Caraviello, S. J. Bertics, M. H. Ramos, R. R. Grummer, (2007, 2007Supplemental Choline for Prevention and Alleviation of Fatty Liver in Dairy Cattle.Journal of Dairy Science, 905May 2007, 241324180022-0302

11. S. Davidson, B. A. Hopkins, J. Odle, C. Brownie, V. Fellner, L. W. Whitlow, (2008, 2008Supplementing limited methionine diets with rumen-protected methionine, betaine, and choline in early lactation holstein cows. Journal of Dairy Science, 91Vol. 91, 4April 2008), 15521559 0022-0302

12. R. M. C. Dawson, D. W. Grime, D. B. Lindsay, (1981, 1981On the insensitivity of sheep to the almost complete microbial destruction of dietary choline before alimentary-tract absorption.Biochemical Journal, 1962May 1981), 4995040000-1470

13. K. N. Deuchler, L. S. Piperova, R. A. Erdman, (1998, 1998Milk choline secretion as an indirect indicator of postruminal choline supply.Journal of Dairy Science, 811January 1998), 2382420022-0302

14. R. N. Dilger, T. A. Garrow, D. H. Baker, 2007Betaine can partially spare choline in chicks but only when added to diets containing a minimal level of choline.Journal of Nutrition, 13710October 2007), 222422280022-3166

15. S. S. Donkin, . July, 201, Rumen, 2011Rumen-Protected Choline, 11.07.2011, Available from http://www.extension.org/26158 rumen-protected-choline#

16. P. Elek, J. R. Newbold, T. Gaal, L. Wagner, F. Husveth, 2008Effects of rumen-protected choline supplementation on milk production and choline supply of periparturient dairy cowsAnimal211November 2008), 15951601

17. B. Emmanuel, J. J. Kennelly, (1984, 1984Kinetics of methionine and choline and their incorporation into plasma lipids and milk components in lactating goats.Journal of Dairy Science, 679September 1984), 191219180000-0022

18. R. A. Erdman, B. K. Sharma, (1991, 1991Effect of dietary rumen-protected choline in lactating dairy cows.Journal of Dairy Science, 745May 1991), 164116470022-0302

19. R. A. Erdman, 1994Production responses in field study herds fed rumen protected choline. Journal of Dairy Science, 77Suppl. 1), 186Abstr.), 0022-0302

20. R. A. Erdman, R. D. Shaver, J. H. Vandersall, 1984Dietary choline for the lactating cow: possible effects on milk fat synthesis.Journal of Dairy Science672February 1984), 4104150022-0302

21. Food and Nutrition Board (1998), Institute of Medicine. Choline, In Dietary Reference Intakes: Thiamin, Riboflavin, Niacin, Vitamin B-6, Vitamin B012, Pantothenic Acid, Biotin, and Choline, National Academy of Sciences, 039065542390422 , Washington, D.C., US

22. C. L. Girard, A. Preynat, B. Graulet, H. Lapierre, 2010Could'minor' nutrients such as B vitamins alter major metabolic pathways in lactating dairy cows?, In: Energy and protein metabolism and

nutrition, G.M. Crovetto (Ed.), 235243Wageningen Academic Publishers (EAAP publication N°127), 978-9-08686-153-8Wageningen. The Netherlands

23. O. W. Griffith, 1987Mammalian sulfur amino acid metabolism: an overview.Methods Enzymology, 1433663760076-6879

24. D. Gruffat, D. Durand, B. Graulet, D. Bauchart, (1996, 1996Regulation of VLDL synthesis and secretion in the liver.Reproduction Nutrition Development, 3643753891297-9708

25. R. R. Grummer, 1993Etiology of lipid related metabolic disorders in periparturient dairy cattle. Journal of Dairy Science, 7612December 1993), 388238960022-0302

26. R. R. Grummer, 2006Etiology, pathophysiology of fatty liver in dairy cows.In Production diseases in farm animals, Joshi, N.P. & Herdt, T.H. (Ed.), 141153Wageningen Academic Publishers,9-07699-857-4The Netherlands

27. R. R. . Grummer, July 2011Managing the Transition Cow-Emphasis on Ketosis and Fatty Liver Syndrome, 15.07.2011, Available fromhttp://www.vtdairy.dasc.vt.edu/pdf/cow_college/2011/18%20grummer-paper.pdf

28. J. R. Hartwell, M. J. Cecava, S. S. Donkin, 2000Impact of dietary rumen undegradable protein and rumen-protected choline on intake, peripartum liver triacylglyceride, plasma metabolites and milk production in transition dairy cowsJournal of Dairy Science12December 2000), 290729170022-0302

29. K. J. Harvatine, D. E. Bauman, 2006SREBP1 and thyroid hormone responsive spot 14 (S14) are involved in the regulation of bovine mammary lipid synthesis during diet-induced milk fat depression and treatment with CLA.Journal of Nutrition 136, 10October 2006), 24682474

30. G. D. Henderson, G. P. Xue, A. M. Snoswell, 1983Carnitine and creatine content of tissues of normal and alloxan-diabetic sheep and ratsComparative Biochemistry and Physiology Part B: Comparative Biochemistry, 76B22952981096-4959

31. P. Holtenius, 1989Plasma lipids in normal cows around partus and in cows with metabolic disorders with and without fatty liver.Acta Veterinaria Scandinavica, 4December 1989), 4414451751-0147

32. Guretzky. N. A. Janovick, D. B. Carlson, J. E. Garrett, J. K. Drackley, 2006Lipid metabolite profiles and milk production for holstein and jersey cows fed rumen-protected choline during the periparturient periodJournal of Dairy Science891January 2006), 1882000022-0302

33. S. A. Kempson, M. H. Montrose, 2004Osmotic regulation of renal betaine transport: transcription and beyond.Euroopean Journal Physiology, 4493December 2004), 2272340000-0031

34. G. D. Kennedy, P. B. Young, S. Kennedy, J. M. Scott, A. M. Molloy, D. G. Weir, J. Price, (1995, 1995Cobalt-vitamin B-12 deficiency and the activity of methylmalonyl CoA mutase and methionine synthase in cattle.International Journal for Vitamin and Nutrition Research, 6542412470300-9831

35. J. E. Kinsella, 1973Preferential labelling of phosphatidylcholine during phospholipid synthesis by bovine mammary tissue.Lipids, 87July 1973), 3934000024-4201

36. A. Kuksis, S. Mookerjea, 1978Choline.Nutrition Reviews,367July 1978), 2012070029-6643

37. D. W. La Count, J. K. Drackley, D. J. Weigel, (1995, 1995Responses of dairy cows during early lactation to ruminal or abomasal administration of L-carnitine. Journal of Dairy Science, 788August 1995), 182418360022-0302

38. D. W. La Count, L. S. Emmert, J. K. Drackley, (1996, 1996Dose response of dairy cows to abomasal administration of four amounts of L-carnitine.Journal of Dairy Science, 794April 1996), 5916020022-0302

39. G. E. Lobley, A. Connell, D. Revell, 1996The importance of transmethylation reactions to methionine metabolism in sheep: effect of supplementation with creatine and choline.British Journal of Nutrition, 751January 1996), 47561475-2662

40. A. L. Lock, M. Rovai, T. A. Gipson, M. J. De Veth, D. E. Bauman, 2008A conjugated linoleic acid supplement containing Trans-10,

Cis-12 Conjugated Linoleic Acid reduces milk fat synthesis in lactating goats. Journal Dairy Science, 91, 9September 2008), 329132990022-0302

41. J. M. Mato, L. Alvarez, F. J. Corrales, M. A. Pajares, (1994, 1994S-adenosylmethionine and the liver. In The Liver: Biology and Pathobiology, I. M. Arias, J. L. Boyer, N. Fausto, W. B. Jakoby, D. A. Schachter, & D. A. Shafritz (Ed.), 461470Raven PressLtd, 0-78170-133-3York, US

42. R. D. Mc Carthy, G. A. Porter, L. C. Griel, Jr , 1968Bovine ketosis and depressed fat test in milk: a problem of methionine metabolism and serum lipoprotein aberration.Journal of Dairy Science516June 1968), 4594620022-0302

43. R. D. Mc Carthy, G. A. Porter, Griel Jr, L. C. , 1968Bovine ketosis and depressed fat test in milk: a problem of methionine metabolism and serum lipoprotein aberration.Journal of Dairy Science513March 1968), 4594620022-0302

44. L. R. Mc Dowell, 1989Choline. In: Vitamins in animal nutrition: comparative aspect to human nutrition, L.R. McDowell (Ed.), New York: Academic Press, Inc., 3473640-12438-372-1

45. A. V. Mc Pherson, B. J. Kitchen, 1983Reviews of the progress of dairy science: the bovine milk fat globule membrane-its formation, composition, structure and behaviour in milk and dairy products. Journal of Dairy Research, 5011071330022-0299

46. J. H. Moore, W. W. Christie, (1981, 1981Lipid metabolism in the mammary gland of ruminant animals. In: Lipid Metabolism in Ruminant Animals.W.W. Christie (Ed.), Pergamon Press, 2272780-08023-789-4UK.

47. A. R. Neill, D. W. Grime, R. M. C. Dawson, 1978Conversion of choline methyl groups through trimethylamine into methane in the rumen.Biochemical Journal, 17035295350264-6021

48. A. R. Neill, D. W. Grime, A. M. Snoswell, A. J. Northrop, D. B. Lindsay, R. M. C. Dawson, 1979The low availability of dietary choline for the nutrition of the sheep.Biochemical Journal, 18035595650264-6021

49. National Research Council2001Vitamins. In: Nutrient Requirements of Dairy cattle, Seventh Revised Edition, Subcommittee on Dairy Cattle Nutrition, Committee on Animal Nutrition, Board on Agriculture and Natural Resource, 162177National Academy Press, 100309069971DC:

50. A. Preynat, H. Lapierre, M. C. Thivierge, M. F. Palin, J. J. Matte, A. Desrochers, C. L. Girard, (2009, 2009Effects of supplements of folic acid, vitamin B12, and rumen-protected methionine on whole body metabolism of methionine and glucose in lactating dairy cows. Journal Dairy Science, 922January 2009), 6776890022-0302

51. M. S. Piepenbrink, T. R. Overton, 2003Liver metabolism and production of cows fed increasing amounts of rumen-protected choline during the periparturient period. Journal Dairy Science, 865May 2003), 172217330000-0022

52. L. Pinotti, A. Baldi, I. Politis, R. Rebucci, L. Sangalli, V. Dell'Orto, 2003Rumen protected choline administration to transition cows: effects on milk production and Vitamin E status. Journal of Veterinary Medicine Series A, 50February 2003) 118210093-1184X

53. L. Pinotti, A. Baldi, V. Dell'Orto, (2002, 2002Comparative mammalian choline metabolism with emphasis on role in ruminants, especially the high yielding dairy cow.Nutrition Research Review, 152December 2002), 315331L0000-0954

54. L. Pinotti, A. Campagnoli, V. Dell'Orto, A. , Baldi(2005Choline: Is there a need in lactating dairy cow? Livestock Production Science, 981-2December 2005), 1491520301-6226

55. L. Pinotti, A. Campagnoli, L. Sangalli, R. Rebucci, V. Dell'Orto, A. Baldi, 2004Metabolism of periparturient dairy cows fed rumen-protected choline. Journal Animal and Feed Science,13Suppl. 1 (August 2004), 5511230-1388

56. L. Pinotti, C. Polidori, A. Campagnoli, V. Dell'Orto, A. Baldi, 2010A meta-analysis of the effects of rumen protected choline supplementation on milk production in dairy cows. In: Energy and protein metabolism and nutrition, G.M. Crovetto (Ed.), 321322Wageningen Academic Publishers (EAAP publication N°127) 978-9-08686-153-8Wageningen, The Netherlands.

57. D. L. Pullen, J. S. Liesman, R. S. Emery, (1990, 1990A species comparison of liver slice synthesis and secretion of triacylglycerol from non-esterified fatty acids in media.Journal of Animal Science, 685May 1990), 139513990000-0021

58. B. S. Robinson, A. M. Snoswell, W. B. Runciman, R. N. Upton, (1984, 1984Uptake and output of various forms of choline by organs of conscious chronically catherized sheep.Biochemical Journal, 21723994080264-6021

59. N. Ruiz, R. D. Miles, R. H. Harms, (1983, 1983Choline, methionine and sulphate interrelationships in poultry nutrition- A review. World's Poultry Science Journal, 391851980043-9339

60. T. Rukkwamsuk, T. Wensing, M. J. H. Geelen, (1999, 1999Effect of fatty liver on hepatic gluconeogenesis in periparturient dairy cows. Journal of Dairy Science, 823March 1999), 5005050022-0302

61. J. Sales, P. Homolkaa, V. Koukolováa, 2010Effect of dietary rumen-protected choline on milk production of dairy cows: A meta-analysis.Journal of Dairy Science, 938August 2010), 374637540022-0302

62. G. Savoini, A. Agazzi, G. Invernizzi, D. Cattaneo, L. Pinotti, A. Baldi, 2010Polyunsaturated fatty acids and choline in dairy goats nutrition: production and health benefitsSmall Ruminants Research, 881-2February 2010), 135 EOF144 EOF0921-4488

63. W. A. Scheer, M. C. Lucy, M. S. Kerley, J. N. Spain, 2002Effects of feeding soybeans and rumen protected choline during late gestation and early lactation on performance of dairy cows. Journal of Dairy Science, 85Suppl.1, 2760022-0302

64. J. M. Scott, 1999Folate and vitamin B12. Proceedings of the Nutrition Society, 582June 1999) 4414480029-6651

65. B. K. Sharma, R. A. Erdman, (1988a, 1988aAbomasal infusion of choline and methionine with or without 2-amino-2-methyl-1-propanol for lactating dairy cows.Journal of Dairy Science, 719September 1998), 240624110022-0302

66. B. K. Sharma, R. A. Erdman, (1988b, 1988bEffects of high amounts of dietary choline supplementation on duodenal choline flow and

production responses of dairy cows.Journal of Dairy Science, 7110October 1988), 267026760022-0302

67. B. K. Sharma, R. A. Erdman, 1989aIn vitro degradation of choline from selected feedstuffs and choline supplements.Journal of Dairy Science7210October 1989), 277127760022-0302

68. B. K. Sharma, R. A. Erdman, 1989bEffects of dietary and abomasally infused choline on milk production responses of lactating dairy cows.Journal of Nutrition, 1192February 1989), 2482540022-3166

69. A. M. Snoswell, G. P. Xue, (1987, 1987Methyl group metabolism in sheep.Comparative Biochemistry and Physiology Part B: Comparative Biochemistry, 8823833941096-4959

70. M. H. Stipanuk, 1986Metabolism of sulfur-containing amino acids.Annual Reviews Nutrition, 61792090199-9885

71. P. M. Ueland, 2011Choline and betaine in health and disease.Journal of Inherited Metabolic Disease, 341February 2011), 3150141-8955

72. A. M. Vuuren, R. M. A. van Goselink, R. Zom, M. J. Veth, J. de Baal, van Goselink, R. M. A., Zom, R., Veth, M. J. de, Baal, J. van (2010Effect of rumen-protected choline on performance and hepatic triacylglycerol concentrations in early-lactating dairy cattle.In: Energy and protein metabolism and nutrition, G.M. Crovetto (Ed.), 493494Wageningen Academic Publishers (EAAP publication N°127) 978-9-08686-153-8Wageningen, The Netherlands.

73. C. C. Whitehead, J. I. Portsmouth, 1989Vitamin requirements and allowances for poultry. In: Recent Advances in Animal Nutrition, Haresign, W. & Cole D. A. J., (Ed.), 3586Butterworths, 0407London, UK.

74. G. Xu, J. Ye, J. Liu, Y. Yu, (2006, 2006Effect of rumen-protected choline addition on milk performance and blood metabolic parameters in transition dairy cows.Asian-Australasian Journal of Animal Sciences, 1933903950000-1011

75. G. P. Xue, A. M. Snoswell, (1985, 1985Regulation of methyl group metabolism in lactating ewes.Biochemistry International, 113September 1989), 3813850158-5231

76. L. C. Zahra, T. F. Duffield, K. E. Leslie, T. R. Overton, D. Putnam, S. J. Le Blanc, (2006, 2006Effects of Rumen-Protected Choline and Monensin on Milk Production and Metabolism of Periparturient Dairy Cows.Journal Dairy Science, 8912December 2006), 480848181811-9743

77. S. H. Zeisel, 1988Vitamin-like" molecules. In: Modern Nutrition and Health and Disease, Shils M. & Young V., (Ed.), 4404521007 81741335Lea & Febiger.

78. S. H. Zeisel, 1992Choline: an important nutrient in brain development, liver function and carcinogenesis. Journal American College of Nutrition, 11 5 473481 ,0731-5724

79. S. Zeisel, H., Costa. K. A. Da, 2009Choline: An Essential Nutrient for Public Health. Nutrition Reviews, 6711November 2009), 6156230029-6643

80. S. H. Zeisel, Costa. K. A. Da, P. D. Franklin, E. A. Alexander, J. T. Lamont, N. F. Sheard, A. Beiser, an. Choline, nutrient. essential, humans. for, F. A. S. E. B. The, Vol. Journal, 209320980892-6638

81. S. H. Zeisel, M. Mar, Z. Zhou, Costa. K. A. Da, 1995Pregnancy and lactation are associated with diminished concentrations of choline and its metabolites in rat liverJournal of Nutrition, 12512304930540022-3166

CHAPTER 10

Use of Yeast Probiotics in Ruminants: Effects and Mechanisms of Action on Rumen pH, Fibre Degradation, and Microbiota According to the Diet

Frédérique Chaucheyras-Durand[1, 2], Eric Chevaux[1], Cécile Martin[3] and Evelyne Forano[2]

[1] Lallemand Animal Nutrition, Blagnac, France

[2] INRA UR 454 Microbiologie, Saint-Genès Champanelle, France

[3] INRA UMR 1213 Herbivores, Saint-Genès Champanelle, France

1. INTRODUCTION

The valorization of fibrous feed sources by ruminants is possible thanks to their unique digestive system involving an intensive preliminary ruminal fermentation step prior to a more classical enzymatic phase. The reticulo-rumen hosts a highly specialized anaerobic microbial community responsible for fibre breakdown, which is influenced by biochemical and microbial characteristics of the rumen environment. In particular, the role of the different microbial species involved in pH regulation and the influence of feed management are presented in section 2. Indeed, intensive farming pratices may disturb the microbial balance due to an excessive high fermentable carbohydrate supply required to sustain high animal performance, and it can turn into metabolic disorders that are likely to impact animal health as reviewed in section 3. This is one area where yeasts probiotics can help the ruminant and the feed nutritionist optimizing the cows nutrition owing to an increasingly well understood proper mode of action. Section 4 reports the positive effects these feed additives, under the form of active dry yeast, have on rumen fermentation, feeding behaviour and feed efficiency, as well as tips to properly assess these effects.

Once the optimal rumen conditions are set up (section 6), fibre will be efficiently digested. It becomes then interesting to dive into the world of the fibrolytic microbiota in section 5 to truly percieve the unicity of the fibre rumen degradation process, bearing in mind that the nature of fibre will impact its digestibility and subsequent animal production response. In addition to its role on rumen pH stabilization that directly affects the fibrolytic microflora, yeast probiotics represent a valuable tool to optimize cow nutrition as detailed in section 7. However, section 8 will emphasize the yeast strain effect and the need of a viable feed additive to be able to offer a comprehensive solution to ruminants' diet formulation. Finally, besides the clearly established benefits on rumen management and fibre degradation, live yeast as probiotics are also currently being assessed in other promising fields of applications (section 9).

2. RUMEN PH: A KEY PARAMETER LINKED TO RUMEN FUNCTION

Due to intense microbial activity, fermentation of feedstuffs in the reticulo-rumen produces a wide range of organic acids. Some of these acids can accumulate and reduce ruminal pH if rumen buffering systems are unable to counteract their impact. Low rumen pH for prolonged periods can negatively affect feed intake, microbial metabolism, and nutrient degradation, and leads to acidosis, inflammation, laminitis, diarrhea and milk fat depression. High yielding dairy cows and fattening beef cattle fed diets rich in readily fermentable starch or sugars at high feed intake levels are particularly susceptible to acidosis, and goats, sheep and other ruminants are also prone to the disease. It is now recognized that subacute ruminal acidosis (SARA) affects from 10% to 40% of dairy cattle in a herd, resulting in large financial losses and major concern for animal welfare reasons. Therefore, rumen pH regulation is a key determinant in the maintenance of an optimal rumen function.

2.1. How To Measure Rumen Ph Accurately

Common field techniques for pH measurement have been relied on collection of samples by rumenocentesis or oral stomach tubing [1,2]. Rumenocentesis has proven to be a more reliable technique for the determination of ruminal pH than oral stomach tubing because saliva contamination is often associated with the stomach tubing technique [3,4]. If rumenocentesis may be done with minimal disturbance [5], frequent sampling raises ethical issues and is not without risk for the animal health. Enemark et al. [2] conducted a study to evaluate the potential of biochemical markers in blood, feces, and urine to predict ruminal pH. They concluded that no peripheral markers could properly predict ruminal pH. A permanent surgical modification, such as rumen cannulation, and the use of an external data logger connected to a pH probe immerged into the rumen [3,6] have been successful in well controlled research studies to monitor rumen pH kinetics, which allow to better characterize microbial fermentations and predict acidosis situations. Recently, telemetric boluses able to measure and record rumen pH in cattle continuously have been developed by different companies. When interrogated by wireless, the bolus transmits the recorded data to an operator standing beside the cow with a receiving station. These rumen pH boluses methods offer a simple, accurate and long lasting measurement of pH in intact cattle [7]. They have been successfully applied in controlled animal studies and offer the opportunity to link pH kinetics to measurements in field situations, but clarifications are still needed about the location of the probes (reticulum, rumen) and thereby the representativeness of the measure, their calibration, long-term measure accuracy, and life time. Moreover, the cost of these systems are still high and the current proposed boluses are not yet applicable to non cannulated small ruminants.

2.1.1. Microbial Mechanisms Which Lead To Ph Modulation And Acidosis

Rumen microbial populations hydrolyze and ferment dietary compounds into volatile fatty acids (VFAs), whose amounts drive pH evolution. Moreover, lactic acid is a common product of carbohydrate fermentation, produced by bacterial species such as *Streptococcus bovis*, *Selenomonas ruminantium*, *Mitsuokella multiacidus*, *Lachnospira multipara* or *Lactobacillus sp. S. bovis* is considered as a major contributor in lactate production from high fermentable diets. Indeed, it is able of very rapid growth, is acid-resistant and produces extracellular and intracellular amylases which hydrolyze raw starch and soluble starch, respectively [8]. Moreover, it has been shown that *S. bovis*produces mainly L-lactate under moderately acidic pH but shifts its metabolism towards D-Lactate production when the pH decreases [9], this latter isoform being more toxic as it is less efficiently re-utilized by the microbiota and the animal tissues. *Megasphaera elsdenii* is considered as the predominant lactate-utilizing bacterial species in the rumen and can be found in large numbers in the rumen of cereal grain-fed cattle [10]. *Selenomonas ruminantium* subsp *lactylitica* is another important lactate-utilizing species. Contrary to *S. ruminantium*, *M. elsdenii* is not submitted to catabolite repression by soluble sugars [11] and ferments lactate to propionate via the acrylate pathway [10]. It exhibits also a lactate racemase activity which is involved in the conversion of D- into L-lactate, which is more easily metabolized. Nevertheless, with high amounts of readily fermentable carbohydrates, or during adaptation from forage to concentrate diets, acid overload of the rumen is possible and may lead to a strong decline in rumen pH, which may trigger acidosis in cattle [1]. Indeed, as rumen pH falls, lactate producers may outnumber lactate utilizers, leading to an accumulation of this metabolite in the rumen. Due to the low pK_a (3.7) of lactic acid compared to the pK_a of the major VFAs (4.8-4.9 for acetate, propionate and butyrate), even low amounts of lactic acid may play a major role on the onset of acidosis. If rumen pH continues to fall, *Lactobacilli* may replace*S. bovis*, initiating a spiraling effect with excessive D-lactate accumulation [9].

Thanks to their capacity to engulf and slowly ferment starch granules into VFAs (particularly butyrate), rumen protozoa can compete with lactate-producing amylolytic bacteria and lactic acid can be actively taken up by entodiniomorphid ciliates [12]. Overall these processes have a beneficial effect on pH stabilization and may participate to limit the severity of acidosis.

2.1.2. Effect Of The Diet On Rumen Microbiota, Microbial Fermentations And Ph Evolution

The effect of a diet shift (from high forage to high concentrate) on the composition of the rumen microbiota has been extensively studied, in particular since the last 10 years because of the development of culture-independent techniques quantifying microbial abundance and assessing population dynamics. Tajima et al. [13] have shown that a diet shift from high forage to high grain in steers induced profound changes in bacterial abundances, an increase in *S. bovis* and *Prevotella ruminicola* 16S *rrs* gene copy numbers and a decline in fibrolytic *Fibrobacter succinogenes* population densities being measured. Using quantitative PCR, Mosoni et al. [14] measured significant decrease in*F. succinogenes, Ruminococcus albus* and *R. flavefaciens* 16S *rrs* gene copy numbers/g of rumen contents in sheep fed 50% concentrate 50% hay, compared with a 100% hay diet. In lambs, the effect of hay *vs* concentrate diet fed at weaning was studied on abundance of different species of the rumen microbiota [15]. Whereas abundance of total bacteria, measured by qPCR, was significantly higher with concentrate diet than with hay diet, the relative abundance of the fibrolytic species *F. succinogenes* and that of methanogens were significantly lowered in the presence of concentrate. *R. flavefaciens* abundance was 2.5-fold lower with the concentrate diet. The rumen microbiome of dairy cows in which subacute ruminal acidosis (SARA) had been induced with either grain or alfalfa pellets has also been analysed [16]. T-RFLP analysis indicated that the most predominant shift during SARA was a decline in Gram-negative *Bacteroidetes* organisms. However, the proportion of *Bacteroidetes*was greater in alfalfa pellet-induced SARA than in mild or severe grain-induced SARA. This shift was also evident from real-time PCR data for *P. albensis, P. brevis,* and *P. ruminicola*, belonging to the phylum *Bacteroidetes*. The real-time PCR analysis also indicated that in severe grain-induced SARA,*S. bovis* and *Escherichia coli* were dominant, *M. elsdenii* dominated in mild grain-induced SARA, and*P. albensis* was abundant in alfalfa pellet-induced SARA. Comparing 16S rRNA gene libraries of hay*vs* high grain-fed beef cattle, Fernando et al. [17] reported significantly higher numbers of bacteria of the phylum *Fibrobacteres* in libraries of hay-fed cattle whereas the libraries of grain-fed animals contained a significantly higher numbers of bacteria of the phylum *Bacteroidetes*. Real-time PCR analysis revealed increases in *M. elsdenii, S. bovis, S. ruminantium*, and *P. bryantii* populations during adaptation to the high-grain diet, whereas the fibre-degrading *Butyrivibrio fibrisolvens* and *F. succinogenes* populations gradually decreased as the animals were adapted to the high-concentrate diet. All together, these studies indicate a negative effect of low pH on cellulolytic bacteria. Indeed, they cannot grow with a low intracellular pH, and an increase in pH gradient leads to an entry of undissociated VFAs in the cells and an accumulation of dissociated anions in the intracellular compartment induces severe toxicity for the bacteria [18].

An increase in the percentage of rapidly degradable starch in the diet generally favors the development of protozoa as soon as the rumen pH is not below 5.5 [19]. The genus *Entodinium* can then represent up to 95% of the total ciliate community. When rumen pH is below 5.5, ciliate protozoa populations are decreased and defaunation can even be observed transiently [20].

A low rumen pH has also a strong impact on rumen fungi. Indeed, the production of zoospores by*Caecomyces* have been sharply decreased *in vitro* at pH 5.5. Zoospore numbers were below 10^3/ml or even not detected in animals fed diets inducing low rumen pH [21]. Moreover, the presence of large amounts of soluble sugars, as with high concentrate diets, may induce saturation of the spore adhesion sites and reduce fungal colonization [22].

Changes in the structure of the rumen microbiota are generally accompanied with modifications of fibrolytic activities. Indeed, compared with a forage diet, cereal grain supplementation induces a decrease in specific and total polysaccharidase activities of the solid-associated microorganisms, whereas the response of glycosidase activities is more variable [19]. A relationship between the decrease in polysaccharidase activities (xylanase, avicelase) of these microorganisms and the decrease in ruminal fibre degradation rate has been found by several authors [23-25]. Low pH seems to be more detrimental to growth and survival of cellulolytic microorganisms than to microbial cellulases whose activities are generally optimal at moderately acidic pH (between 5.5 and 6.0) [18]. However, Martin et al. [23] have quantified cellulase and hemicellulase activities and 16S rRNA of cellulolytic bacteria in rumen contents of cows fed a 40% barley diet, and found that cereal supplementation modified the activity but not the abundance of cellulolytic bacterial community.

Sauvant et al. [26] summarized studies conducted on 14 feedstuffs and showed that a strong relationship exists between rumen pH values induced *in vitro* by each feedstuff's fermentation and its percentage of Dry Matter (DM) degradation (Figure 1), indicating that the nature of the feedstuff impacts on its acidogenic potential. Indeed, rapidly degradable starch (as in barley or wheat) will more strongly impact rumen pH than slowly degradable starch (as in corn or sorghum).

For example, when comparing wheat and corn supplementation in beef steers, mean pH was less and time below pH 6.2 was greater for the wheat based diet than for the corn based diet, which was linked to a higher lactate and VFA concentration [27]. The effect of 3 dietary challenges differing by the nature and degradation rate of their carbohydrates (wheat, corn or beet pulp) was investigated on rumen pH kinetics and fermentation profile in sheep [28]. Mean ruminal pH was significantly less for wheat than for corn and beet pulp at 4.85, 5.61, and

6.09, respectively. This was correlated with a change in the fermentation profile: ruminal lactic acidosis was induced by wheat, whereas butyric and propionic SARA were respectively provoked by corn and beet pulp after the 3 day challenge.

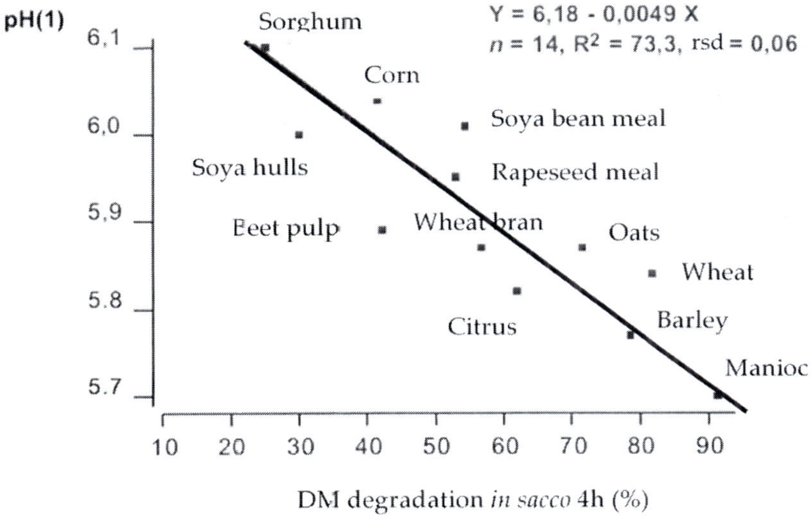

Figure 1. Relationships between acidogenic potential of feedstuffs and their degradation *in sacco*. From [26].

The particle length of forages can greatly affect rumen pH. Indeed, physically effective Neutral Detergent Fibre (peNDF) represents the physical characteristics of fibre by accounting for particle length and NDF content, which promote chewing and the flow of salivary buffers to the rumen [29]. Yang and Beauchemin [30] compared rumen pH response when short (7.9 mm) or long (19 mm) cut alfalfa silage was included in either high or low concentrate diets. They showed that increasing peNDF intake reduced ruminal acidosis; mean ruminal pH and the duration that pH remained below 5.8 were highly correlated to intake of long particles.

3. IMPACT OF A LOWERED RUMEN PH ON RUMEN EFFICIENCY AND ANIMAL PRODUCTIVITY

3.1. Consequences Of A Low Rumen Ph: Acidosis, Inflammation, Rumen Wall Integrity And Impact On Animal Health

Acute acidosis occurs after the consumption of an excessive quantity of readily fermentable carbohydrates that rapidly alters ruminal function and can have irreversible metabolic consequences. Ruminal perturbations include an increased concentration of lactate (up to 100mM) and a decrease in VFA concentration after 8 to 24h, this latter being the result of poor microbial activity and/or of quicker absorption of the VFA from the rumen to the blood in response to pH fall [31]. Rumen pH values can then drop under 5.0 and trigger metabolic acidosis with an accumulation of D-lactate in the bloodstream. SARA is probably more difficult to characterize because biological parameters in the rumen fluctuate within physiological limits and are difficult to maintain [31]. This unstable state may reflect the oscillatory behavior of the ruminal microbial population in response to diet-based fermentative jolts. According to Kleen and Canizzo [32], the exact definition of SARA remains debatable, but it is certain that SARA is present in a large number of dairy herds. SARA is characterized by a drop of ruminal pH to non-physiological levels; pH values of 5.5 and 5.8 and the duration per day below these threshold values are used to define individuals or groups experiencing SARA or being at risk for SARA. SARA is frequent in high producing cattle and has wide-reaching economic consequences, as it has been estimated to cost $1.12 /d per cow in USA [33]. In Europe, field studies data indicate that SARA prevalence would range between 10 and 30% in dairy herds [32]. In these studies, the pH thresholds of 5.5 and 5.8 were generally used, rumenocentesis being the reference method for collecting rumen fluid.

The microbial dysbiosis occurring in the rumen during acidosis may trigger the release of potential harmful molecules which may impact the animal health. Indeed, due to an increase of the death and lysis of Gram-negative bacteria under low pH, free lipopolysaccharide (LPS) concentration is increased in the rumen fluid and translocation of this endotoxin can occur across the rumen mucosa [34]. Endotoxin release can trigger an inflammatory response, with an increase in acute phase protein concentrations in peripheral blood [34-37]. Endotoxin is suggested to be involved in metabolic disorders such as laminitis, abomasal displacement, fatty liver or sudden death syndrome [38].

Moreover, the low pH of rumen digesta may have a negative impact on rumen wall integrity. Repeated aggressions by fermentation acids may cause papillar atrophy, diffuse areas of acute or chronic lesions, scars resulting from severe local rumenitis, perforations and mucormycosis which are at the origin of pain, discomfort, as well as erratic feed intake and alteration of rumen function [39].

Low ruminal pH is often associated with increased occurrence of bloat, which is characterized by an accumulation of gas in the rumen and reticulum. Indeed, frothy bloat is caused by entrapment of gas produced from fermentation of readily digestible feeds (high digestible legumes or cereals). Bloat can impair both digestive and respiratory function, and can occur both in cattle raised on pasture or in confinement [40]. Abscessed livers are generally considered to be associated with both acute and subacute ruminal acidosis. Ulcerative lesions, hairs, and other foreign objects that become embedded in the ruminal epithelium can provide routes of entry into the portal blood for microbes that cause liver abscesses [41]. *Fusobacterium necrophorum* (and/or *F. funduliforme*), a commensal rumen Gram-negative species, has been identified as a causative agent of liver abscess; as it is able to use lactate as its major substrate, and its population increases in the rumen of cattle fed high-grain diets [42]. Diarrhea has been very frequently associated with ruminal acidosis and microbial dysbiosis [1]. Changes in fecal consistency, color, brightness, and odour are generally observed; presence of undigested whole grains and large size particles is also a sign of rumen dysfunction [43].This phenomenon may be linked to excessive hindgut fermentation because too much readily fermentable carbohydrates reach the post-ruminal compartments [36] but also the increase in osmolarity of the digesta would lead to soften the fecal mass [43].

Under low rumen pH conditions, erratic feed intake is generally observed but a decrease in intake, mostly on acidogenic feed, has also been reported [44]. In fattening bulls fed high concentrate diets, it has been observed that animals change their feeding behavior to counteract acidosis by spreading their meals over the day [45]. A 10-30% increase in water intake was observed in sheep submitted to acidotic challenges [46]. Water intake could represent a means to dilute acidity but also to reduce rumen fluid viscosity. An increase in salt licking has been also measured in the same study and in goats fed with high concentrate diets [47]. Licking would favor salivary bicarbonate production. Animals under acidosis would also be able to modify their dietary choice to optimize their digestive comfort. Acidosis and low rumen pH conditions may also have consequences on social behavior. For example, sheep undergoing successive acidotic challenges were more active and more aggressive towards each other, spent more time standing, adopted alarm postures more often, and reacted more slowly to hot stimulus during the acidosis bouts [46]. These discomfort signs would not be only linked to rumen pH evolution but to the set up of an inflammatory status in the rumen triggered by changes in microbiota balance.

3.2. Effect Of Rumen Ph On Milk Yield And Quality

From a dietary standpoint, rumen pH is a function of the dry matter intake (DMI) where it becomes below 6 when DMI exceeds 3.8% body weight, i.e. high producing animals with elevated nutritional requirements are more at risk [26]. The quality of the ingested feed directly matters too where pH turns out below 6 when the rumen digested starch accounts for more than 40% of the diet DM [26].

Cows fed high-concentrate diet (nadir 75:25 concentrate:forage ratio) will have a lower ruminal pH, acetate, and butyrate concentrations, whereas propionate concentration will go up. When the rumen acidity is alleviated with a buffer, total VFA production increases, and so does milk production and milk fat content, especially for high concentrate fed cows. Milk fatty acid profile gives also a good insight of what happened in the rumen and more trans 10-11 C18:1 is well correlated to a depressed milk fat due to its inhibitory effect on de novo fatty acids synthesis in the mammary gland [48]. In addition, the stage of lactation may modulate the animal sensitivity to high-concentrate diet with a better resistance to less optimal rumen fermentation conditions for late lactation cows [49]. However, not only the forage:concentrate ratio matters on rumen pH but the nature or technological process of the grains [50] and the frequency of distribution of the concentrate [51] also do.

High fibre diets will not sustain an elevated production of propionate that will negatively impact the milk lactose synthesis and overall milk yield. The cow will thus mobilize her body fat reserves (ketone bodies metabolized in the liver from butyrate) to compensate for this lack of energy.

4. BENEFITS OF USING YEAST PROBIOTICS TO CONTROL PH STABILITY

4.1. Targets

pH evolution is the result of impaired microbial balance and animal compensation mechanisms. Strategies aiming to induce beneficial effects on the balance of the rumen microbiota and thereby stabilize rumen pH can represent interesting means to reduce the risk of acidosis. This may be achieved by targeting microbial populations involved in massive release of fermentation acids, and/or those implicated in lactic acid removal.

4.2. How Best Measuring A Probiotic Effect On Animal Performance?

Two types of experimental design are basically available to the scientist: contemporaneous or crossover. Parallel designs (i) can be completely randomized design with only one explanatory variable or (ii) randomized complete block design in presence of 2 factors where the experimenter divides animals into subgroups called blocks (eg. sex, origin, size…) such that the variability within blocks is less than the variability between blocks. In crossover design, each experimental unit receives two or more treatments through time, and as the comparison of treatments is made within subjects, each subject acts as its own control which increases statistical power to detect a direct treatment effect [52] and makes it more efficient than the randomized complete block design. However, there are limitations important to bear in mind amongst with a carryover effect is likely to occur between periods, the latter being able to vary between treatments.

The particular nature of probiotics as live microorganisms impacting the rumen flora balance and fermentations make their comparative assessment critical when using experimental design encompassing a carry-over effect. The inclusion of a washout period between successive treatments is a good way of minimizing the remanent treatment effect over time, but there is good evidence suggesting that the 15-28 days usually applied are not long enough.

Indeed, in a complete rumen content transfer study between two cows, Weimer et al. [53] showed that it could last up to 65d for the bacterial community composition to reach back its original profile. A measurement of methanogens population dynamics over time [54] indicated that 4 weeks were not enough to adapt from the dietary shift of grazing to concentrate. These recent microbial studies support questioning about the relevance of crossover type of designs in assessing probiotics effect on rumen parameters [55]. However, it would not be fair omitting to report studies where such a design allowed displaying significant probiotic effects, but the inconsistence or absence of response with a latin-square design may also be due to the tested probiotic strains themselves or to the too short adaptation period.

4.3. Experimental Proofs

Stabilization of ruminal pH in the presence of yeast probiotics has been reported by several authors [56-59]. In a meta-analysis, Sauvant et al. [26] concluded that yeast supplementation increased ($P<0.05$) rumen pH *in vitro*, but did not find any significant *in vivo* effect neither on pH, nor on VFAs or lactate. However, the authors admitted that the studies selected for the meta-analysis had used different strains of *S. cerevisiae*, or yeast culture which is defined to be mainly composed by dead cells and fermentation products. More than an increase in mean rumen pH, reductions in duration within a day under a certain pH threshold, as well as in area under the pH curve have been measured in the presence of live yeast probiotics [56, 59]. A recent study conducted in a commercial dairy herd [60] compared sodium bicarbonate and live yeast supplementation in 2 pens of 60 cows on milk production and feed efficiency and rumen pH was monitored every 5 min during 5 weeks in 4 cows equipped with a pH probe. Sodium bicarbonate is very often used as an efficient buffer to overcome pH fall in dairy cows. Mean pH remained consistently higher for the live yeast supplemented cows when compared to the control group cows (6.22 vs 6.03). In addition, live yeast supplemented cows spent less time below a pH threshold of 5.6.

4.4. Modes Of Action On Rumen Microbiota and Lactate Accumulation

Effects of live yeasts have been studied on lactate-metabolizing bacteria. *In vitro*, one strain of *S. cerevisiae* was able to outcompete *S. bovis* for the utilization of sugars; due to a higher affinity of the yeast cells for sugars, the reduction in quantity of fermentable substrate available for the bacterial growth consequently limited the amount of lactate produced [61]. Dead cells had no effect on lactate production. Moreover, stimulation of growth and metabolism of lactate-utilizing bacteria, such as *M. elsdenii* or *S. ruminantium,* was observed *in vitro* in the presence of different live yeasts [61-64] through a supply of different growth factors such as amino acids, peptides, vitamins, and organic acids, essential for the lactate-fermenting bacteria. The impact of yeast probiotics on ruminal lactate concentration has been confirmed in *in vivo* studies. In sheep receiving a live yeast product during their adaptation to a high-concentrate diet, ruminal lactate concentration was significantly lower compared to control animals. Consequently, rumen pH was maintained at values compatible with an efficient rumen function, as shown by higher fibrolytic activities in the rumen of the supplemented animals [24,65]. In dairy cows, reductions in ruminal lactate concentrations have also been observed with yeast probiotics [66-67].

According to the composition of the diet, the fermentation pattern can be shifted to butyric orientated acidosis [28]. Brossard et al. [6,12] reported the pH stabilising effect of one strain of *S. cerevisiae* in sheep fed a high-wheat diet under a butyric latent acidosis. Authors suggested that this strain could act by stimulating ciliate Entodiniomorphid protozoa, which are known to engulf starch granules very rapidly and thus compete effectively with amylolytic bacteria for their substrate [68]. In addition, starch is fermented by protozoa at a slower rate than by amylolytic bacteria and the main end-products of fermentation are VFAs rather than lactate, which may explain why these ciliates had a stabilizing effect in the rumen by delaying fermentation.

When ruminants encounter successive acidotic bouts, it is not well known whether live yeast supplementation could alter rumen microbiota and fermentations. Indeed, the severity of acidosis may change with repeated challenges, partly because of modifications in feeding behavior [69], and because of possible shifts in rumen microbial communities leading to selection of the most acid resistant species. Studies in sheep submitted to acidotic challenges showed that cellulolytic bacterial culturable population was greatly decreased after a first acidotic challenge but that after 3 challenges, the level of population came back to normal [70]. However, it is probable that this population, enumerated in a filter paper-based medium, had encountered profound changes in its structure and/or diversity. In this study, with repeated challenges, a positive evolution of rumen pH parameters were observed in live yeast supplemented animals which was accompanied with decreased numbers of lactate producing bacteria and a beneficial effect on bacterial diversity which was maintained at a higher level [71].

Provided an adequate balance between soluble nitrogen and carbohydrate supply, it is likely that live yeast probiotics can enhance microbial growth; indeed, more digested carbohydrates would be incorporated into microbial mass thanks to an optimized fermentation coupling and not "wasted" under the form of VFAs, thereby the risk of acidosis would be reduced [72].

4.5. Beneficial Consequences Of Yeast Probiotics On Rumen Fermentations, Feeding Behavior, Feed Efficiency, And Animal Production

Bach et al. [56] reported that the supplementation of live yeast increased average rumen pH and average maximum pH by 0.5 units, and average minimum pH by 0.3 units in loose-housed lactating cows (Figure 2). In this study, a significant change was observed in the eating behavior of the animals. Cows supplemented with live yeast had a shorter inter-meal interval (3.32h) than unsupplemented cows (4.32h). This change in feeding behavior could help in rumen pH recovery, or the beneficial effect of live yeast on pH stabilization could induce a change in eating behavior.

A meta-analysis conducted on all types of yeast (including live yeast and yeast culture) and all types of dairy ruminants (cows, goats, ewes) [58] concluded that the addition of yeast improved milk yield by 1.2 g/kg body weight. In their multi-analysis reporting data collected from 14 dairy cow trials fed the same live yeast strain, De Ondarza et al. [73] found that live yeast improved ($P < 0.0001$) milk yield by 1.15 kg/day. The effect was slightly greater for cows in early lactation (<100 Days In Milk, DIM) than for cows >100 DIM, suggesting that animal performance is improved when the acidosis risk is high, notably at critical periods of the lactation cycle.

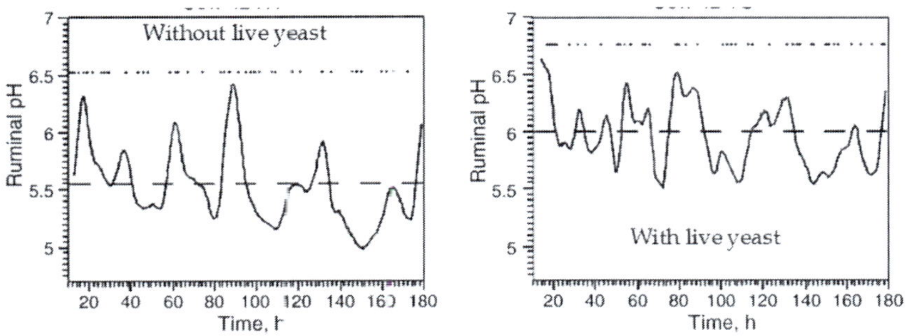

Figure 2. Ruminal pH pattern (solid line) during the 8 days of sampling as affected by live yeast supplementation.The dashed line depicts average ruminal pH. The dots indicate the beginning of a meal. From [56], example shown with one cow.

The effect of yeast probiotics on DM intake shows either no effect [73] or a significant increase in DMI [58]. Live yeast supplementation seems to have an effect on intake pattern rather than on intake *per se*[56]. As a result, feed efficiency is generally improved in the presence of live yeast [73,74]. Milk composition is generally not or only slightly affected by yeast supplementation. Milk fat and protein percentages have been found to be slightly but significantly lower in the presence of live yeast [73], but due to the increase in milk yield, yields of milk fat and true protein were higher than in control cows.

5. FIBRE DIGESTION IN THE RUMEN: A KEY PROCESS IN RUMINANT NUTRITION

By symbiosis with specific micro-organisms, ruminants possess a unique ability to use plant cell wall components as energy and nutrient sources and thereby convert plant biomass into milk, meat, wool and hides. A large proportion of energy intake of ruminant comes in the form of structural complex polysaccharides (cellulose, hemicelluloses, pectins), which are mainly present in the plant cell walls. Indeed, the rumen harbors an abundant and diversified community of bacteria, fungi and protozoa able to thoroughly hydrolyze plant cell wall polysaccharides. Effective degradation is the result of microbial adhesion to plant tissue and production of active enzymatic machinery well adapted to plant cell wall breakdown.

5.1. Relation Between Fibre Digestion And Intake And Productivity

Digestion of fibre is the result of the competition between rates of passage and degradation and the ruminal passage rate (%/h) depends on fibre particles size and digestibility [75]. Reducing particle size will increase DMI but the effect on total digested fibre is also related to the quality of the roughage and its nature: legumes NDF is quicker digested than perennial grass NDF despite a higher lignification, but less resistance to breakdown [76]. Particle size also affects the reticulo-omasal passage kinetics along with the intrinsic fragility of the fibre, its density and shape. The importance of particle size on forage rumen degradation has been recently highlighted [77] as the adjustment parameter to increase the available surface area for attachment of ruminal fibrolytic bacteria and protozoa without negatively affecting cellulolytic activity and other fermentation processes in the rumen.

Fibre occupies space and limit intake by filling the rumen as they are hollow and therefore fill a bigger volume than their mass indicates. In addition, a fraction of the dietary fibre will remain undigested or slowly degraded and will accelerate the rumen filling [78] reducing thus the entrance of other important ingredients to meet the animal nutritional requirements. Knowing that feed intake is the main predictive variable of milk yield [79], the increase of dietary forage will lead to a milk yield reduction besides isonitrogenous rations [80]. Rinne et al. [81] also concluded to a linear decrease of milk yield when the corn silage NDF content increased due to later harvest.

5.2. How To Measure Fibre Digestion

Different methods can be used to measure fibre digestion in the rumen. This compartment is mostly targeted because in general the proportion of fibre which is digested in the hindgut is small. However, the contribution of the large intestine to plant cell wall digestion may increase with the proportion of cereal in the diet [82].

Degradation of dry matter, and NDF fraction of raw materials or more complex mixture of ingredients can be assessed with various *in vitro* techniques requiring mixed rumen contents [83,84], *in situ* (nylon bags) kinetics [82,85] or rumen evacuation [86] in rumen cannulated animals, or in non cannulated ruminants (total fecal collection). The measurement of particle sizes in the fecal material using the Penn State forage and total mixed ration particle separator can be of interest to estimate fibre digestibility [60].

Fibre degrading functional groups can be enumerated on complex culture media in which a source of polysaccharide is added as sole energy source. Measurement of fibrolytic activities can be performed on pure cultures as well as on rumen contents samples. After extraction of ruminal microbial enzymes, activities are measured against various polysaccharides and the concentration of reducing sugars released after enzyme action is determined [19]. PCR-based techniques using specific primer sets are powerful to quantify absolute or relative abundance of targeted fibrolytic species within a complex sample [14,87,88], or to specifically detect and quantify *in vivo* the expression of cellulase or hemicellulase genes from selected microorganisms [89].

5.3. Microbial Communities Involved In Fibre Degradation In The Rumen

In the rumen, degradation and fermentation of plant cell wall polysaccharides is achieved by bacteria, protozoa and fungi. The different fibrolytic species, or even strains, are specialized to a various extent in the degradation of specific substrates. The overall effective degradation is the result of these different capacities, related to substrate composition and to interactions existing between these communities and also between the fibrolytic and the non-fibrolytic microorganisms within the ecosystem.

In the Bacteria domain, the cellulolytic function is covered by a very limited number of cultivated species. These species are established a few days after birth in the newborn ruminant, although no solid feed penetrates into the rumen [90]. Indeed, from one week of age, the size of the cellulolytic bacterial community is close to that found in adult animals. Cellulolytic bacteria are unable to properly colonize the rumen in absence of a complex and diversified bacterial fermentative community [91,92]. In young lambs kept without contact with their dams or other adults, cellulolytic bacteria were not detected in the rumen during three months after birth, which suggests the essential role of newborn-dam contacts in the transmission of rumen microbiota and rumen maturation [92].

The concentration of fibrolytic bacteria is generally close to 10^9 culturable cells/g of rumen content. Quantitative PCR studies have shown that the main cellulolytic species *Fibrobacter succinogenes*, *Ruminococcus flavefaciens* and *Ruminococcus albus* represent 1-5% of the total bacteria [14, 93] but recent data suggest that these bacteria account for about 50% of the total active cellulolytic bacteria [94]. *F. succinogenes* is very active on crystalline cellulose and hemicelluloses (xylans). However, it is only able to use products of cellulose hydrolysis [94]. *R. albus* and *R. flavefaciens* are active on cellulose, xylans and pectins. Other species are considered as secondary fibrolytic species such as*Butyrivibrio fibrisolvens* and *P. ruminicola*, because they are not able to breakdown the cellulose polymer. However, they possess high carboxymethylcellulose-, xylan- and pectin-degrading activities and probably play an important role in overall fibre digestion [95,96].

The enzymatic equipment of the three main cellulolytic species has been well studied since the last 20 years. In the database CAZy (Carbohydrate Active enZymes, http://www.cazy.org ; [97]) are referred protein sequences involved in carbohydrate binding and hydrolysis. The recent whole genome sequencing programs confirm that a huge number of genes is involved in fibre breakdown in each bacterial cell, demonstrating great functional redondancy, which is essential for the good functionning of the ecosystem. Genome sequences of strains belonging *to F. succinogenes, R. flavefaciens, R. albus, P. ruminicola,* and *P. bryantii* are now available. From these genome sequences, 183 putative CAZymes have been found for *F. succinogenes*, and more than 140 for *R. flavefaciens* and *R. albus*[98].

Efficacy of fibrolytic bacteria to degrade plant cell wall components are explained by their adhesion capacities and the production of a well adapted enzymatic equipment. Bacteria use different strategies to colonize plant material: for example, *Ruminococci* exhibit several structures on their cell surface, such as type IV pili and components of glycocalyx. Moreover, they produce an elaborate cellulosomal enzyme complex that is anchored to the bacterial cell wall [99,100]. In *F. succinogenes*, attachment to the substrate is mediated by fibro-slime proteins and type IV pilin structures attached to the outer membrane; 13 cellulose binding proteins anchored on the outer membrane seem to be important in effective adhesion to crystalline cellulose [101].

Ciliate protozoa also participate to fibre degradation. Characterization of their ability to directly process plant material have been addressed by diverse strategies, such as direct, biochemical detection of specific fibrolytic enzymes in extracts derived from individual protozoan species [102], or by molecular cloning studies to directly identify protozoal genes encoding enzymes capable of degrading cellulose or hemicellulose [103]. Among protozoa, only Entodiniomorphs (*Polyplastron, Eudiplodinium,Epidinium)* are considered as cellulolytic. Their abundance is between 10^4 and 10^6 cells/g of rumen content. Ciliates are able to engulf whole plant particles, and digest plant polymers in digestive vacuoles. They synthesize a well adapted enzymatic equipment composed of cellulases and hemicellulases [104,105]. Up to now, about a dozen of fibrolytic genes have been identified in the various protozoa species. An activity-based metagenomic study of a bovine ruminal protozoan-enriched cDNA expression library identified four novel genes possibly involved in cellulose and xylan degradation [106]. Several studies have reported that defaunation, i.e. removal of protozoa, can have a negative effect on fibre degradation in the rumen [107,108]. Mosoni et al. [88] showed that long term defaunation had rather a beneficial effect on the abundance of fibrolytic bacterial species *R. flavefaciens* and *R. albus*, quantified by qPCR, but not on that of *F. succinogenes,* which is the most efficient in low digestible plant cell wall degradation, which could explain at least in part, the observed negative effect on fibre digestion.

Anaerobic fungi are also involved in digestion of plant material. They represent a very homogenous phylogenetic group (phylum *Neocallimasticota*) and a very specialized functional group as all species are fibrolytic [109]. The fungal biomass is estimated to represent between 5 and 10% of the total microbial mass. During their life cycle, flagelatted zoospores alternate with filamentous sporangia which are tightly attached to plant tissues, thanks to their cellulosome-like complexes [110]. Rumen fungi produce a very efficient set of cellulases and hemicellulases, whose specific activities are higher than that of bacteria [111]. They also possess esterase activities which contribute to the cleavage of ester bridges which link phenolic compounds of lignin to structural carbohydrates [112,113]. Moreover, thanks to the development of a rhizoidal network they are able to weaken and even disrupt plant tissue which enhances accessibility to digestible structures [114]. Studies carried out with gnotoxenic lambs harbouring or not fungi confirmed their important role in fibre breakdown in the rumen [115].

6. LIMITING FACTORS IN FIBRE DIGESTION

6.1. Animal Characteristics

A cow chews during eating and rumination to reduce feed (forage) particle sizes and allow the best fermentation process possible via a better distribution of feedstuff and bacteria in the rumen as well through rumen pH maintenance (high buffer capacity of the saliva). Indeed, this first step of the digestive process stimulates saliva production (274 ml/ min chewing and 6g sodium bicarbonate/ liter of saliva) and rumen motility. With an average daily time spent eating, ruminating and resting of 1/3, a production of up to 150 l of saliva per day is achieved. However, about half of the saliva will be produced during rumination, whereas eating will account for 20% and resting 30% [116].

The chewing responses to forage fragility and digestibility have been described [117]: at equal particle size, a low NDF Digestibility (NDFD) rate and less fragile forage increase by about 30 min/day the chewing time when compared to a high NDFD and fragile hay, whereas fragility appears less related to chewing when forage NDFD is similar. These results suggest that increased dietary physically effective NDF may affect chewing activity either through prolonging chewing time or increasing chewing rate. In addition, longer particle size will promote salivation and thus a shorter time with rumen pH<5.8 [118].

From a species standpoint, chewing activity is highly related to the intake capacity and body weight. Animals with a greater intake capacity seem to chew feed more efficiently (i.e. goat, sheep), while heavier animals (cows) can cope with relatively more fibre, because rumination capacity is in line with body size [119].

6.2. Composition Of The Diet And Structure Of Fibre

Many biotic and abiotic factors may limit the efficacy of fibre degradation in the rumen which may be driven by changes in fibre colonization efficacy. For example, the chemical composition of the plant material modulates the rate and extent of fibre digestion [120]. Digestibility of forage fibre (cell walls) has long been known to be negatively associated with lignin concentration. This relationship between lignin and fibre digestibility is very strong for a same forage compared according to different maturity stages, but it is less clear when comparing different forages harvested at a similar maturity stage, so with similar lignin concentrations [121]. To explain the observed variation in fibre digestibility of forages with similar lignin concentrations, composition of lignin and chemical cross-linking of lignin to cell wall polysaccharides have been suggested as involved additional factors. For example, cross-linking of lignin and arabinoxylans may limit cell wall digestibility by placing lignin in very close proximity to the polysaccharides and preventing physical access by hydrolytic microbial enzymes [120]. The slow entrance of microbial cells into some plant cell tissues such as sclerenchyma and also their slow diffusion capacities down the lumina represent also an important limitation factor for totally efficient fibre digestion [122].

Several studies have shown that the feed particle size may influence the degradation rate of fibre fractions as well as the bacterial colonization of the feed particles. Witzig et al. [123] investigated the effect of the forage source and particle size on the composition of the ruminal *Firmicutes* community assessed by qPCR and Fluorescent In Situ Hybridization *in vitro*. They found that *Ruminococcus albus*was more abundant on short particle size of forage, whereas the xylanolytic *Roseburia* sp. was favored by coarse particle grass silage based diets, and that abundance of *Clostridium* cluster XIV was higher with increasing grass silage proportion in the diet.

6.3. Characteristics Of The Rumen Environment

As described earlier in this chapter, it has been demonstrated that a diet rich in readily fermentable carbohydrates can adversely alter the structure and/or activities of fibre-degrading community, because of a decline in ruminal pH and acidosis occurrence. As a consequence, ruminal digestion of NDF is decreased [124] (Figure 3).

It is generally admitted that most of fibre-degrading microorganisms are sensitive to oxygen because most of them lack detoxification enzymes necessary for removal of reactive oxygen species. The presence of dissolved oxygen in the rumen ecosystem has been demonstrated [125,126] and oxygen regularly enters the rumen due to feed and water uptake and mastication, which can be illustrated by a greater post-feeding redox potential as measured in dairy cows by Marden et al. [57,127]. Newbold et al. [128] measured the concentration of cellulolytic bacteria in Rusitec in which either normal or low O_2 concentrations had been maintained. Oxygen concentration significantly influenced cellulolytic bacteria, whose numbers were increased by almost 15-fold when low O_2 concentrations were applied in the fermenters. Adhesion of cellulolytic bacteria to cellulose has been shown to be inhibited in the presence of oxygen *in vitro* [129].

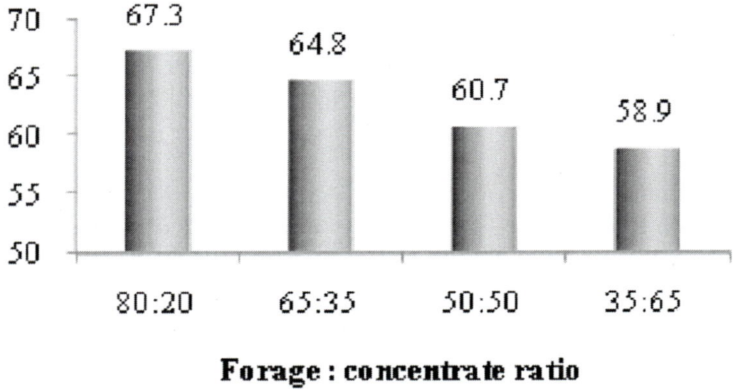

Figure 3. Effect of forage:concentrate ratio on apparent rumen NDF digestibility (%) in cows. From [124].

6.4. Physiology Of Fibrolytic Microorganisms and Microbial Interactions

Among biotic factors, the existence of a complex set of interactions between fibrolytic microbes and the other actors of feed digestion does impact fibre degradation. For example, synergistic cross feeding interactions have been described between cellulolytic and non cellulolytic species which lead to a global improvement in degradation [130]. A relevant example is the interaction between proteolytic bacteria and cellulolytic bacteria, the former releasing ammonia, used as preferential nitrogen source for the latter, and the latter releasing soluble sugars from cellulolysis, which will be metabolized by proteolytic bacteria. Moreover, hydrogen transfer between fibre degrading organisms and hydrogen consuming methanogens is necessary for an optimal functioning of fibre degradation mechanisms. Indeed, methanogens help to reduce the hydrogen partial pressure and thereby avoid the inhibition of ferredoxine oxidoreductase which has an essential role on NADH re-oxidation [130]. The result of this interaction is a gain in energy for both partners and an increase in fibre digestion. On the opposite, competition mechanisms have been described between cellulolytic bacterial species for adhesion on cellulose [131,132]. Secretion of inhibitory peptides by *Ruminococcus* strains have been shown *in vitro*to impact growth of rumen fungi [133]. Finally, the physiology of the microorganisms plays also an essential role on overall fibre digestion. Indeed, there are great differences between species regarding their preference and affinity for substrates, their energy requirements, or their capacity to resist to environmental stresses.

7. BENEFITS OF USING YEAST PROBIOTICS TO PROMOTE FIBRE DIGESTION

7.1. Targets

To optimize fibre digestion, there is a need to minimize the indigestible fibre fraction, maximize rate of fibre digestion, and maintain a ruminal environment that promotes the population of fibre-digesting bacteria. The indigestible fibre in forages (iNDF) is related to lignin concentration, but also contains structural carbohydrates (cellulose and hemicellulose) which are 'trapped' with lignin. Whereas lignin, of which biochemical degradation process involves oxidative pathways, is considered not digested in the animal gastro-intestinal tract, the release of the carbohydrates bound to lignin would be interesting in terms of increasing feed value of the forage.

To achieve these goals with probiotics, several strategies may be developed depending on the dietary conditions of the animals. Indeed, indirect or direct effects can be sought. Indirect benefits could be mediated through pH stabilization effects (see section 4), or modification of the environment of the microbiota which will definitely sustain or promote fibre-degrading microbiota and their action on plant cell walls. Direct effect of probiotics on fibrolytic microorganisms can also be wished to exist, as nutritional requirements for peptides, amino acids, ammonia, organic acids or branched chain fatty acids have been described for bacteria and fungi and the supply of these components might be achieved through the use of probiotics.

7.2. Experimental Proofs

Using different methods, it has been reported that live yeast supplementation improves rumen fibre digestion *in vivo* [85,134-137], although this has not always been observed [138].

7.3. Modes Of Action On Rumen Microbiota

In vitro, the potential of probiotic yeasts to enhance growth and activity of fibre-degrading rumen microorganisms has been established. Fungal zoospore germination and cellulose degradation were increased in the presence of a strain of *S. cerevisiae* [139]; the authors suggested that yeasts could enhance fungal colonization of plant cell walls, which was confirmed recently [136]. The effectiveness of some yeast strains to stimulate growth or/and activities of fibrolytic bacteria has also been demonstrated. *In vitro*, a *S. cerevisiae* strain stimulated growth of *Fibrobacter succinogenes* S85 and reduced the lag time for growth of *Ruminococcus albus* 7, *Ruminococcus flavefaciens* FD1, and*Butyrivibrio fibrisolvens* D1 [140]. Callaway and Martin [141] showed that the same yeast could accelerate the rate, but not the extent, of cellulose filter paper degradation by *F. succinogenes* S85 and*R. flavefaciens* FD1. *In vivo*, in gnotoxenic lambs harbouring three species of bacteria (*F. succinogenes, R. albus,* and *R. flavefaciens*) as sole cellulolytic organisms, cellulolytic bacteria became established earlier and remained at a high and stable level even after a stressful period (lambs were fitted with a rumen cannula) in the lambs receiving a probiotic yeast daily [137]. Ciliate protozoa, which are not able to establish unless bacterial communities have previously colonized the rumen [142], appeared more rapidly in the rumen of conventional lambs in the presence of live yeasts [143].This supports the hypothesis that live yeast supplementation accelerates maturation of the rumen microbial ecosystem. Fibre degradation processes would thereby be set up more efficiently in the early age of the animal, as shown by the increase in polysaccharidase and glycoside-hydrolase activities in the presence of yeast in the rumen of gnotoxenic lambs [137].

There are some evidence that live yeast additives indirectly promote fibre degradation or fibrolytic microbial activities by stabilizing rumen pH in case of ruminal acidosis (see section 4). Greater polysaccharide-degrading activities of the solid-associated bacterial fraction in rumen-cannulated adult sheep fed a high-concentrate diet were measured in the presence of yeasts [144]. The proportions of 16S rRNA of *F. succinogenes, R. albus,* and *R. flavefaciens* have been shown to increase in the rumen of sheep receiving another yeast product [145]. A 2 to 4-fold increase in the number of 16S rRNA gene copies of *R. albus* and *R. flavefaciens* was also measured with real-time PCR in rumen contents of sheep receiving a high-concentrate diet and a live yeast probiotic [14].

Guedes et al. [85] reported that a live yeast strain increased NDF degradation of different corn silage samples incubated *in sacco*. In their study, cows were fed with grass silage-corn silage based diet and the rumen pH was not indicative of SARA situation. However, it is noteworthy that a yeast effect was observed on pH and lactate concentration but the authors suggested that the yeast efficacy was not only attributable to a pH stabilization effect. Using the same technique, Chaucheyras Durand et al. [136, unpublished] have studied the effect of the same yeast strain on fibre degradation of different substrates and followed the kinetics of colonization by fibre-degrading bacteria and fungi using qPCR in rumen cannulated cows. In this study, the diet offered to the cows was composed of grass silage and hay and was not at risk regarding SARA. Results showed that the supplementation of 10^{10} cfu/day/cow of the yeast additive promoted colonization of fibrous substrates by cellulolytic bacteria (*F.succinogenes, R.flavefaciens, B.fibrisolvens*) and fungi but that the degree of stimulation was depending on the nature of the substrate, and on the microbial species targeted. It was noticed that feedstuffs with highest levels of lignin and thereby with less easily accessible digestible carbohydrates were better degraded in the presence of yeast, suggesting a particularly marked impact on the microbial breakdown of lignin-polysaccharide linkeages. The same strain of *S. cerevisiae* significantly improved NDF degradation of 40 corn silages samples incubated *in sacco* in rumen cannulated cows, with differences in the degree of improvement according to the degradability of the corn silage [85]. Indeed, the yeast probiotic increased NDF degradation of the low digestible corn silages more strongly than that of the high digestible corn silages (Figure 4). These results suggest that live yeast could help to reduce indigestible NDF by promoting the action of bacteria and fungi involved in the hydrolysis of lignin-polyholoside bonds (Figure 5).

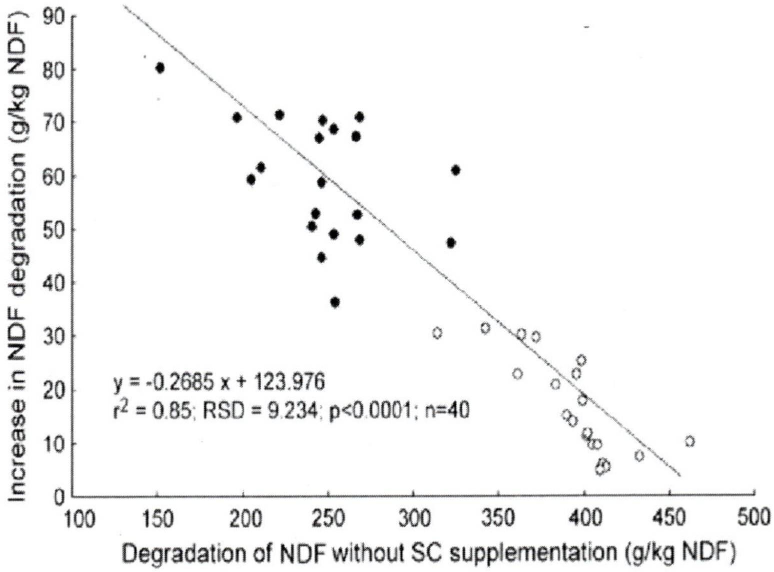

Figure 4. Effects of supplementation with a yeast probiotic (*Saccharomyces cerevisiae* CNCM I-1077) on fibre (NDF) degradation of maize silages after 36h of incubation in the rumen of cows: open circles, high fibre degradation group, full circles, low fibre degradation group. From [85].

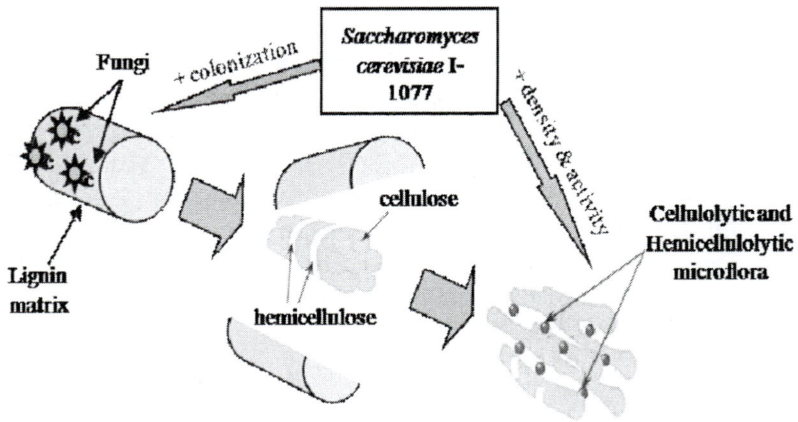

Figure 5. A proposed scheme for mode of action of *Saccharomyces cerevisiae* CNCM I-1077 on fibre degrading communities.

In the study of Chaucheyras-Durand et al. [136, unpublished], a positive effect of live yeast was demonstrated for the first time on *Butyrivibrio fibrisolvens* abundance on fibrous substrates. The hemicellulose fraction of forages consumed by ruminants consists mainly in xylan substituted with acetyl, arabinosyl, and glucuronyl residues. Xylan is also cross-linked via ferulic and p-coumaric acids which are esterified to the arabinose side chains. It is supposed that the ester linkages between these phenolic acids and polysaccharides provide a steric hindrance to the degradation of fibre by rumen microbiota. Consequently, the promotion of *B. fibrisolvens*, that possesses ferulic and p-coumaric acid esterases which hydrolyze these ester linkages [146] appears particularly interesting.

One of the main factors implicated in the beneficial effect of live yeasts on fibre-degrading bacteria is probably the capacity of yeast cells to scavenge oxygen. Indeed, although the rumen environment is known to be strictly anaerobic, dissolved oxygen can be detectable *in situ*; as high as 16 liters of oxygen can enter an ovine rumen daily during feed and water intake, rumination or salivation [147]. Most of ruminal microorganisms are considered to be highly sensitive to oxygen, but this is particularly true for fibre-degrading organisms. Respiratory-deficient mutants of *S. cerevisiae* were unable to stimulate bacterial numbers in rumen-simulating fermenters, whereas the wild-type parent strains, able to consume oxygen, did effectively stimulate bacterial activities [128]. Other studies have reported that redox potential of rumen fluid was lowered in the presence of live yeasts in lambs [143], in sheep [148] and in cows [57] suggesting that live yeast cells could create more favorable environmental conditions for growth and activities of the cellulolytic microbiota. Due to the fact that live yeasts could release vitamins or other growth factors to closely associated bacterial cells [149], yeast impact could also be mediated through the interplay between different bacterial species (i.e. non cellulolytic species) and would not only be explained by a direct effect on oxygen consumption.

7.4. Consequences On Rumen Fermentations, Feed Efficiency, and Animal Production

The beneficial effects on fibre digestion can be partly at the origin of the increase in dry matter intake often observed with yeast supplementation [149], but more generally a better fibre digestion is recognized to benefit the animal rumen health and its function by improvement of feed efficiency. The study carried out by Bitencourt et al. [150] did support this assumption with cows fed a corn silage, soybean meal, citrus pulp and steam-flaked corn based TMR. The diet NDF digestibility was improved by 11.3% in presence of 10^{10} cfu/day of the live yeast and the milk production tended to be improved by 0.9 kg/d. Cows were not in SARA situation ($6.43<pH<6.5$). In De Ondarza et al. multi-analysis [73], live yeast effect was particularly strong in low yielding cows. In addition, feed efficiency of the supplemented animals was improved which illustrates a better use of the diet. When targeting the cows fed diet above 30% NDF, feed efficiency was higher than the overall mean and the live yeast treated animals gained an extra 40g of milk per kg DMI. The shorter intervals between meals of live yeast fed cows reported in [56] strongly suggests the fact that the TMR digestibility was improved as the meal size and length were not affected by the treatment. As mentioned earlier, improvement of rumen pH for the cows receiving the live yeast at the same dose than the previously cited studies would also support a higher activity of the cellulolytic flora and thus explain the higher meal frequency.

8. IMPORTANCE OF YEAST VIABILITY AND STRAIN SELECTION

A better understanding of the modes of action of live yeast probiotics is important to further select of new yeast strains acting on specific key target microorganisms and areas of ruminal fermentation. Therefore, strain selection process is obviously critical in terms of safety; chosen organisms should be on the GRAS (Generally Recognized As Safe) list, or sufficient evidence would have to be provided to guarantee their innocuity for the animal, consumer and environment. Moreover, strain selection is important as different probiotics clearly exhibit markedly different effects on digestive microbiota of the same targeted organism. Dose response effects have also been reported for a same strain within the same experiment [63,85], suggesting that an optimal concentration of live cells has to be defined precisely according to the product application.

Efficacy of probiotics is strongly related to cell viability and metabolic activity [151], therefore, stability within the rumen is also an important consideration. Although yeast strains cannot properly colonize the rumen for a long period of time, certain strains can remain metabolically active in rumen fluid for more than 24 h [152] and live cells may be recovered from the faeces of treated animals up to several days after their initial incorporation in the diet. One objective when selecting a new probiotic strain will then be to assess its capacity to persist for a long time at a significant concentration in the targeted digestive compartment. Production, storage, and delivery protocols for yeast products should be designed to maintain yeast cell viability. High temperature storage, or in the presence of components such as minerals acting as oxidizing agents, may compromise viability [153]. The most common and officially recognised method for quantification of viable yeast probiotics is the colony forming unit (CFU) plate counting technique. Although it is perfectly adapted to take into account cells which have the capacity to multiply in optimal environmental conditions, it has long been recognized that microbial cells may exist in a latent state, in which they will not form colonies on nutrient media but may have other measurable activity [154]. For example, throughout alcoholic fermentation, *Saccharomyces cerevisiae* cells have to cope with stress conditions that could affect their viability and thereby enter into a Viable But Not Culturable (VBNC) state [155,156]. Further methodological developments would be necessary in order to take into account this status, which would improve our understanding on adaptive responses of probiotic yeasts to digestive conditions.

9. CONCLUSIONS AND FUTURE WORK

Yeast probiotics benefit from a natural and well-accepted image by the consumer, as they are not involved in health disorders and do not have any detrimental impact on environment. Moreover, yeasts have been used for a long time in human nutrition. More and more well controlled research studies indicate that they can be useful to positively balance the rumen microbiota, stabilize rumen pH, and promote microbial degradation of plant cell walls. Thanks to their action, improvement in animal production and health can be obtained and in that sense one can expect a promising future for these additives in ruminant nutrition. As particularly shown for fibre degradation, the nature of dietary ingredients has a great influence in the rumen response to yeast probiotics. More research is needed to enlarge the efficacy data base using various diets and raw materials, which in term would lead to elaboration of predictive tools applicable on farms.

In the context of a high feed cost, fermentation aids such as live yeast represent a valuable nutritional tool which allows increasing the forage portion of the diet and consequently limiting the costly sources of energy. In addition, current intensive farming practices require high levels of fermentable carbohydrates which put the animal at risk of developing metabolic disorders. In that sense, yeast probiotics become even more relevant when the digestive microbiota is challenged, for example during a feed transition (weaning, grazing, step up feeding programs) or during periods of stress (hot temperature, transportation). In these particular conditions, higher yeast doses appear to better support rumen challenges. As differences have been reported in terms of response of the ruminal microbiota to different yeast additives (strain and capacity to retain metabolic activity), it is important to focus on the way the yeast strain is selected. Future research will also need to address the behavior of the yeast cells in the digestive environment. Indeed, identification of specific metabolic and physiologic characteristics exhibited by the yeast strains would allow a better understanding of their interactions within the animal gut and will help to further select more targeted additives for improved benefits in ruminant nutrition.

During plant cell wall breakdown and fermentation, most of cellulolytic bacteria, with the exception of*Fibrobacter succinogenes*, produce a lot of hydrogen, which is used to reduce carbon dioxide by*Archaea* methanogens to produce methane. This hydrogen transfer is important for a good functioning of the rumen ecosystem, but at the same time methane formation represents a loss of energy (10-12% of the metabolizable energy of the host animal) and this gas being a potent greenhouse gas, it should be decreased [157]. Studies with gnotobiotically-reared lambs have shown that animals inoculated with *F. succinogenes* were less prone to produce methane than lambs inoculated with *Ruminococci* and fungi, without significant modifications of rumen fibre degradability and volatile fatty acid concentrations [158]. The use of microbial solutions to promote *F. succinogenes* would then appear interesting to be able to mitigate methane emissions by cattle.

It is noteworthy that the increase in feed efficiency reported in presence of yeast probiotics has already an indirect effect on polluting outputs as it will decrease the amount of output/kg of milk/meat produced, but targeting microorganisms directly involved in these fermentative processes may be of interest.

Biohydrogenation mechanisms would also be a good target as they appear to be involved in milk fat depression which is very commonly observed in high-yielding cows, at risk for SARA. Under certain conditions, rumen microbial biohydrogenation results in the formation of fatty acids that are potent inhibitors of milk fat synthesis, i.e. trans10,cis12-CLA, and of possibly related intermediates from linolenic acid and other polyunsaturated fatty acids [48]. It has been shown that *Butyrivibrio sp.* is able to produce mainly trans-11,vaccenic acid via cis9, trans11-CLA instead of trans10,cis12-CLA from linolenic acid. By increasing the *Butyrivibrio* sp. population so that they utilize more linolenic acid at the expense of the organisms which form the detrimental isomer trans10,cis12 CLA, the potential exists to avoid a decrease in milk fat content. Stabilising ruminal pH through the addition of live yeasts should be beneficial for improved growth of these organisms which are sensitive to low pH. Moreover, promising data have been recently obtained that show a stimulation of *B. fibrisolvens* colonization on plant cell walls.

Yeast probiotics which have a good survival beyond the rumen may have interesting effects on intestinal homeostasis, and could thereby positively influence immune system and animal health. Indeed, certain strains of *Saccharomyces* may reduce pathogen load or their effects through competitive exclusion, cell binding or degradation of the toxins produced by intestinal pathogens. The beneficial effect that live yeast can have on pH regulation could also limit the release of inflammatory molecules, such as lipopolysaccharide or biogenic amines, and counteract the set up of acid-resistance mechanisms which may increase the virulence of certain pathogens. It has been reported that acidification of the rumen environment may increase mycotoxin absorption at low pH and decrease microbial detoxication mechanisms [159], so a better control of rumen pH by probiotic yeast may also aid in decreasing mycotoxin animal exposure.

REFERENCES

1. Nocek JE. Bovine acidosis: implications on laminitis.Journal of Dairy Science 1997801005

2. Enemark JM, Jorgensen RJ, Kristensen NB.An evaluation of parameters for the detection of subclinical rumen acidosis in dairy herdsVeterinary Research Communications200428687

3. Nocek JE, Kautz WP, Leedle JA, Allman JG.Ruminal supplementation of direct-fed microbials on diurnal pH variation and in situ digestion in dairy cattle.Journal of Dairy Science 2002852429433

4. Duffield TF.Monitoring strategies for metabolic diseases in transition dairy cows. Médecin Vétérinaire du Québec 200434 EOF35 EOF

5. Deiss. V. Mialon, S. Andanson, F. Anglard, I. Veissier, Andanson S, Anglard F, Veissier I. An assessment of the impact of rumenocentesis on pain and stress in cattle and the effect of local anaesthesiaThe Veterinary Journal2012

6. L. Brossard, C. Martin, F. Chaucheyras-Durand, B. Michalet-Doreau, Protozoa involved in butyric rather than lactic fermentative pattern during latent acidosis in sheep.ReproductionNutrition Development 200444195

7. Mottram T, Lowe J, McGowan M, Phillips N. Technical note: A wireless telemetric method of monitoring clinical acidosis in dairy cows. Computers and Electronics in Agriculture 2008;64 45-48.

8. Stewart CS, Flint HJ, Bryant MP. The rumen bacteria. In : Hobson PN, Stewart CS (eds.) The rumen microbial ecosystem.London: Chapman & Hall; 19971072

9. J. B. Russell, T. Hino, Regulation of lactate production in Streptococcus bovis: a spiraling effect that contributes to rumen acidosis.Journal of Dairy Science1985681712

10. Counotte GHM, Prins RA, Janssen RHA, Deie MJA.Role of Megasphaera elsdenii in the fermentation of DL-[2-C13]lactate in the rumen of dairy cattle. Applied and Environmental Microbiology 1981424649655

11. Russell JB, Baldwin RL.Substrate preferences in rumen bacteria: evidence of catabolite regulatory mechanismsApplied and Environmental Microbiology 1978362319329

12. L. Brossard, F. Chaucheyras-Durand, B. Michalet-Doreau, C. Martin, Dose effect of live yeasts on rumen microbial communities and fermentations during butyric latent acidosis in sheep: new type of interactionAnimal Science 2006821

13. K. Tajima, R. I. Aminov, T. Nagamine, H. Matsui, M. Nakamura, Y. Benno, Diet-dependent shifts in the bacterial population of the rumen revealed with real-time PCR.Applied and Environmental Microbiology2001672766

14. P. Mosoni, F. Chaucheyras-Durand, C. Béra-Maillet, E. Forano, Quantification by real-time PCR of cellulolytic bacteria in the rumen of sheep after supplementation of a forage diet with readily fermentable carbohydrates. Effect of a yeast additive. Journal of Applied Microbiology20071032676

15. D. R. Yáñez-Ruiz, B. Macías, E. Pinloche, C. J. Newbold, The persistence of bacterial and methanogenic archaeal communities residing in the rumen of young lambsFEMS Microbiology Ecology 2010722272278

16. E. Khafipour, S. Li, J. C. Plaizier, Krause, Rumen microbiome composition determined using two nutritional models of subacute ruminal acidosis.Applied and Environmental Microbiology 2009 752271157124

17. S. C. Fernando, H. T. Purvis, Najar. F. Z. 2nd, L. O. Sukharnikov, C. R. Krehbiel, T. G. Nagaraja, Desilva. U. Roe, Rumen microbial population dynamics during adaptation to a high-grain dietApplied and Environmental Microbiology2010762274827490

18. Russell JB, Wilson DB.Why are ruminal cellulolytic bacteria unable to digest cellulose at low pH? Journal of Dairy Science 1996791503

19. C. Martin, G. Fonty, B. Michalet-Doreau, Factors affecting the fibrolytic activity of the digestive microbial ecosystems in ruminants. In: Martin SA (ed.) Gastrointestinal Microbiology in Animals. Trivandrum: Research Signpost; 2002117

20. Goad DW, Goad CL, Nagaraja TG.Ruminal microbial and fermentative changes associated with experimentally induced subacute acidosis in steers.Journal of Animal Science199876234

21. E. Grenet, A. Breton, P. Barry, G. Fonty, Rumen anaerobic fungi and plant substrates colonization as affected by diet composition. Animal Feed Science and Technology19892655

22. G. Fonty, E. Grenet, Effects of diet on the fungal population of the digestive tract of ruminants. In: Mountfort DO, Orpin CG (eds.) Anaerobic fungi: biology, ecology and function. New York: Marcel Dekker; 1994229239

23. C. Martin, L. Millet, G. Fonty, B. Michalet-Doreau, Cereal supplementation modified the fibrolytic activity but not the structure

of the cellulolytic bacterial community associated with rumen solid digesta.ReproductionNutrition Development 200141413

24. B. Michalet-Doreau, D. Morand, C. Martin, Effect of the microbial additive Levucell SC CNCM I-1077 on microbial activity in the rumen during the stepwise adaptation of sheep to high concentrate diet. Reproduction Nutrition Development 1997supplEE5 81.

25. P. Nozière, J. M. Besle, B. Michalet-Doreau, Effect of barley supplement on microbial fibrolytic enzyme activities and cell wall degradation rate in the rumenJournal of Science of Food and Agriculture 199672235

26. D. Sauvant, Le contrôle de l'acidose ruminale latenteINRA Productions Animales 20061926978

27. C. Philippeau, C. Martin, B. Michalet-Doreau, Influence of grain source on ruminal characteristics and rate, site, and extent of digestion in beef steers.Journal of Animal Science1999771587

28. A. Lettat, P. Nozière, M. Silberberg, D. P. Morgavi, C. Berger, C. Martin, Experimental feed induction of ruminal lactic, propionic, or butyric acidosis in sheepJournal of Animal Science20108 89304 13046

29. Mertens DR.Creating a system for meeting the fiber requirements of dairy cows.Journal of Dairy Science199780714631481

30. Yang WZ, Beauchemin KA.Altering physically effective fiber intake through forage proportion and particle length: digestion and milk production.Journal of Dairy Science200790734103421

31. C. Martin, L. Brossard, M. Doreau, Mécanismes d'apparition de l'acidose ruminale latente et conséquences physiopathologiques et zootechniques. INRA Productions Animales 20061993

32. Kleen JL, Cannizzo C. Incidence, prevalence and impact of SARA in dairy herds. Animal Feed Science and Technology 2012;172 4-8.

33. Stone WC.The effect of subclinical acidosis on milk components. Cornell Nutrition conference for feed manufacturers. Cornell University, Ithaca NY 19994046

34. D. G. Emmanuel, A. Jafari, K. A. Beauchemin, J. A. Leedle, B. N. Ametaj, Feeding live cultures of Enterococcus faecium and

Saccharomyces cerevisiae induces an inflammatory response in feedlot steers.Journal of An:mal Science200785233

35. Gozho GN, Krause DO, Plaizier JC.Ruminal lipopolysaccharide concentration and inflammatory response during grain-induced subacute ruminal acidosis in dairy cows.Journal of Dairy Science2007902856866

36. Plaizier JC, Krause DO, Gozho GN, McBride BW.Subacute ruminal acidosis in dairy cowsthe physiological causes, incidence and consequences. Veterinary Journal 200817612131

37. Khafipour E, Krause DO, Plaizier JC. A grain-based subacute ruminal acidosis challenge causes translocation of lipopolysaccharide and triggers inflammation. Journal of Dairy Science 2009;92(3) 1060-1070.

38. Q. Zebeli, B. U. Metzler-Zebeli, Interplay between rumen digestive disorders and diet-induced inflammation in dairy cattle. Research in Veterinary Science 2012

39. P. Thompson, A. Hentzen, W. Schultheiss, The effect of rumen lesions in feedlot calves: which lesions really affect growth? In: Proceedings from the 4th Schering Plough Ruminant day, 2006University of Pretoria, Pretoria, South Africa, 2327

40. Y. Wang, W. Majak, T. A. Mc Allister, Frothy bloat in ruminants: Cause, occurrence, and mitigation strategiesAnimal Feed Science and Technology2012103 EOF114 EOF

41. Vasconcelos JT, Galyean ML. ASAS centennial paper: Contributions in theJournal of Animal Science to understanding cattle metabolic and digestive disorders. 20082008868171117 21

42. Tadepalli S, Narayanan SK, Stewart GC, Chengappa MM, Nagaraja TG. Fusobacterium necrophorum: A ruminal bacterium that invades liver to cause abscesses in cattle. Anaerobe 2009;15(1/2) 36-43.

43. J. L. Kleen, G. A. Hooijer, J. Rehage, J. P. Noordhuizen, Subacute ruminal acidosis (SARA): a reviewJournal of Verterinary Medecine A : Physiology, Pathology, Clinical Medicine 2003508406414

44. L. Commun, Olivera. L. Alves de, L'acidose subclinique chez les ruminants. Conséquences comportementales et indicateurs

physiologiques périphériques. Journées Nationales des GTV Nantes 200910911100

45. Martin. C. Mialon, F. Garcia, J. B. Menassol, H. Dubroeucq, I. Veissier, D. Micol, Effects of the forage-to-concentrate ratio of the diet on feeding behaviour in young Blond d'Aquitaine bulls Animal200821682

46. L. Commun, M. Silberberg, Martin. C. Mialon, I. Veissier, Behavioral adaptations of sheep to repeated acidosis challenges. Animal 2012In press.

47. M. Desnoyers, C. Duvaux-Ponter, K. Rigalma, S. Roussel, C. Martin, S. Giger-Reverdin, Effect of concentrate percentage on ruminal pH and time-budget in dairy goatsAnimal200821802

48. J. J. Kennelly, B. Robinson, G. R. Khorasani, Influence of carbohydrate sourca and buffer on rumen fermentation characteristics, milk yield, and milk composition in early-lactation holstein cows. Journal of Dairy Science 1999822486

49. Khorasani GR, Kennelly JJ.Influence of carbohydrate source and buffer on rumen fermentation characteristics, milk yield, and milk composition in late-lactation holstein cows.Journal of Dairy Science2001841707

50. A. Offner, A. Bach, D. Sauvant, Quantitative review of in situ starch degradation in the rumenAnimal Feed Science and Technology 200381 EOF

51. Yang CM, Varga GA.Effect of three concentrate feeding frequencies on rumen protozoa, rumen digesta kinetics and milk yield in dairy cows.Journal of Dairy Science198972950

52. Uriarte. R. Diaz, Incorrect analysis of crossover trials in animal behaviour researchAnimal Behavior 2002634815822

53. Weimer PJ, Stevenson DM, Mantovani HC, Man SLC.Host specificity of the ruminal bacterial community in the dairy cow following near-total exchange of ruminal contents.Journal of dairy Science2010935902

54. Williams YJ, Popovski S, Rea SM, Skillman LC, Toovey AF, Northwood KS, Wright ADG. A vaccine against rumen methanogens

can alter the composition of archaeal populations. Applied Environment Microbiology 2009;75(7) 1860-1866.

55. K. A. Beauchemin, W. Z. Yang, D. P. Morgavi, G. R. Ghorbani, W. Kautz, J. A. Z. Leedle, Effects of bacterial direct-fed microbials and yeast on site and extent of digestion, blood chemistry, and subclinical ruminal acidosis in feedlot cattle. Journal of Animal Science 2003811628

56. Bach A, Iglesias C, Devant M. Daily rumen pH pattern of loose-housed dairy cattle as affected by feeding pattern and live yeast supplementation. Animal Feed Science and Technology 2007;136 156-163.

57. J. P. Marden, C. Julien, V. Monteils, E. Auclair, R. Moncoulon, C. Bayourthe, How does live yeast differ from sodium bicarbonate to stabilize ruminal pH in high-yielding dairy cows? Journal of Dairy Science 200891935283535

58. M. Desnoyers, S. Giger-Reverdin, G. Bertin, C. Duvaux-Ponter, D. Sauvant, Meta-analysis of the influence of Saccharomyces cerevisiae supplementation on ruminal parameters and milk production of ruminantsJournal of Dairy Science2009921620

59. M. Thrune, A. Bach, M. Ruiz-Moreno, Linn. J. G. Stern, Effects of Saccharomyces cerevisiae on ruminal pH and microbial fermentation in lactating dairy cows. Journal of Dairy Science 2007Suppl. 1) 172.

60. M. B. De Ondarza, T. Hall, J. Sullivan, E. Chevaux, Effect of live yeast supplementation on milk yield, milk components, and rumen pH in dairy cows. Journal of Dairy Science 2012E-suppl. In press.

61. F. Chaucheyras, G. Fonty, G. Bertin, J. M. Salmon, P. Gouet, Effects of a strain of Saccharomyces cerevisiae (Levucell SC), a microbial additive for ruminants, on lactate metabolism in vitro. Canadian Journal of Microbiology 199642927

62. Nisbet DJ, Martin SA.Effect of a Saccharomyces cerevisiae culture on lactate utilization by the ruminal bacterium Selenomonas ruminantium.Journal of Animal Science1991694628

63. Newbold CJ, McIntosh FM, Wallace RJ.Changes in the microbial population of a rumen-simulating fermenter in response to yeast cultureCanadian Journal of Animal Science199878241

64. F. Rossi, A. D. Luccia, D. Vincenti, P. S. Cocconcelli, Effects of peptidic fractions from Saccharomyces cerevisiae culture on growth and metabolism of the ruminal bacteria Megasphaera elsdeniiAnimal Research200453177

65. B. Michalet-Doreau, D. Morand, Effect of yeast culture, Saccharomyces cerevisiae CNCM I-1077, on ruminal fermentation during adaptation to high-concentrate feeding. In: 4èmes Rencontres autour des Recherches sur les Ruminants, Paris. 1997121

66. Williams PEV, Tait CA, Innes GM, Newbold CJ.Effects of the inclusion of yeast culture (Saccharomyces cerevisiae plus growth medium) in the diet of dairy cows on milk yield and forage degradation and fermentation patterns in the rumen of steers.Journal of Animal Science1991693016

67. R. S. Marsola, M. G. Favoreto, F. T. Silvestre, J. C. Shin, N. Walker, A. Adesogan, C. R. Staples, J. E. P. Santos, Effect of feeding live yeast on performance of holstein dairy cows during summer. Journal of Dairy Science 2010E-Suppl1 432.

68. Owens FN, Secrist DS, Hill WJ, Gill DR. Acidosis in cattle: a review.Journal of Animal Science 1998761275286

69. F. Dohme, T. J. De Vries, K. A. Beauchemin, Repeated ruminal acidosis challenges in lactating dairy cows at high and low risk for developing acidosis: ruminal pHJournal of Dairy Science 2008 913354

70. F. Chaucheyras-Durand, M. Silberberg, L. Commun, C. Martin, D. P. Morgavi, Repeated ruminal acidotic challenges in sheep: effects on pH and microbial ecosystem and influence of active dry yeasts. Microbial Ecology 200957564

71. M. Silberberg, F. Chaucheyras-Durand, L. Commun, Martin. C. Richard-Mialon, D. P. Morgavi, Repeated ruminal acidotic challenges in sheep: effects on pH and microbial ecosystem and influence of active dry yeasts. Journal of Animal Science 2009E-Suppl) 280.

72. F. Chaucheyras-Durand, N. D. Walker, A. Bach, Effects of active dry yeasts on the rumen microbial ecosystem: past, present and futureAnimal Feed Science and Technology20081455

73. De Ondarza MB, Sniffen CJ, Dussert L, Chevaux E, Sullivan J, Walker ND. Case study: Multiple-Study analysis of the effect of live yeast on milk yield, milk component content and yield, and feed efficiency. The Professional Animal Scientist 2010;26 661-666.

74. U. Moallem, H. Lehrer, L. Livshitz, M. Zachut, S. Yakoby, The effects of live yeast supplementation to dairy cows during the hot season on production, feed efficiency, and digestibilityJournal of Dairy Science200992343

75. P. Huhtanen, U. Asikainen, M. Arkkila, S. Jaakkola, Cell wall digestion and passage kinetics estimated by marker and in situ methods or by rumen evacuations in cattle fed hay 2 or 18 times dailyAnimal Feed Science and Technology2007206 EOF227 EOF

76. M. Oba, Allen, Evaluation of the importance of the digestibility of neutral detergent fiber from forage: Effects on dry matter intake and milk yield of dairy cows.Journal of Dairy Science1999823589596

77. Zebeli Q, Aschenbach JR, Tafaj M, Boguhn J, Ametaj BN, Drochner W. Invited review: Role of physically effective fiber and estimation of dietary fiber adequacy in high-producing dairy cattle. Journal of Dairy Science 2012;95(3) 1041-1056.

78. Mertens DR.Challenges in measuring insoluble dietary fiber.Journal of Animal Science2003811232333249

79. Hristov AN, Price WJ, B. Shaffi B. A meta-alaysis examining the relationship among dietary factors, dry matter intake, and milk and milk protein yield in dairy cows. Journal of Dairy Science 2004;87(7) 2184-2196.

80. West JW, Mandebvu P, Hill GM, Gates RN. Intake, milk yield, and digestion by dairy cows fed diets with increasing fiber content from bermudagrass hay or silage. Journal of Dairy Science 1998;81(6) 1599-1607.

81. M. Rinne, P. Huhtanen, S. Jaakkola, Digestive processes of dairy cows fed silages harvested at four stages of grass maturity.Journal of Animal Science200280719861998

82. C. Martin, C. Philippeau, B. Michalet-Doreau, Effect of wheat and corn variety on fiber digestion in beef steers fed high-grain diets.Journal of Animal Science1999772269

83. Hall MB, Mertens DR.In vitro fermentation vessel type and method alter fiber digestibility estimatesJournal of Dairy Science201291301

84. Spanghero M, Berzaghi P, Fortina R, Masoero F, Rapetti L, Zanfi C, Tassone S, Gallo A, Colombini S, Ferlito JC. Technical note: precision and accuracy of in vitro digestion of neutral detergent fiber and predicted net energy of lactation content of fibrous feeds. Journal of Dairy Science 2010;93(10) 4855-4859.

85. C. M. Guedes, D. Gonçalves, M. A. M. Rodrigues, A. Dias-da-Silva, Effect of yeast Saccharomyces cerevisiae on ruminal fermentation and fiber degradation of maize silage in cows. Animal Feed Science and Technology 200814527

86. G. Towne, T. G. Nagaraja, C. Owensby, D. Harmon, Ruminal evacuation's effect on microbial activity and ruminal function. Journal of Animal Science198662783

87. S. Koike, Y. Kobayashi, Development and use of competitive PCR assays for the rumen cellulolytic bacteria: Fibrobacter succinogenes, Ruminococcus albus and Ruminococcus flavefaciens.FEMS Microbiology Letters2001204361

88. P. Mosoni, C. Martin, E. Forano, D. P. Morgavi, Long-term defaunation increases the abundance of cellulolytic ruminococci and methanogens but does not affect the bacterial and methanogen diversity in the rumen of sheepJournal of Animal Science2011893783791

89. C. Béra-Maillet, P. Mosoni, A. Kwasiborski, F. Suau, Y. Ribot, E. Forano, Development of a RT-qPCR method for the quantification of Fibrobacter succinogenes S85 glycoside hydrolase transcripts in the rumen content of gnotobiotic and conventional sheepJournal of Microbiological Methods2009771816

90. G. Fonty, F. Chaucheyras-Durand, E. Forano, Structuration de l'écosystème ruminal chez le nouveau-né: influence de facteurs écologiques. Journées Nationales GTV:conference proceedings, May 13-15, 2009Nantes, France. 205214

91. G. Fonty, P. Gouet, J. P. Jouany, J. Senaud, Ecological factors determining establishment of cellulolytic bacteria and protozoa in the rumen of meroxenic lambs.Journal of General Microbiology 1983129213

92. G. Fonty, J. Senaud, J. P. Jouany, P. Gouet, Establishment of the microflora and anaerobic fungi in the rumen of lambsJournal of General Microbiology19871331835

93. Weimer PJ, Waghorn GC, Odt CL, Mertens DR.Effect of diet on populations of three species of ruminal cellulolytic bacteria in lactating dairy cows.Journal of Dairy Science1999821122134

94. Y. Kong, Y. Xia, R. Seviour, M. He, T. Mc Allister, R. Forster, In situ identification of carboxymethyl cellulose-digesting bacteria in the rumen of cattle fed alfalfa or triticale.FEMS Microbiology Ecology2012801159167

95. G. Suen, P. J. Weimer, D. M. Stevenson, F. O. Aylward, J. Boyum, J. Deneke, C. Drinkwater, N. N. Ivanova, N. Mikhailova, O. Chertkov, L. A. Goodwin, C. R. Currie, D. Mead, P. J. Brumm, The complete genome sequence of Fibrobacter succinogenes S85 reveals a cellulolytic and metabolic specialist PLoS One 2011

96. Dodd D, Mackie RI, Cann IK. Xylan degradation, a metabolic property shared by rumen and human colonic Bacteroidetes. Molecular Microbiology 2011;79(2) 292-304.

97. B. Henrissat, G. J. Davies, Glycoside hydrolases and glycosyltransferases. Families, modules, and implications for genomicsPlant Physiology 200012441515 1519

98. H. J. Flint, K. P. Scott, S. H. Duncan, P. Louis, E. Forano, Microbial degradation of complex carbohydrates in the gut.Gut Microbes20123418

99. Miron J, Ben-Ghedalia D, Morrison M. Invited review: adhesion mechanisms of rumen cellulolytic bacteria. Journal of Dairy Science 2001;84(6) 1294-1309.

100. Flint HJ, Bayer EA.Plant cell wall breakdown by anaerobic microorganisms from the Mammalian digestive tractAnnals of New York Academy of Sciences 20081125280

101. Ransom-Jones E, Jones DL, McCarthy AJ, McDonald JE. The Fibrobacteres: an mportant phylum of cellulose-degrading bacteria. Microbial Ecology 2012;63(2) 267-281.

102. Williams AG, Coleman GS.Hemicellulose-degrading enzyme in rumen ciliate protozoa.Current Microbiology19851228590

103. Devillard E, Newbold CJ, Scott KP, Forano E, Wallace RJ, Jouany JP, Flint HJ.A xylanase produced by the rumen anaerobic protozoan Polyplastron multivesiculatum shows close sequence similarity to family 11 xylanases from gram-positive bacteria. FEMS Microbiology Letters 1999;181(1) 145-152.

104. C. Béra-Maillet, E. Devillard, M. Cezette, J. P. Jouany, E. Forano, Xylanases and carboxymethylcellulases of the rumen protozoa Polyplastron multivesiculatum, Eudiplodinium maggii and Entodinium sp.FEMS Microbiology Letters20052441149156

105. Devillard E, Béra-Maillet C, Flint HJ, Scott P, Newbold CJ, Wallace RJ, Jouany JP, Forano E. Characterization of XYN10B, a modular xylanase from the ruminal protozoan Polyplastron multivesiculatum, with a family 22 carbohydrate-binding module that binds o cellulose. Biochemical Journal 2003;373 495-503.

106. S. D. Findley, M. R. Mormile, A. Sommer-Hurley, X. C. Zhang, P. Tipton, K. Arnett, J. H. Porter, M. Kerley, G. Stacey, Activity-based metagenomic screening and biochemical characterization of bovine ruminal protozoan glycoside hydrolasesApplied and Environmental Microbiology2011772281068113

107. J. P. Jouany, D. I. Demeyer, J. Grain, Effect of defaunating the rumenAnimal Feed Science and Technology198821229

108. M. Eugène, H. Archimède, D. Sauvant, Quantitative meta-analysis on the effects of defaunation of the rumen on growth, intake and digestion in ruminantsLivestock Production Science20048581

109. Orpin CG, Joblin KN.The rumen anaerobic fungi. In : Hobson PN, Stewart CS (eds.) The rumen microbial ecosystem. London, Chapman & Hall;1997140195

110. T. Nagy, R. B. Tunnicliffe, L. D. Higgins, C. Walters, H. J. Gilbert, M. P. Williamson, Characterization of a double dockerin from the cellulosome of the anaerobic fungus Piromyces equiJournal of Molecular Biology20073733612622

111. Akin DE, Borneman WS.Role of rumen fungi in fiber degradation.Journal of Dairy Science1990731030233032

112. Ljungdahl LG.The cellulase/hemicellulase system of the anaerobic fungus Orpinomyces PC-2 and aspects of its applied useAnnals of New York Academy of Sciences 20081125308

113. M. Qi, P. Wang, L. B. Selinger, L. J. Yanke, R. J. Forster, T. A. Mc Allister, Isolation and characterization of a ferulic acid esterase (Fae1A) from the rumen fungus Anaeromyces mucronatus.Journal of Applied Microbiology2011110513411350

114. G. Fonty, M. Chavarot, J. Lepetit, J. Canistro, R. Favier, Mechanical resistance of wheat straw after incubation in cultures of ruminal cellulolytic microorganismsAnimal Feed Science and Technology1999297 EOF

115. G. Fonty, A. G. Williams, F. Bonnemoy, S. E. Withers, P. Gouet, Effect of anaerobic fungi on glycoside hydrolase and polysaccharide depolymerase activities, in sacco straw degradation and volatile fatty acid concentrations in the rumen of gnotobiotically-reared lambs.ReproductionNutrition Development 199535329

116. Maekawa M, Beauchemin KA, Christensen DA. Chewing activity, saliva production, and ruminal pH of primiparous and multiparous lactating dairy cows. Journal of Dairy Science 2002;85(5) 1176-1182.

117. Grant R. Forage fragility, fibre digestibility and chewing response in dairy cattle. Proceedings of 2010 Tri-State Dairy Nutrition Conference, Fort-Wayne, Indiana, USA, 20-21 April. 22pp.

118. Beauchemin KA, Yang WZ.Effects of physically effective fiber on intake, chewing activity,and ruminal acidosis for dairy cows fed diets based on corn silage.Journal of Dairy Science200588621172129

119. De Boever JL, Andries JI, De Brabander DL, Cottyn BG, Buysse FX.Chewing activity of ruminants as a measure of physical structure- A review of factors affecting it. Animal Feed Science and Technology 1990274281291

120. Varga GA, Kolver ES.Microbial and animal limitations to fiber digestion and utilization.Journal of Nutrition 1997Suppl) 819S EOF823S EOF

121. Jung HG, Mertens DR, Phillips RL.Effect of reduced ferulate-mediated ignin/arabinoxylan cross-linking in corn silage on feed intake, digestibility, and milk production. Journal of Dairy Science 2011941051245137

122. Weimer PJ.Why don't ruminal bacteria digest cellulose faster? Journal of Dairy Science 199679814961502

123. M. Witzig, J. Boguhn, S. Kleinsteuber, I. Fetzer, M. Rodehutscord, Influence of the maize silage to grass silage ratio and feed particle size of rations for ruminants on the community structure of ruminal Firmicutes in vitro.Journal of Applied Microbiology 2010109619982010

124. Moorby JM, Dewhurst RJ, Evans RT, Danelón JL.Effects of dairy cow diet forage proportion on duodenal nutrient supply and urinary purine derivative excretionournal of Dairy Science 200689935523562

125. R. I. Scott, N. Yarlett, K. Hillman, T. N. Williams, A. G. Williams, D. Lloyd, The presence of oxygen in rumen liquor and its effects on methanogenesisJournal of Applied Bacteriology 198355143

126. K. Hillman, D. Lloyd, A. G. Williams, Use of a portable quadrupole mass spectrometer or the measurement of dissolved gas concentrations in ovine rumen liquor in situ.Current Microbiology198512335

127. Marden JP, Bayourthe C, Enjalbert F, Moncoulon R. A new device for measuring kinetics of ruminal pH and redox potential in dairy cattle. Journal of Dairy Science 2005; 88(1) 277-281.

128. Newbold CJ, Wallace RJ, McIntosh FM.Mode of action of the yeast Saccharomyces cerevisiae as a feed additive for ruminants.British Journal of Nutrition 1996762249261

129. V. Roger, G. Fonty, S. Komisarczuk-Bony, P. Gouet, Effects of physicochemical factors on the adhesion to cellulose Avicel of the rumen bacteria Ruminicoccus ftavefaciens and Fibrobacter succinogenes subsp. succinogenes. Applied and Environmental Microbiology 1990563081

130. G. Fonty, E. Forano, Les interactions microbiennes impliquées dans la cellulolyse ruminale. Comptes Rendus de l'Académie d'Agriculture de France 1998841135148

131. J. Chen, P. Weimer, Competition among three predominant ruminal cellulolytic bacteria n the absence or presence of non-cellulolytic bacteria. Microbiology 200114721

132. P. Mosoni, G. Fonty, P. Gouet, Competition between ruminal cellulolytic bacteria for adhesion to cellulose.Current Microbiology 19973514447

133. A. Bernalier, G. Fonty, F. Bonnemoy, P. Gouet, Inhibition of the cellulolytic activity of Neocallimastix frontalis by Ruminococcus flavefaciens.Journal of General Microbiology19931394873880

134. Plata PF, Mendoza MGD, Barcena-Gama JR, Gonzalez MS.Effect of a yeast culture Saccharomyces cerevisiae) on neutral detergent fiber digestion in steers fed oat straw based diets.Animal Feed Science and Technology199449203

135. R. L. A. Miranda, M. G. D. Mendoza, J. R. Barcena-Gama, Ferrara. R. Gonzalez, C. M. E. Ortega, P. M. A. Cobos, Effect of Saccharomyces cerevisiae and Aspergillus oryzae cultures and NDF level on parameters of ruminal fermentation. Animal Feed Science and Technology199663289

136. F. Chaucheyras-Durand, A. Ameilbonne, N. D. Walker, P. Mosoni, E. Forano, Effect of a ive yeast, Saccharomyces cerevisiae I-1077 on in situ ruminal degradation of alfalfa hay and fibre-associated microorganisms. Journal of Animal Science 2010E-Suppl. 2) 145.

137. F. Chaucheyras-Durand, G. Fonty, Establishment of cellulolytic bacteria and development of fermentative activities in the rumen of gnotobiotically-reared lambs receiving the microbial additive Saccharomyces cerevisiae CNCM I-1077.ReproductionNutrition Development 20014157

138. Angeles SC, Mendoza GD, Cobos MA, Crosby MM, Castrejon FA.Comparison of two commercial yeast cultures (Saccharomyces cerevisiae) on ruminal fermentation and digestion in sheep fed on corn-stover dietSmall Ruminant Research 19983145

139. F. Chaucheyras, G. Fonty, G. Bertin, P. Gouet, Effects of live Saccharomyces cerevisiae cells on zoospore germination, growth, and cellulolytic activity of the rumen anaerobic ungus, Neocallimastix frontalis MCH3.Current Microbiology199531201

140. Girard ID, Dawson KA.Effect of a yeast culture on growth characteristics of representative ruminal bacteria Journal of Animal Science 1995

141. Callaway TS, Martin SA.Effects of a Saccharomyces cerevisiae culture on ruminal bacteria that utilize lactate and digest cellulose.Journal of Dairy Science1997802035

142. G. Fonty, J. Senaud, J. P. Jouany, P. Gouet, Establishment of ciliate protozoa in the rumen of conventional and conventionalized lambs: influence of diet and management conditions.Canadian Journal of Microbiology198834235

143. F. Chaucheyras-Durand, G. Fonty, Influence of a probiotic yeast (Saccharomyces cerevisiae CNCM I-1077) on microbial colonization and fermentation in the rumen of newborn ambs. Microbial Ecology in Health and Disease 20021430

144. J. P. Jouany, F. Mathieu, J. Senaud, J. Bohatier, G. Bertin, M. Mercier, The effect of Saccharomyces cerevisiae and Aspergillus oryzae on the digestion of the cell wall fraction of a mixed diet in defaunated and refaunated sheep rumen.ReproductionNutrition Development 199838401

145. F. Chaucheyras, L. Millet, Michalet-Doreau, Effect of the addition of Levucell SC on the rumen microflora of sheep during adaptation to high starch diets. Reproduction Nutrition Development 1997; EE 5 (suppl) 82.

146. C. S. Mc Sweeney, A. Dulieu, R. Bunch, Butyrivibrio, and other xylanolytic microorganisms from the rumen have cinnamoyl esterase activity. Anaerobe1998415765

147. Newbold CJ.Microbial feed additives for ruminants. In: Wallace RJ, Chesson A (eds.) Biotechnology in animal feeds and animal feeding. Weinheim: VCH;1995259278

148. F. Mathieu, J. P. Jouany, J. Sénaud, J. Bohatier, G. Bertin, M. Mercier, The effect of Saccharomyces cerevisiae and Aspergillus

oryzae on fermentations in the rumen of faunated and defaunated sheep; protozoal and probiotic interactions.ReproductionNutrition Development 1996363271287

149. Jouany JP.Optimizing rumen functions in the close-up transition period and early actation to drive dry matter intake and energy balance in cows. Animal Reproduction Science 200696250

150. L. L. Bitencourt, M. N. Pereira, B. M. L. de Oliveira, J. R. M. Silva, Júnior. G. S. Dias, F. Lopes, R. C. M. de Melo, Júnior. S. Siécola, Response of lactating cows to the supplementation with ive yeast. Journal of Dairy Science 2008E-suppl1) 264.

151. F. Chaucheyras-Durand, G. Fonty, Effects and modes of action of live yeasts in the rumenBiologiaBratislava) 2006616741750

152. F. Durand-Chaucheyras, G. Fonty, G. Bertin, M. Theveniot, P. Gouet, Fate of Levucell SC-1077 yeast additive during digestive transit in lambs. Reproduction Nutrition Development 199838275

153. Sullivan ML, Bradford BJ.Viable cell yield from active dry yeast products and effects of storage temperature and diluent on yeast cell viability.Journal of Dairy Science2011941526531

154. Davey HM. Life, death and in between: meanings and methods in microbiology Applied and Environmental Microbiology 2011771655715576

155. A. Zuzuarregui, L. Monteoliva, C. Gil, M. L. del Olmo, Transcriptomic and proteomic approach for understanding the molecular basis of adaptation of Saccharomyces cerevisiae o wine fermentation. Applied and Environmental Microbiology 20067 21836847

156. I. Andorra, B. Esteve-Zarzoso, J. M. Guillamon, A. Mas, Determination of viable wine yeast using DNA binding dyes and quantitative PCRInternational Journal of Food Microbiology 20101442257262

157. D. P. Morgavi, E. Forano, C. Martin, C. J. Newbold, Microbial ecosystem and methanogenesis in ruminantsAnimal 201047 1024 1036

158. F. Chaucheyras-Durand, S. Masséglia, G. Fonty, E. Forano, Influence of the composition of the cellulolytic flora on the

development of hydrogenotrophic microorganisms, hydrogen utilization, and methane production in the rumens of gnotobiotically reared ambs. Applied and Environmental Microbiology 2010762 479317937

159. H. Boudra, an. Mycotoxins, menacing. insidious, for. factor, quality. the, forages. of, performances. the, the. of, ruminants, Fourrages 2009199265

Index